Globalizing South China

RGS-IBG Book Series

The *Royal Geographical Society (with the Institute of British Geographers) Book Series* provides a forum for scholarly monographs and edited collections of academic papers at the leading edge of research in human and physical geography. The volumes are intended to make significant contributions to the field in which they lie, and to be written in a manner accessible to the wider community of academic geographers. Some volumes will disseminate current geographical research reported at conferences or sessions convened by Research Groups of the Society. Some will be edited or authored by scholars from beyond the UK. All are designed to have an international readership and to both reflect and stimulate the best current research within geography.

The books will stand out in terms of:

- the quality of research
- their contribution to their research field
- their likelihood to stimulate other research
- being scholarly but accessible.

Published

Geomorphological Processes and Landscape Change: Britain in the Last 1000 Years
David L. Higgitt and E. Mark Lee (eds)

Globalizing South China
Carolyn Cartier

Globalizing South China

Carolyn Cartier

First published 2001

2 4 6 8 10 9 7 5 3 1

Blackwell Publishers Ltd
108 Cowley Road
Oxford OX4 1JF
UK

Blackwell Publishers Inc
350 Main Street
Malden, Massachusetts 02148
USA

British Library Cataloguing in Publication Data

A CIP catalogue record for this book is available from the British Library.

Library of Congress Cataloging-in-Publication Data

Cartier, Carolyn
Globalizing South China / Carolyn Cartier.
p. cm. — (RGS-IBG book series)
Includes bibliographical references and index.
ISBN 1-55786-887-5 (hbk : alk. paper) — ISBN 1-55786-888-3 (pbk : alk. paper)
1. China, Southeast—Economic conditions. 2. Globalization—Economic aspects—China, Southeast. 3. Globalization—Social aspects—China, Southeast. 4. China, Southeast—Foreign economic relations. 5. China, Southeast—Commerce. I. Title. II. Series.

HC428 .S74 .C37 2001
337.51'2—dc21

2001037468

This book is printed on acid-free paper.

Contents

Plates

Figures

Table

Maps

Series Editors' Preface

The RGS-IBG Book Series publishes the highest quality of research and scholarship across the broad disciplinary spectrum of geography. Addressing the vibrant agenda of theoretical debates and issues that characterize the contemporary discipline, contributions will provide a synthesis of research, teaching, theory and practice that both reflects and stimulates cutting-edge research. The series seeks to engage an international readership through the provision of scholarly, vivid and accessible texts.

Nick Henry and Jon Sadler
Series Editors

Preface

> [G]eographical knowledge is too broad and too important to be left to geographers. Its reconstruction as a preparation for a civilized life and its synthesis as an endpoint of human understandings depends on overcoming the old Kantian distinctions between history (narration) and geography (spatial ordering) and between geography (the outer world of the objective material conditions) and anthropology (the inner world of subjectivities). It would probably require the reconstitution of some new structure of knowledge.
> (David Harvey, 2000, p. 558)

This book is an inquiry into the possibilities of research on an urban and industrializing transboundary region, the south China coast, which has experienced some of the most rapid economic growth in world history. Numerous accounts about the region have examined its contemporary economic transformations. This analysis instead is based in critical perspectives on regional scholarship in geography and China area studies, and departs from existing work on regional economies in Asia. Alternatively, the book approaches regional analysis from the perspective of "intellectual globalization," the debates about the organization of knowledge, in universities and scholarly research, and their adequacy for negotiating societal complexities in the contemporary era.[1] How can organizations of knowledge suited to a cosmopolitan geographical ethic intersect in innovative ways that transcend the problems of (inter)disciplinary studies, and reinvigorate diverse worldviews, replacing partial knowledges with whole and situated accounts? The book that follows presents one answer to this epistemological problem through the region in south China, by foregrounding debates about its processes of formation and by replacing portrayals of rapid economic growth with critical perspectives about regional transformation. Thus "globalizing" south China means first working to open up a cosmopolitan knowledge about the region, in ways that recognize both historical processes at the basis of regional formation and how places in the contemporary region have experienced the transformations wrought by internationalized transboundary activity.

Because, in Fredric Jameson's (1998b, p. 55) words, "globalization is a communicational concept, which alternately masks and transmits cultural or economic meanings," entering the arena of scholarship about globalization requires some cautionary remarks. While the idea of globalization became popular in the late twentieth century, what may be "globalized" must be understood in the context of historic cultural practices, political activities, and economic processes, now transformed. Similarly, in this project globalization does not focus on the currently typical debates in political and economic studies, about the effects of globalizing processes on economic activity or nation-state power, because those debates, despite their important contributions, have tended not to sustain engagements with the emplaced conditions of regional formations. This project instead concerns how to understand the complexities of regional formations as situated intersections of cultural, political, and economic forms. In the same vein, what is globalization is not global in a landed geographical sense, but global as unprecedented possibilities for rapid transference and new combinations of resources, especially capital, labor, information, and ideas. Regional formations, where these and other resources originate, touch down, accumulate, recirculate, and generate new social relations, are especially important sites of globalizing processes. The largest cities of the south China coast have been the centers of transboundary regional activity and are a primary focus of the study.

While studies of "the local and the global" proliferated in geography and related fields through the 1980s and 1990s, the phrase often appeared more as a metaphorical signifier, substituting recognition of transnational and transboundary processes for examination of how such processes actually work out and how they connect diverse places and regions. The problems of the "local and the global" also effectively elided the region, whereas contemporary theorizations of dynamic scale relations help us to think through spatialities that constitute and interrelate places, regions, states, and other territories. The conceptualization of transboundary regional activity through relations of scale is one of the most important ways of moving toward seeing the realities of globalized activity. A focused study on scale relations in China would form its own subject of investigation, and so this book introduces concepts of scale and depends on a dialectical scale perspective as a framework for analysis.

The methodological orientation of this project reflects late twentieth-century shifts in Anglo-American geography, when geographers began to adopt approaches reflecting the broader poststructural theoretical shift in the academy. The approach adopted in this study retains a material grounding yet understands constructivist knowledges in order to examine both realities of the south China coast and changing representations about the region over time. The analysis recognizes too the instability of human categories, and the importance of understanding "difference" in individuals and within cultural groups, in issues of class, race, ethnicity, gender, sex, and sexuality. This stance is not one that reflects methodological trends, but rather one that finds how theory has finally caught up with practice and empirical reality. Maintenance of a relatively fixed and unchanging identity is, on a world scale, often a privileged position of power and an elite and male-gendered opportunity. Feminist scholarship has allowed us to understand that the "normative" perspective is regularly a patriarchal position, and the analysis depends on feminist critiques to understand a range of issues, including embodied subjectivity and gendered economic activity.

The book is also a product of the era of *Orientalism*, Edward Said's (1978) exposition of the practices and representations of European orientalisms. It uses the basic precepts of the *Orientalism* project as a tool to intervene in normative social scientific conceptualizations of Asia and China. I assume in this book that what Said achieved for humanistic scholarship, in conceptualizing how the West invented "the Orient" through myriad representations, has not been reliably achieved for social scientific analyses of Asia, and especially economic accounts. Because social scientific acceptance of the problem of representations has been incomplete if it has begun at all, this project challenges continuing Western-centrisms retained by many political economic analyses about China, and the larger Asian region. Organization of knowledge for a truly cosmopolitan ethic could not begin to proceed any other way.

Similarly, the critical interventions in Western political economic perspectives about China are not wholly indebted to the legacies of structural Marxism, not least because that mode has tended to presuppose places and processes of Western industrialized geographies. This book contends, as one of its central arguments, that Han Chinese cultural understandings of place and space, human–spatial relations, and some economic reworkings of the space economy in the industrial era have their own logic, and that they reflect a long history of cultural complexes and dynamic spatial relations in Chinese society, and organizational strategies of the Chinese state. One book can only begin to explore these issues, and for this reason alone, the approach must be historical. The critical edges do belong to understanding that globalization, in certain institutional contexts, is deployed as a euphemistic disguise for the neoliberal regime and its problems, a contemporary legacy of the Enlightenment project and diverse forms of imperialism. Still, treating globalization as the property of the West denies how ideas and experiences of globalization have global contexts and localized meanings. Ultimately, the critical interventions belong to concern for the problems of uneven development and ethical accountings of political economic activity, at all scales.

An important observation about neoliberal capitalism at the turn of the millennium is the set of experiential contradictions that have emerged as economic activity works its course and leaves new extremes in its wake: "it appears both to include and to marginalize in unanticipated ways; to produce desire and expectation on a global scale yet to decrease the certainty of work or the security of persons; to magnify class differences but to undercut class consciousness; above all, to offer up vast, almost instantaneous riches to those who master its spectral technologies – and simultaneously, to threaten the very existence of those who do not" (Comaroff and Comaroff, 2000, p. 298). In this complexity of extremes, the world economy promotes the learning of certain kinds of knowledges and the jettisoning of others. The contemporary south China coast has been a region of riches, and the stories repeated about it have been partial accounts that reflect the decades of rapid accumulation that ricocheted across the Pacific in the second half of the twentieth century. This book assembles regional knowledges that have not been reliably told.

Several chapters of this book draw on material published (or about to be published) in different versions elsewhere. A short section in chapter 1 is taken from "Origins and evolution of a geographical idea: the macroregion in China," *Modern China*, 28(1), 2002. The discussion of the Bukit China movement in chapter 5 was originally published in "The dead, place/space, and social activism: constructing the nationscape in historic Melaka," *Environment and Planning D: Society and Space*, 15(5), 1997. A longer version of chapter 7 appeared as "'Zone fever', the arable land debate, and real estate speculation: China's evolving land use regime and its geographical contradictions," *Journal of Contemporary China*, 10(28), 2001. Chapter 8 includes short sections from "Cosmopolitics and the maritime world city," which appeared in *The Geographical Review*, 89(2), 1999, and "The state, property development, and symbolic landscape in high rise Hong Kong," *Landscape Research*, 24(2), 1999. I thank the editors and referees of these pieces for their helpful comments.

Over five years the conceptualization and materialization of this book spanned three colleges and universities in the USA, and work on three continents, from China and Southeast Asia, to the USA, and the UK, and I am especially grateful to the people and institutions who supported the project throughout. I owe a special debt to John Davey, who originally contracted the book for Blackwell, and to my editor, Sarah Falkus, who has followed the manuscript to completion. Joanna Pyke, the editorial controller, Brian Johnson, who managed the graphics, and John Taylor, my copy editor, were especially helpful during the publication process. I am grateful to the reviewers for the press who encouraged a leaner and restructured manuscript, and especially to Larry Ma, whose detailed suggestions were subtle, precise, and brilliantly insightful. Jane Sinclair, who produced the line illustrations, has been more a collaborator than a cartographer and created a set of distinctive contemporary illustrations to accompany the text. David Hooson read an earlier draft and staunchly supported my rewriting regional geography before it became fashionable again to do so, and Roland Greene read the manuscript from the perspective of a literary critic and encouraged development of the ideas about regionalism and geographical representations.

Diverse institutes and sources of funding supported research and writing related to the book: the Association of Asian Studies, the Centre for Advanced Studies at the National University of Singapore, the Chiang Ching-kuo Foundation, the Freeman Foundation, the Luce Foundation, the Nanyang Research Institute at Xiamen University, the University of California Humanities Research Institute at UC Irvine, the University of Oregon Department of Geography, and the University of Oregon Humanities Institute. I am deeply grateful to many individuals in China, Hong Kong, and Malaysia who granted interviews, some repeatedly, for material that appears in chapters 5, 7, and 8 of the book.

The process of producing this book has also been richer for conversations and correspondences with many colleagues and friends, especially John Agnew, Tony Bebbington, Bill Barron, Karen DeBres, Cynthia Brokaw, Bob Burnett, Bob Cartier, Cezanne Cartier, Kam Wing Chan, Josephine Chua, Lily Chua, Stephen Durrant, Maram Epstein, Mary Erbaugh, Cindy Fan, Susanne Freidberg, Katherine Gibson, Bryna Goodman, David S. G. Goodman, Julie Graham, Wang Gungwu, David Harvey, Laura Hess, Lisa Hopkinson, Richie Howitt, Denise Humphreys, Jean K. M. Hung, Nayna Jhaveri, John Paul Jones, Ron Knapp, Dick Kraus, Wendy Larson, Marcia Levenson, Alan Lew, Martin Lewis, Christine Loh, Li Si-Ming, George Lin, Larry Ma, Mary MacDonald, Janet Momsen, Dick Peet, Tim Oakes, Kim Rhody, Sang Sze-lan, Elizabeth Sinn, Victor Sit, Helen Siu, Chris Smith, Paul Starrs, Laura Steinman, Wing-shing Tang, Greg Veeck, Billy Villet, Peter Walker, Kären Wigen, Anthony Yeh, You-tien Hsing, and Yue-man Yeung. The China Specialty Group of the Association of American Geographers has served as a forum for initial presentation of some of the ideas in the book, and I am thankful to Cindy Fan especially for encouraging my central participation in the group.

The seeds of this book were in a dissertation I wrote at Berkeley in geography, when that department was experiencing the chaotic processes of a paradigm shift. Only after I left that milieu could I produce scholarship that transcended the political–intellectual boundaries that otherwise were entrenched there at the time. I am grateful to the Berkeley geography department for support over three degrees, and especially to David Hooson and Jay Vance, and also to Dick Walker. David Hooson and Jay Vance were unstintingly supportive, and Dick Walker coached me through postdoctoral academic politics. I am also appreciative to Michael Watts, for being the leading model of professionalism, Allan Pred, whose boundary breaking theoretical work has been an inspiration, and Bob Reed, who steered me toward Southeast Asian studies. I learned to think critically about Chinese historiography from courses with Frederic Wakeman in ways that have proved particularly sustaining over time.

The origins of this book more properly lie in a course I taught at Vassar College in 1993–4, "China and the Chinese Overseas." I am thankful to the students of that class for apparently finding the subject matter about as interesting as I did. From Vassar I remember gratefully the collegialty of Gabrielle Cody, Miriam Cohen, Harvey Flad, Donald Gillin, Brian Godfrey, Lucy Johnson, David Kennett, Jannay Morrow, and Cindy Wall, none of whom bear any responsibility for my failure to acculturate to the east coast.

I sought to return to the west coast – from where Asia simply never could have been the "Far East" – to build an alternative basis for the kind of scholarship I wanted to produce. While I was at the University of Oregon, the university's exchange with Xiamen University allowed me to spend the summer of 1995 in Xiamen, after a research period in Suzhou. The following summer I was at the Stanford Center for Chinese language study in Taipei, and the next summer at the National University of Singapore. These experiences proved foundational in my work to rethink the region, and I am grateful to the people who made them possible, especially Stephen Durrant, Cynthia Brokaw, Cui Gonghao, Liao Shaolian, Su Zixing, Kris Olds, and Peggy Teo. Wendy Larson, Bryna Goodman, and Cynthia Brokaw formed the faculty research group at Oregon, "Gender in Historical and Transnational China," which sustained my intellectual interests there. Cynthia Brokaw, Bryna Goodman, Stephen Durrant, and Dick Kraus read parts of the first draft of the manuscript and offered valuable suggestions. My graduate students in geography at Oregon provided inspiration and some research assistance, especially Jeff Baldwin, Becky Mansfield, and Jessica Rothenberg-Alami. My research assistant, Xingli Zhang, a doctoral student in international relations at the University of Southern California, scrupulously compiled the index.

In a variation on a well turned phrase, this project could only have come together in Los Angeles: my warmest professional thanks are for my colleagues at the University of Southern California, Bernie Bauer,

Michael Dear, Rod McKenzie, Stephanie Pincetl, Laura Pulido, Doug Sherman, Billie Shotlow, Chris Williamson, John Wilson, and Jennifer Wolch, who make the geography department an eminently collegial environment, without equal in my experience. I am especially thankful to Jennifer Wolch and Michael Dear for ensuring that I had the three resources at the basis of scholarly production – space, time, and money – and the best professional counsel. Jennifer and Michael also introduced me to Los Angeles in ways that have made me feel incredibly fortunate. Jennifer particularly makes the department an energized place of the highest standards, and I have been grateful for her presence and mentorship.

The book is dedicated to the memory of my parents, Helen D. Boller Cartier and Raymond E. Cartier, whose ends of life have marked this project like monuments to historical eras. After prolonged illnesses, my father passed away before I left for my first journey to China and Southeast Asia, and my mother passed away while I was in middle of the final revisions of the manuscript for the press. I have much to thank them for, in propelling me into the international world, thinking the world of me, and in their various activities and passions, for reminding me of the limits of scholarship.

NOTE

1 On this subject see, in addition to Harvey (2000), ideas of Masao Miyoshi (1998) and Bill Readings (1996).

1

Negotiating Geographical Knowledges

> As the last remaining socialist country with perhaps the fastest economic growth in the world today, China presents a challenge to critical thinking about globalization. It is imperative that the question of alternatives and other possibilities and potentialities be raised in any attempt at theorizing or conceptualizing the process of globalization. Globalization is generally perceived as the result of the collapse of Soviet-style socialism, as well as the unprecedented expansion of transnational capitalism. While avowedly Eurocentric in its hegemonic formations, globalization also sets up an indispensable structural context for analyzing what happens in the world today. Therefore, globalization must be grasped as a dialectical process: it refers at once to an idea, or an ideology – that is, capitalism disguised as a triumphant, universal globalism – and a concrete historical condition by which various ideas, including capitalism in its present guise, must be measured. China's challenge to globalization can be perceived in both senses, first to global capitalism as an ideology and then to the "new world order," or "world-system," as an accepted reality. China has become increasingly integrated into the global economic system, yet retains its ideological and political self-identity as a third-world, socialist country. Will China offer an alternative?
> (Liu Kang, 1998, p. 164)

Globalization has emerged as a common term, yet is an unwieldy conceptual idea used in diverse contexts and to signal, or disguise, a variety of different cultural, economic, and political positions. It is fundamentally associated with the increasing internationalization of capitalist practices, through firms and transnational corporate activities in the world economy, backed and challenged by political forces, accompanied by cultural forms, and mediated by local resistances. In the contemporary geographical imaginary, coastal south China is one area of the world whose economic processes and social relations, in dialectical formation with the world economy, have contributed to contemporary understandings about globalization. South China's rise has also destabilized the national order of things on the Chinese domestic scene. These events, though, were not entirely new to the late twentieth century. For most of its history, the south China coast has been an internationalized transboundary region, the primary zone of contact between the larger empire of which it has been a part and the world economy. It has also been a region of social activism and revolution. Despite the totalizing qualities of some globalization narratives, regions, like coastal south China, continue to be distinctive in particular ways. The focus of this analysis is that basic geographical problem – the tension between the forces of globalization and the production of local and regional difference.

The book that follows has had several points of intellectual origin. One of those points was the recognition that much of post-Second World War scholarship about the south China coast, and especially about the period of the nineteenth century, when China faced Western demands for free trade (how much has changed in a century?), bore all the marks of Cold War era politicized debate. In the 1990s the political rhetoric of the Cold War era was refashioned into discourses of neoliberalism, which continue to promote Western political economic goals in new "globalized" ways. Another pivotal concern was the set of methodological disjunctures between research paradigms in China area studies and contemporary theoretical approaches in geography and related fields, and resulting gaps in knowledge about regional formation in China. The interrogation of prevailing paradigms and the formation of other modes of explanation have been in order. The most important concern was understanding that most of the material processes that constitute a regional formation in coastal south China coast are transboundary, transnational, and, in a few significant ways, simultaneously transhistorical in nature. Coming to terms with these perspectives meant that the analysis had to treat regional space as a set of dynamic, scaled processes, which would frame a globalizing regional formation in relation to the territorial coherence of the dynastic era and twentieth-century state-making project in China. What follows is a course of unbounding south China, to raise complex questions about the implications of historical geographies for understanding contemporary regional formations, and their imaginations; how regional formations, materially and discursively, are responses to other territorial transformations and processes of globalization; the ways in which regional formations emerge in contexts of economic restructuring; whether articulating regional formations may lead beyond the problems of statist paradigms and nationalisms; and how in a transhistorical maritime region, oceans connect rather than divide.

This chapter introduces these issues to set forth larger-scale contexts of understanding China in the contemporary world order, and also to begin to establish how what we may know about a country and its regions are regularly partial and shifting accounts of more complex processes. My strategy is to place an unfolding geography about south China in the history of its scholarship, and at the same time, to call into question some of the ways in which that scholarship has been written. This is a critical and contexualist approach, which combines theoretical orientations with understandings of regional realities, and seeks to mediate between ideas about the conceptual space of flows in transboundary space economies, located geographies of production and exchange, and cultural spheres of agency and symbolic meaning. The first half of the book concerns historical geographies of the south China coast and their contested representations, in the sense of what Felix Driver (1992, p. 35) has assessed less as "a prop for the present," and more as a means of articulating between geographical realities and understandings of the past, and their conditions and representations in the present. Instead of a linear accounting of the regional past, the historical analysis recalls how history matters in situated and lived geographies – that is, explorations at the intersection of place/space, time, narrative, and body – and in doing so substitutes for progressivist history a regional geography formed of diverse places and landscapes of disruption and discontinuity. These geographical reorientations establish the means for understanding causal relations at the basis of many questions about the contemporary regional formation, the rise of the south China coast under reform. The next sections begin with an empirical account of contemporary regional transformation, but with the recognition that the apparently factual description is a partial view of what must be assessed in a broader and historicized theoretical analysis.

Region of Reform

The south China coast erupted on the world economic map in the final quarter of the twentieth century, compelling widespread interest in special economic zones, capital flows, the global shift in low-wage manufacturing industries, Chinese overseas business networks, and the rise of China as a potential economic superpower. The process of economic reform was formally initiated at the Third Plenum of the Eleventh Party Congress Central Committee in Beijing in December 1978 and has unfolded incrementally in diverse market-oriented economic policies. China's economic transformation has been wide-ranging and complex, and while central state policy has appeared to drive reform, innovative economic practices undertaken by local and regional officials have also substantially led reform initiatives.[1]

The geographical foundation of the export-oriented sector of reform was the "open policy" and its system of special economic zones, open cities, and open development regions, all established by the state to concentrate foreign investment and export-oriented manufacturing in coastal China. China's leadership established the four original special economic zones (SEZs) in Guangdong and Fujian provinces, which are the two homeland provinces of the majority of Chinese overseas. At the time of their selection, the first four SEZs were not important locations in China's existing administrative system, but were strategically linked to historic trading economies or Chinese overseas communities, or both. Two of the cities, Shantou and Xiamen, are centers of historic trade and emigration and had also been open ports under the treaty system.[2] The other two cities, Shenzhen and Zhuhai, were border frontiers with Hong Kong and Macao, respectively. The geographic specificity of reform wove the economies of Hong Kong and Macao into Guangdong province fifteen to twenty years ahead of the scheduled repatriation of the two colonies. Through the 1980s, economic relations between Hong Kong, Taiwan, and the coastal zones of Guangdong and Fujian provinces became so closely tied that the greater part of Hong Kong's former manufacturing industry relocated to Guangdong, and the majority of external investment in Fujian had come from Taiwan (Luo and Howe, 1993). By 1997, Hong Kong had already served as the major source and conduit of capital and manufacturing expertise for southern China.

The successful establishment of the four SEZs led to a series of special open cities and development zones which enlarged the spatial scope of reform. As Dali Yang (1997, p. 30) has assessed, "These special zones grew at a torrid pace and prompted Deng Xiaoping to urge . . . that more coastal cities be given various special policies." In 1984, Beijing announced the opening of fourteen coastal port cities to trade and foreign investment. As economic activity grew beyond the special zones, the state kept pace by designating open counties and open regions. In 1985, the state established three open economic regions, which formed areas of concentrated economic transformation: the Zhujiang delta in Guangdong, the Minnan delta region in southern Fujian, and the Yangzi delta region encompassing Shanghai and its hinterland in the Su'nan area of Jiangsu province (map 1.1). As a result of this geography of export-oriented reform, Guangdong and Fujian rapidly changed places in the hierarchy of provincial significance, from middling and low economic importance, respectively, to become the first provinces in China under reform to receive foreign investment.[3] In 1988, the State Council named Hainan Island the fifth SEZ and the thirty-first province of China, and declared all coastal provinces open to foreign investment. In 1990, the State Council finally granted Shanghai its own special zone, the gargantuan 350 km^2 Pudong New Area across the Huangpu River from central Shanghai. The geographical nature of reform put the south China coast at the center of domestic economic planning for the first time in Chinese

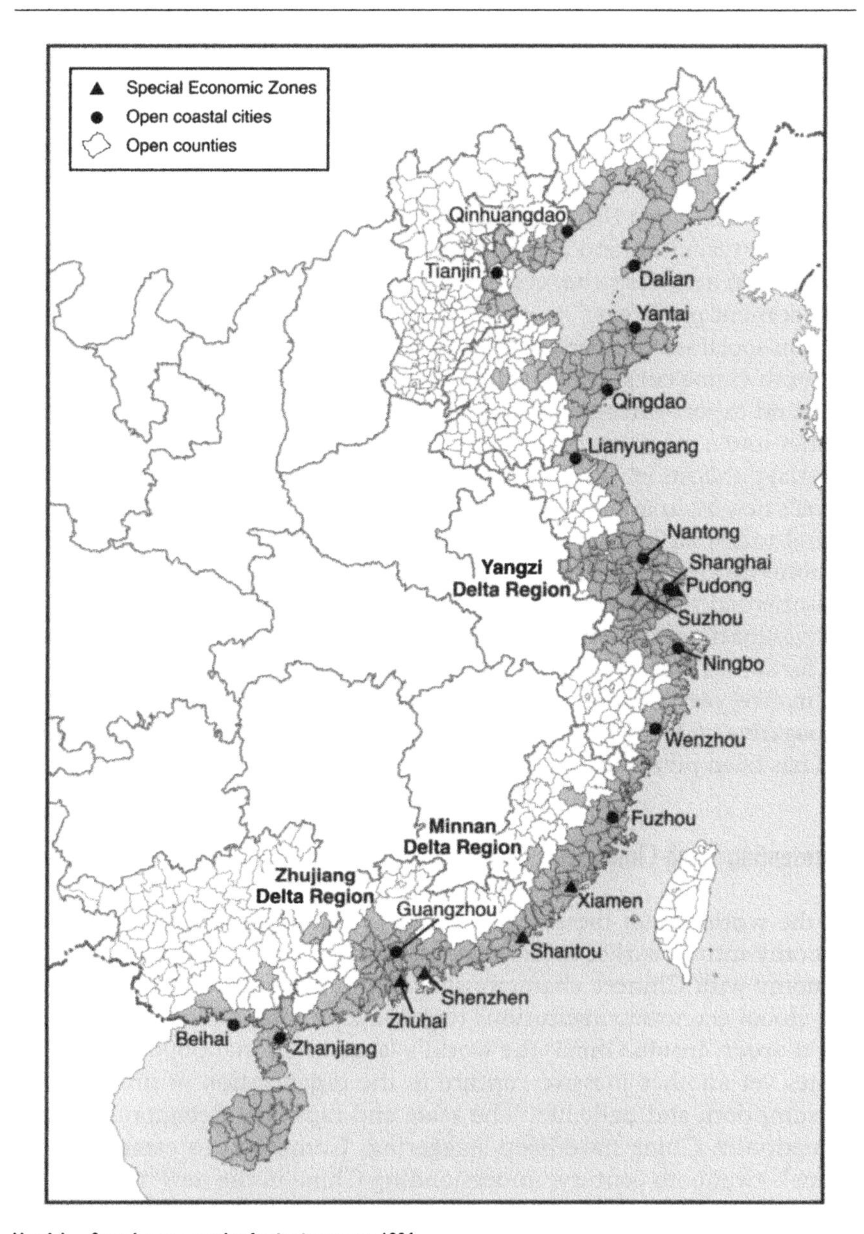

Map 1.1 Coastal areas opened to foreign investment, 1996
Source: Zhongguo duiwai jingji maoyi nianjian (1996/7); line work by Jane Sinclair.

history. In the process, the south coast between Shanghai and Hainan transformed from a relatively peripheral Chinese region into a series of port city-based boom towns tied to the world economy. The state continued to open coastal counties to foreign investment so that by the middle of the 1990s open areas formed a continuous sub-provincial open coastal zone. As the SEZs in Guangdong and Fujian shed their experimental status, cities and provinces across China, especially in the interior, began to press the central government for their own special development privileges.[4]

From social science perspectives on world economic activity, the rise of the south China coast has appeared as evidence about how foreign investment and export-oriented development can turn a once remote maritime frontier into a magnetic center of regional change. The World Bank's influential publication *The East Asian Miracle* confirmed the role of SEZs in China's new internationalized economy: "An export-push strategy has been central to China's rapid development since the government opened the economy to the outside world in 1978. Mechanisms have included export-oriented special economic zones (SEZs) and open cities; export incentives for domestic enterprises and foreign investors in targeted sectors, and for some firms, mandatory export targets. Success has been spectacular: in five years, exports grew nearly tenfold to $72 billion in 1991" (Panagariya, in World Bank, 1993, p. 144). Enough literature in the same vein has been published on this subject to stock a small library.

Representing South China

On the world scale, the transformation of the largest central planned economy into a market economy – in China's terms, a socialist market economy with Chinese characteristics – has challenged other countries and global economic institutions to reconceptualize China's role in the world order. Inside China, the world's largest national population negotiates yet another massive rupture in the organization of production, consumption, and daily life. The scale and rapidity of economic transformation in China have been staggering. Compared to cataclysms in China's twentieth century, understanding China in the new millennium is a relatively reasonable project, since information about China's changing condition is now widely available. Yet in the intensive interest to publish materials about China's transformation as rapidly as it has unfolded, accounts of the reform experience have often lacked contextual and space–time dimensions and have typically disregarded relevant geographies and regional histories. The goal here is to examine some of the scholarly perspectives on rapid growth in south China as an exercise in representations, and to clear the ideological ground in order to build a different kind of account of the transboundary region. Critical assessment of two reform era representations of China, the SEZ "experiment" and the "miracle"economy approach, suggest ways of seeing beyond normative political economic discourses.

Special zone "experiment" meets "miracle" development

Deng Xiaoping, recently emerged from a power struggle to capture the leadership of the Communist Party in the wake of the death of former Chairman Mao Zedong, introduced the geographical component of China's export-oriented reforms as an "experiment." In 1979, a Communist Party document named Shenzhen, Zhuhai, and Shantou and Xiamen "experimental special economic zones" (*Selected Works of Deng Xiaoping*, 1984, p. 416, n.113). The language eased in the new policies and helped quiet opponents of reform.[5] Yet after just months of SEZ implementation, notions of experimentation faded as realities of new economic activity began to result in completely new geographies of production and accumulation, and consequent new geometries of power. Deng Xiaoping promoted the success of the SEZ experiment by highly publicized site visits. In 1984, during a tour of Shenzhen, Zhuhai, and Xiamen, Deng encouraged the use of international capital and expertise, and invited Chinese overseas investment. In 1992, at a critical juncture in the second decade of the reform era, after a period of high inflation following the Tiananmen crisis, Deng purposefully conducted another southern tour of the SEZs and exhorted officials to stay the course of reform and intensify rapid growth. As the new economic practices taking place in SEZs were sanctioned at the highest levels, they became geographical symbols of nationalist reform ideology (Crane, 1996). What was once SEZ "exceptionalism" became normative practice, as cities and towns across China implemented the special zone concept, often without official permission.

The success of the reform program has typically been measured in terms of economic growth. From 1979 to 1999, China's economy grew at an average annual rate of 9.7 percent. In the first decade of reform, China's economy grew at an average annual rate of 9 percent. China's economy slowed especially from 1989 to 1991, and began to improve again after 1992. From 1993 to 1997, China's economy again maintained a relatively high growth rate around 9 percent. Even in 1998 and 1999, after the economic downturn in the Asian region, the annual growth rate maintained between 7 and 8 percent (*ZGTJNJ*, 2000). Throughout, the high national growth rate was achieved by higher than average growth rates in the southern coastal provinces, especially in the first decade of reform. Coastal provinces regularly led the country with double-digit growth rates, and at its peaks in Guangdong, in 1985 and 1992, the rate was as high as over 20 percent per annum (*ZGTJNJ*, 1986, 1993). But the use of aggregate economic figures abstracts and homogenizes space, as well as a vast array of diverse and transforming conditions under reform. Economic accounting measures, even as they record and legitimize economic policies, absorb and mask particular kinds of differences in the economic landscape. Separated from historical and geographical context, these contemporary measures of China's economic growth have presented a new China as another "miracle" economy.

The miracle economy position originated in economic analyses about Hong Kong, Taiwan, Singapore, and South Korea, during the period from the 1970s to 1997, and was a collective product of writings by social scientists, the World Bank, and the media, in their attempts to forge general explanations for high-growth economic conditions in Asia.[6] These four Asian economies, variously termed the NIEs (newly industrialized economies), the NICs (newly industrialized countries), the Four Tigers, or the Four Dragons, became in certain ways models for China's export-oriented economy.[7] The miracle position has been intensively debated, and was only fundamentally sidelined by the events of the so-called Asian financial crisis that set off in Thailand in 1997. Proponents of the miracle position have attributed rapid regional growth in Asia to a set of economic policies associated with the neoliberal regime, including privatization and free market policies, and a diminished role of the state. But Robert Wade (1990, 1993a, b) and other regional specialists have cautioned against the totalizing quality of the miracle position, which has not considered differences among Asian countries, and the fact that across Asia the state has actively intervened in economic planning and articulated industrial policy. Nevertheless, part economic theory, part ideological platform, perspectives derived from the neoliberal regime, especially in its US-based worldview, have been powerful determinants of which research topics scholars privilege and what conclusions they find.

The miracle account would also understand China's decision to open to the world economy as evidence of the failures of communism, the global success of Western economic systems, and a vindication of the entire Cold War project. But the leading Asian account of China's policy shift reflects Chinese leaders' recognition of rapid development on non-Western terms in the NIEs, and while on initial state visits in the region after 1976 (Shirk, 1994; Yabuki, 1995). The four NIEs share with China a Confucian cultural base, and Taiwan, Hong Kong, and Singapore are all majority Chinese populations. In China's enduring historic perspective, Hong Kong, Taiwan, and Singapore are peripheral islands of Chinese immigrants that developed well beyond the conditions of the motherland, a frank upset to the ideological remnants of the Chinese world order. In reform planning, China adapted policies of the NIEs and invigorated these historic connections to tap capital flows for SEZ development. Thus the Asian account of China's economic transformation understands a regional cultural economy in which establishing the SEZs was one event in the articulation of a regionally based "Confucian capitalism," an alternative, albeit in many ways a discursive one, to the hegemony of Western forms. At this juncture we would do well to keep in mind Arjun Appadurai's (2000, p. 13) recognition that "actors in different regions now have elaborate interests and capabilities in constructing world pictures whose very interaction affects global processes. Thus the world may consist of regions (seen processurally), but regions also imagine their own worlds."

Area studies debates

The disjunctures suggested by these different methodological approaches have also played out in Asian area studies fields. By contrast to economistic approaches, organizations of knowledge generated in area studies arenas have more dependably maintained historical and cultural perspectives. They have also maintained greater distance from prevailing theory in disciplinary fields, and, in the need to evolve culturally appropriate approaches, have tended to question methods derived from the Western societal experience. In China area studies, research on the arrival of the West, especially the period of the nineteenth century when European powers forced China into a "semi-colonial" status, necessarily pierced the nation-state boundaries of the area studies paradigm and engendered a first significant wave of transnational research. The historiography of this period has stirred some of the field's most contentious debates. The following sections assess the area studies debates through changes in the China field to establish how the globalization of knowledge is pressing scholars to rethink their approaches to international research.

China's "response to the West"

For China scholars concerned with the relationship between China and the West, John King Fairbank's account of foreign trade under the treaties established a research paradigm that would endure for two decades after its initial appearance in the 1950s: what was the nature of China's "response to the West" (see Teng and Fairbank, 1954)? This theme influenced many scholars in the middle decades of the twentieth century, including Rhoads Murphey, who was among the final doctoral students in geography at Harvard University.[8] Murphey (1953, 1970) focused on the conditions of Shanghai and general questions about the roles and impacts of the "treaty ports" in China. The response to the West perspective treated open ports as nodes of Western practices, and assessed how Chinese society reacted to Western forms of knowledge, economic activities, technology applications, and religious beliefs. Based on such perspectives, the treaty port appeared to have begun its existence in the nineteenth century as a unique type of city. As Paul Cohen (1984, p. 9) evaluated the "response to the West" perspective, "This conceptual framework rested on the assumption that, for much of the nineteenth century, the confrontation with the West was the most significant influence on events in China." In the early 1970s China scholars began to openly question these Western-oriented approaches, and Joseph Esherick (1972) challenged the Fairbank school for its tendency to construe China as a nation that reacted to Western policies and institutions. Esherick renamed this preoccupation with China's response to the West the "impact-response" school of Chinese historiography. Against this backdrop, Cohen (1984) called for a "China-centered" view of Chinese history. Partly as a result of these shifts in historiographic method, China scholarship has moved toward localized studies of social and economic history.[9] These debates, however, did not foreground the complexities in China area studies scholarship engendered by China's opening to the world economy.

In the face of rapid regional development and similarly intensive needs to account for it, few social scientists have paused to consider larger epistemological questions about *how* to understand Asian political economy. In other terms, what Edward Said's *Orientalism* achieved for the humanities, in demonstrating how Western writers constructed partial and problematic ideas about Asia, has not substantially influenced social scientific analysis. One alternative reading has emerged from André Gunder Frank (1998) in a new account of world economic history. Contrary to widely accepted views of European economic hegemony from the time of the Renaissance to the middle of the twentieth century, Frank has argued that the Asian region, with China at its center, really dominated the evolving world economy until less than two centuries ago. By this account, only in the early nineteenth century did China cede central world economic position to Europe and the West. The gross fallacy in scholarly analysis that led to the misunderstanding of Asia's position in world history, according to Frank, has been widespread dependence on Western theory and philosophy. Scholarly analysis based in Western

European social thought, from all points on the political spectrum – the "Marx–Weber" complex – has constructed a hegemonic world view of Western exceptionalism, based on assumed superiority of "rationality, institutions, entrepreneurship, technology, geniality, in a word – race" (Frank, 1998, pp. 4, 20). Frank's position underscores the socially constructed bias obtainable in Western accounts, in which race as "whiteness" is practically synonymous with Western society, however rarely the subject is foregrounded. According to this argument, Europe's early modern position in the world economy is better understood as having tapped into accumulation strategies of existing Asian markets and trade, rather than having developed them. On these terms, the rise of Asian regional economic power in the late twentieth century represents a return to historical conditions rather than a sea change.

The critiques mounted by both Cohen and Frank share a certain antipathy toward Western political and economic theory. Yet as Arif Dirlik (1996) has pointed out, the complication of pledging allegiance to a China-centered approach has denied the significance of approaches originating in the West and appropriately applied or adapted in the Chinese and larger regional contexts. For example, the concept of modernization has taken on new contexts and meanings as deployed by the state and the intelligentsia in China and the NIEs. Rapid economic growth in Asia partially undid the Western teleological narrative of modernization theory, and led to ideas about different cultures of capitalism, Chinese capitalism, and Confucian capitalism in "Greater China." While some China scholars called for an approbation of the application of Western models and Western world views to China, their views, once appropriate in leading the paradigm shift from imperialist orientations to less ethnocentric scholarship, have been in part swamped by forces of intellectual globalization, in which the flow of ideas represents not one point of origin, linear flow, and singular interpretation, but mutual influence, and new confidences about reinterpretations to suit specific cultural and regional circumstances. In the contemporary context, Chinese adaptation of once Western models has been the very result of the economic reforms that propelled China into the world economy after 1978.

Restructuring area studies

In its reflection of the Cold War world, area studies research served to organize knowledge about countries and continents for international security analysis. On the other hand, area studies has been the professional arena of internationalists who have eschewed Western models and pursued area studies on culturally appropriate terms. This latter point creates its own problems on the intellectual high ground of theory. Bruce Cumings (1997b, p. 8), an active critic of the area studies debates, wrote about Asian area studies as "an opera-bouffe which goes as follows: China (or Japan, etc.) is an 'area'. Area studies, as we all know, are a-theoretical. Therefore 'area studies' should be abolished – except for that 'area' known as America, which is far too idiosyncratic and complex to yield to abstract theory, and, of course, no foreigner can really understand it either."[10] The conventions of area studies practice have tended to keep fields circumscribed, organized around Asia, Africa, Latin America, and so on, and, in distancing from theoretical approaches, distinct from the disciplines and scholarly debates over globalization. By contrast, scholars based in disciplinary perspectives have tended to emphasize current method and theory over complex area analysis. Similarly, theorizing globalization has arguably been dominated by scholars who do not reliably maintain comparative area expertise (see Featherstone, 1990; Jameson and Miyoshi, 1998). The area studies–theory divide is also reflected in epistemological divisions between national and transnational positions. The organization of area studies research in the national "container" of the nation-state prevailed through most of the twentieth century. As Vincente Rafael (1994, p. 91) has written, "by privileging the nation-state as the elementary unit of analysis, area studies conceives 'areas' as if they were the natural – or at least the historically necessary – formations for the containment of differences within and between cultures." The contemporary emergence of research on transnational processes has broken down the traditional spatial biases of area studies, but has not dependably bridged divides between national and transnational approaches.

In addition to the area studies–theory divide, area studies has also been plagued by the epistemological separation of classical from modern research fields, so that many area specialists divide regional history into distinct "classical" and "modern" periods of study. A related topical division is the separation of cultural from economic subjects. Such historical divisions have the effect of organizing academic work by assigning the classical period and cultural subjects to the humanities, and the modern era and economic subjects to the social sciences. This observation holds true for the majority of economic analyses of China's reform era, which regularly do not consider cultural contexts, or relevant events of the Maoist era, the rest of the twentieth century, the Qing dynasty, or any other period. This particular (dis)organization of knowledge reflects the problems of the methodological divide between the humanities and social sciences in the academy, arguably undergirded by the legacy of modernization theory (Rostow, 1960). Modernization theory proposed a linear trajectory of societal organization from traditional to modern stages of evolution, and based on the territorial unit of the nation-state and the experience of the industrialized West. In applications of modernization theory, and in addition to the problems of Western ethnocentrism and imperialism embedded in the model, the evolutionary stage perspective tended to be rendered dualistically, which divided the Chinese past and Chinese historiography into a traditional era before Western contact, and a modern era of significant contact with the West.

In the United States, the Social Science Research Council (SSRC) and American Council of Learned Societies (ACLS) hold power to influence research directions in area studies scholarship. By the later 1990s, SSRC committees began advocating funding research on globalization, local–global relations, and how such work should move beyond "existing political boundaries, which limit how problems or questions should be framed" (Abraham and Kassimir, 1997, p. 24). While such new directions in area studies emphasize transnational and transboundary issues, the shifts may still be read in the context of dominant US political interests. As Cumings (1997a, p. 9) would see it, the SSRC and ACLS initiated this restructuring only when the world focus of political economic power shifted after the end of the Cold War. Cumings points out that expectations of area studies experts have shifted from subjects like analyses of Communist political strategies to "informed judgements on 'Chinese economic reforms'," which leads straight back to where we started, with special economic zones.

China area studies and the "macroregion"

The conventions of area studies practice left a legacy of research frameworks whose dualistic epistemologies, in Western impacts and local responses, tradition and modernity, culture and economy, challenged scholars to evolve more complex approaches. In the search for alternatives, one model China scholars widely embraced is a regional approach called the "macroregion." The macroregion model is based on a geography of watersheds, and was derived from location theory and regional systems theory, which were popular traditional methods in economic geography and related fields in middle of the twentieth century. After the 1970s, location theory and regional science lost influence in geographical analysis, but the macroregion continued to be used in China area studies without substantial modification or replacement (Cartier, 2002). Most regional research in China area studies has used the macroregion concept, but with time, increasingly less as an analytical tool and more as a locational device (cf. e.g. Schoppa, 1982; Rowe, 1984,

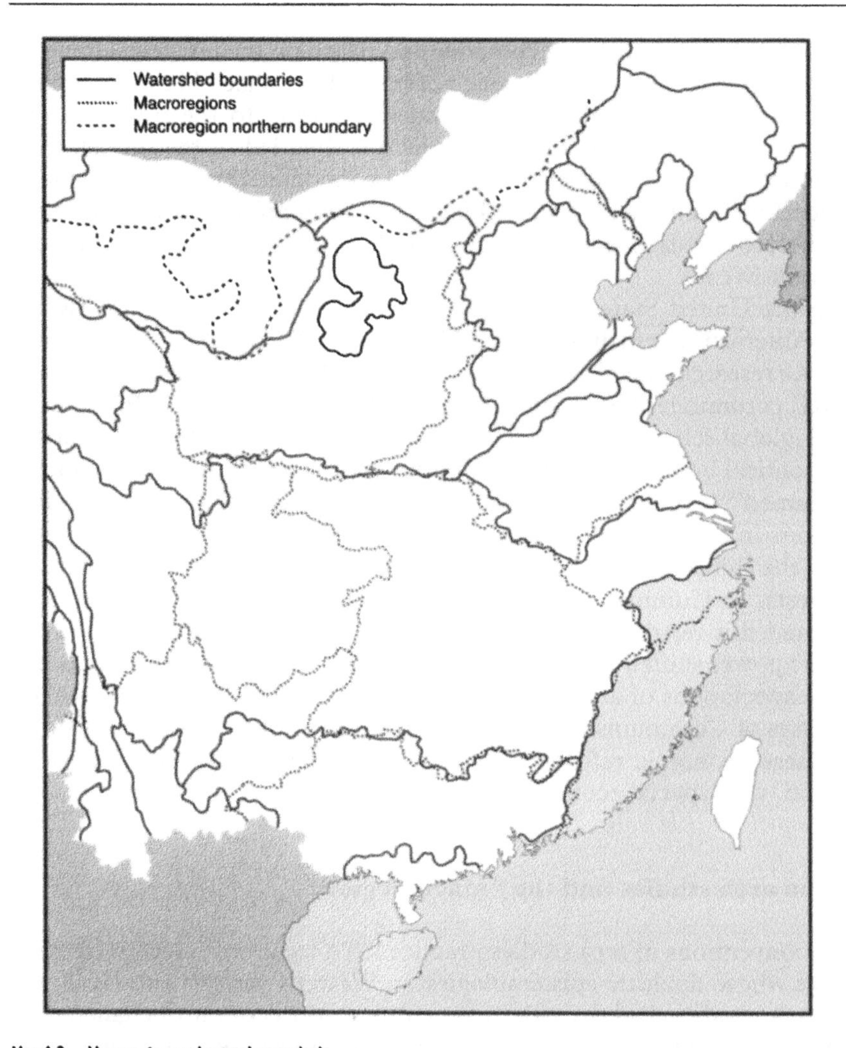

Map 1.2 Macroregions and natural watersheds.
Source: Skinner (1977b) and The Conservation Atlas of China (1990); line work by Jane Sinclair.

pp. 8–9; Naquin and Rawski, 1987; Esherick and Rankin, 1990, pp. 17–19; Spence, 1990, pp. 91–3; Dean, 1993, pp. 21–3; Leong, 1997, p. 19; Wigen, 1999, p. 1185). Continued reference to the macroregion model in the face of the decline of location theory in geography must reflect a lack of engagement between China area studies and advances in geography, and the area studies–disciplinary divide in the academy. Diverse approaches in regional analysis have characterized geography since the era of regional systems theory but have not appeared in China area studies. The following discussion summarizes some of the critical issues about the macroregion approach and its derivation from methods in disciplinary geography.[11]

From the 1970s and through the 1990s, most urban and regional research on late imperial China was influenced by the work of G. William Skinner and two approaches he evolved for studying the urban hierarchy and regional economies in China: the "macroregion", and its derivative framework, the "marketing systems" model. The marketing systems model was based on central place theory, a type of location theory, and a study of marketing towns in Sichuan province.[12] Central place theory accounts for the size and distribution of settlements within an urban system, and, in his assessment of Sichuan, Skinner (1964, 1965a, b) found the pattern of towns to represent the classic hexagonal pattern of the central place model. The macroregion, a neologism, was the more popular of the two approaches, and combined the idea of a system of marketing towns with concepts from core–periphery models and regional systems theory. Based on recognition of regional variations in economy and urbanization rates in Han China, Skinner (1977a, b) posited the existence of nine macroregions whose areas corresponded to drainage basins or watersheds (map 1.2). Skinner promoted general use of the models, and his prominence in the field of China area studies lent considerably to their popularity among area studies scholars.

Skinner's focus on marketing towns distinguished the existence of important settlements in productive agricultural areas, and the origins of the towns in increasing local diversification of the agricultural economy. This insight demonstrated how local level settlements that arose from the agricultural economy were integrated into the larger Chinese urban system but, importantly, that their existence was fundamentally economic, the result of local retail economies, and not owed to the establishment of an administrative center by the imperial order. This realization, while apparently a relatively simple contrast, helped to lead the move away from the dominance of imperial history from the perspective of the capital to locally and regionally based social and economic studies of Chinese society. Understanding an integrated system of economic settlements also helped to break down the rigid stereotypical assumptions about social life in China being divided into rural and urban realms, and notions that the rise of a system of economic settlements must be tied to the development of industrial capitalism.[13]

Skinner readily borrowed from the geographical literature to develop the models, but did not heed the many critical analyses of their limitations. He also did not present a data analysis of marketing towns in Sichuan, but nevertheless found there the classical hexagonal geometry of settlement distribution predicted by central place theory. Because marketing town distribution differs empirically in shape and arrangement, Richard Szymanski and John Agnew (1981, p. 39) critically evaluated the application of central place theory to Sichuan province as "diagrammatic and without mathematical basis." The macroregion was also presented as a theoretical model and was never systematically applied to the nine macroregions of Han China, with the exception of the southeast coast macroregion (Skinner, 1985). This was a problematic approach to research design, however, since location theory does not theorize long distance trade, which has been the basis of the regional economy on the south coast. Similarly problematic, location theory is inherently ahistorical, and does not conceptualize the origins and evolution of social and economic activity (see Smith, 1989). Location theories are based on principles of neoclassical economics, in least cost locations as a function of transportation, and have their origins in economic landscapes of industrializing Europe, which raises questions about their suitably for a "China-centered" approach.

In his application of the macroregion model to the southeast coast, Skinner (1985) also historicized the model through the concept of cycles of regional development, a related systems model. The historical analysis for the southeast coast defined the level of economic activity to be high during the Yuan dynasty of Mongol rule when the port of Quanzhou, just inland from Xiamen in Fujian, functioned as the center of lucrative long distance trade. After this period, Skinner held that the coastal economy fell into precipitous decline as a result of imperial trade proscriptions during the early Ming period. The arrival of the Portuguese and Spanish to the coast of China in the sixteenth century led an economic resurgence in the region from about 1520 to 1640, explained Skinner, but another round of trade prohibitions from the middle of the seventeenth to the middle of the eighteenth centuries plunged the regional economy into decline once again. About the coastal economy at the end of the imperial period, Skinner (1977a, p. 279) concluded, "This dark age for the Southeast Coast was ended only in the 1840s, when Fuzhou and Xiamen were opened as treaty ports."

Skinner explained growth and decline of regional trade activity from the perspective of imperial trade policy, based on accounts of the region from the perspective of the capital. Local histories, by contrast, widely noted how mariners regularly circumvented trade bans. Based on analysis local and official histories, Ng Chin-keong's (1983, p. 53) definitive study of the maritime economy of the south China coast from 1683 to 1735 demonstrated that imperial trade bans did not end the lucrative coastal trade, and "moreover, the more restrictive the law was, the more lucrative the trade became." Dian Murray (1987, p. 10), in her work on coastal piracy, observed "the junk trade was at its height during the late eighteenth and early nineteenth centuries, when Chinese merchants monopolized the exchange of 'Straits produce' from Southeast Asia – rattan, seaweed, and pepper – and thwarted European attempts to supply these goods to China." Sarasin Viraphol (1977, p. 7), writing about the long distance trade with Siam, concluded that the junk trade was at its height at the end of the eighteenth century and during the early decades of the nineteenth century. After exhaustive study of the tea trade on the southeast coast, Robert Gardella (1994, p. 33) concluded that "Skinner greatly overstates the case of an alledged economic depression along the littoral in the early to mid-Qing era, and neglects the impact of interregional and international trade upon the interior or peripheral zones of Southeast China over the same span of time." Indeed, the macroregion analysis did not distinguish between the activities of Western mercantilists and the regionally important junk trade with Southeast Asia, Taiwan, the Liuqius, and Japan. By attributing the return of trade to Western mercantile powers, Skinner also foregrounded the significance of the treaty ports and the tradition and modernity dualisms that characterized the so-called impact-response school of Chinese historiography.

The macroregion approach reflected basic problems in both traditional economic geography and traditional regional geography. Traditional regional geography assessed diverse regional characteristics in the context of bounded political regions or physical regions. Physical regions include drainage basins or watersheds, and regional differentiation based on watersheds has been a particularly common approach in China, where major rivers effectively act as latitudinal divides. But traditional regional geography largely described patterns of regional phenomena, and identified regions as if they are "natural" and "out there" rather than socially produced (Pudup, 1988). Among a range of limitations, absent from traditional regional geography and traditional economic geography were concerns about causal processes and transboundary activities. China is a country of diverse, contending, and ultimately coherent regions, yet adherence to the macroregion approach maintained focus on patterns of economic activities and settlement distribution in the context of prescribed regions. As Martin Heijdra (1995, p. 31) has observed, the macroregion and marketing systems models have been "adopted as a whole by historians with only the most minimal revisions, to such a point that direct spatial investigations of phenomena are excluded from most current research."[14] Reduced to a mapped location, the macroregion is the physical region of a watershed and another "research container."

Transboundary Cultural Economy

By comparison to geography's traditional concern with spatial patterns, theoretical invigorations of human geography in the late twentieth century advanced analysis of geographical processes: historicized processes that produce patterns of human activity (over descriptions of patterned activity); factors of scale in how local, regional, national, and global processes are mutually influential and transformative (as opposed to static conditions captured at singular scales); aspects of human agency and difference among and within culture groups (as opposed to the ascription of human impacts to generalized characteristics of principal culture groups). The concern with process prioritizes examination of the conditions of geographical formation and transformation and dynamic events in space–time relations. The emphasis on scaled processes breaks open the conventional boundaries of nation-states and political administrative geographies of provinces and planning regions to examine questions about transboundary processes and their causal roles in regional formation. The idea of transboundary processes also reflects caution around the "transnational," in the way that the term explicitly codes the national scale. A transboundary perspective, while subject to being read as a reification of bounded space, instead recognizes the existence of boundaries and bounded territorial space, the processes that have given rise to them, the many processes that transcend them, and, critically, how dynamic regional processes are commonly based in transboundary activities. It also signals the possibility of different scales of activity and still recognizes the embeddedness of regions in political geographies. In south China, the transboundary perspective elides questions over national sovereignty between China and Taiwan.

The transboundary region in south China is also a regional cultural economy. In economic geography, the idea of a regional cultural economy has emerged from a rethinking of political economy so that it "must employ cultural terms like symbol, imaginary, and rationality if it is to understand crucial economic processes such as commodification, industrialization, and development" (Peet, 2000, p. 1215). The regional cultural economy also recognizes, after Appadurai (2000, p. 13), how "regions also imagine their own worlds." In the case of south China, alternative regional discursive projects, about rapid economic growth or located cultural complexes, have left us with partial and often ideological accounts of regional processes. This transboundary region is simultaneously China's maritime frontier, the hearth of the Chinese diaspora, a zone of miracle development, a place of global-relative poverty, as evidenced by the rise of "container migration" at the turn of the millennium, and anchored by China's two largest and most internationalized cities, Shanghai and Hong Kong. The legacy of the culture–economy split in social thought, and the parallel problem in traditional regional geography, about systematic or functional regions, based on economic activity, versus unique regions, based on distinctions of local culture and ways of life, has divided these topics (see Entrikin, 1991). Thinking about the transboundary regional economy works to overcome such divides and compels a central perspective for analysis, by bringing together cultural and economic subjects and concerns about both the regional conditions of place and the spatial processes of regional economy. In the regional cultural economy, "economic imagination derives from the cultural history of a people" (Peet, 1997, p. 38), which means that regional imaginations take social forms and contribute to the invention of particular economic strategies. These imaginations and social forms need to be theoretically emplaced.

The transboundary cultural economy insists on a cultural economy perspective to establish a regionality that interlinks culture, as a basis of economic organization, with transboundary economic activity. This concept of region is a mediating spatiality, which both supports and fundamentally questions tendencies of the processes of globalization and their historic forms. This alternative conceptualization borrows ideas from the *Rethinking the Region* project, in which John Allen, Doreen Massey, Alan Cochrane (1998), and others have assessed England's southeast, the Greater London Metropolitan Region, as an actual and discursive region produced under the neoliberal regime inherited from the Thatcher government. *Rethinking the Region* examines a "growth region" by focusing on the characteristics of growth mechanisms, but it does not measure or map growth. Instead the authors recognize how a high growth region is also a "discontinuous region" of locally uneven development. In the discontinuous region spatial disjunctures in patterns of characteristics necessarily emerge, rather than being hidden in only hypothetically homogeneous space. The discontinuous region is both a reality and a methodological strategy which signals how places and localized economies within regions may differ and still intersect to constitute a known regional entity. The analysis understands place and region as constituted out of spatialized social relations, and questions and narratives about them, which then serve as a basis for rewriting regional geographies, and, in turn, contribute to reshaping identities and how they are represented. The study also acknowledges that there can be "no complete 'portrait of a region'" (Allen et al., 1998, p. 2), which compels us to face how studies of places and regions are undertaken for particular purposes – "whether theoretical, political, cultural" – and to acknowledge what those purposes are.

Where literatures of geography and globalization intersect, Neil Smith (1997, p 182) has interpreted globalization as an ideological growth complex, and finds in globalization the reinvention of modernization theory, "the latest stage of uneven development," and "an increasingly pure form of imperialism." Smith's perspective also suggests the difficulties of writing about globalization and its spatialities, because to yield to alternative and "local" readings of globalization would appear to back away from rigorous critical analysis of the problems of neoliberalism. Yet that is precisely the opportunity to engage, the conceptual open space between the realities of the neoliberal regime as the contemporary paradigm of would-be global development, and combinations of cultural, political, economic – and historical – ideas and events whose causal processes are resulting in new regionalities, which simultaneously support and deny globalizing processes. It is important for particular reasons to recognize the significance and effects of economic growth and the ideological complexes which support growth, but unlike most of the scholarship about south China, the present project does not focus on the nature and causes of growth – because those discourses, in turn, however unwillingly or unwittingly, continue to privilege growth. As J. K. Gibson-Graham (1996) has argued, constant privileging of capitalism's hegemonic characteristics, like the growth imperative, results in marginalizing or making invisible a range of important economic subjects. Globalization's effects are highly uneven and so its material geographies must be located, place-based, and regional. Regionality in this sense is a concept for alternative dialogues about would-be global processes.

Region, Place/Space, and Scale

Contemporary ideas about regions have coalesced around several themes: the emergence of the region as the geographical sphere most suited to framing interactions of complex social processes in an era of globalization, the cultural conditions of economic regions, spatial unfixing of regions and regional identities, the significance of interrogating regional representations, and how understanding scale relations contributes to the possibilities of all this (Cartier, 2001b). Regions, whether administrative and bounded like Guangdong province, or imaginary and unmappable by conventional means, like the transboundary economic region at the basis of Greater China, are social and political constructions, and exhibit, and are products of, scale relations. Regional formations are also constituted through emplaced cultural practices. This section develops perspectives on the connections between region, place, space, and scale in order to think through how to emplace the regional cultural economy, and as a basis for identity formation.

The concept of place must inform contemporary understandings of regional formations, yet different intellectual lineages have characterized the scholarship about place. Differences between a social theoretical treatment of place and a phenomenological view of place may be reasonably bridged if we understand place through contemporary theorizations in both geographical and philosophical literatures. Alan Pred (1984, p. 280; 1986, pp. 6–7) theorized place "as historically contingent process that emphasizes institutional and individual practices as well as the structural features with which those practices are interwoven." Pred's theorization foregrounded the importance of human agency in the formation of place, and distinctively, in ideas about the importance of biography and in the limits of contextualized power relations. Doreen Massey's (1994, 1995) concept of place as the "spatial reach of social relations" has focused on the social relations of production that influence the experience of place and create the conditions in which places form and are embedded. These perspectives on social relations necessarily concern scaled social processses, from local to regional, national, and global, so that place can no longer be thought of in its traditional sense as local, bounded, and fixed in character. Rather, place must be dynamic, contested, and multiple in its representative identity positions, as places are buffeted by political and economic events which are also negotiated and resisted through activities of local agents (Massey, 1994, p. 5). Both Pred and Massey have distinctively treated place formation as bound up in multiple social processes.

Recent philosophical treatments of place – as the ontological basis of human existence – in the work of Jeffrey Malpas (1999) and Ed Casey (1996, 1997) have further established the significance of the concept of place in humanistic fields of inquiry. Malpas's (1999, p. 176) emphasis on an ontological inquiry into place finds identity formation and its basis in human subjectivity as "necessarily embedded in place, and in spatalised, embodied activity." This view of place depends on understanding an interplay of interconnected concepts, including agency, spatiality, and experience, in which embodiment is "one's extended, differentiated location in space . . . [and] essential to the possibility of agency and so to experience and thought" (Malpas, p. 133). In other words, to be embodied is to be emplaced. The embeddedness of subjectivity in place finds expression in memory and narrative, which serve to structure representations of place and identity (see also Schama, 1995). These relations are necessarily historically constituted: "To have a sense of the past is always, then, to have a sense of the way in which present and future conditions are embedded within a complex 'history' that is articulated only with respect to particular individuals and concrete objects as they interact within specific spaces and with respect to particular locations" (Malpas, p. 180). Malpas's view then, even as it is tied to a phenomenological legacy, opens up to a material grounding, and, in his own words, "need not be viewed as incompatible with other projects that attempt to fill out more particular, especially socio-cultural, features of our relation to place" (Malpas, p. 197).

Casey's (1997, p. 239) philosophy of place traces the concept of place in the history of Western philosophy and ultimately depends on understanding the body as "the very vehicle of emplacement." This recognition of the situatedness of the body resonates with theorizations of scale and subjectivity in feminist geography (Rose, 1993; McDowell, 1999). Conceptualizing the relation between place and body establishes a critical distinction between place and region, which serves as a corrective to earlier work that has simply treated region as larger-scale place, in apparent detours around the problems of minimally developed concepts in regional geography (see Entrikin, 1991; Johnston, 1991). Casey's (1996, p. 46) view of place also depends on place relationality (rather than place uniqueness): "It is undeniable that the concreteness of place has its own mode of abstractness: that is, in its relationality (there is never a *single* place existing in utter isolation) and in its inherent regionality (whereby a plurality of places are grouped together)." In Casey's terms, regions "affiliate" places, which is not a definition of geographical contiguity but rather a recognition of the interplay of diverse and sometimes disparate actual places in regional formation. Thinking about place as complex social relations and through embodied subjectivity yields approaches for understanding the emplaced contexts of social relations in a transboundary regional economy.

Henri Lefebvre's writings on space provide important approaches to understanding the relationship between the concepts of space and place, and, specifically, how in the course of capitalist development space has been abstracted from place (Lefebvre, 1991). The Lefebvrian project gives more emphasis to political and economic processes, and inherits its concern for a historical and dialectical approach from Hegelian dialectics, by contrast to the Heideggerian legacy in the work of Malpas and Casey. Lefebvre set forth a "conceptual triad" to account for the production of space, or all the ways in which social processes carve out particular kinds of spaces through spatial practices (perceived), representations of space (conceived), and representational space (lived) or spaces of representation (after Stewart, 1995, p. 610). Spatial practices are the located and embodied human activities of production and reproduction and ritualized activities of daily life, which are characteristic of particular societies. Spaces of representation are lived spaces of cultural systems, produced through common use and practice. Representations of space, by contrast, are conceived and abstract spaces, typically born of the activities of modern planning and land ownership. Lefebvre located the difference between actual lived space and conceptualized space at the heart of understanding how societies appropriate space from place and produce new spatial forms. Through these concepts, and different categories of space (e.g. abstract space, social space, absolute space), Lefebvre evolved a dialectical understanding of the relationship between space-in-the-abstract and the cultural embeddedness of place. This Lefebvrian matrix allows a critical negotiation of the place/space dilemma, which goes far to overcome the problems of dualistic approaches (Merrifield, 1993) and other epistemological divides. The Lefebvrian approach also shares common ground with the recent philosophical treatments in its concern for embodied spatial practices at the basis of place-based experiences, and, with Massey, in concepts of scale.

Scaling social processes

In a discussion about scale, John Agnew (1994; see also 1989, 1993) proposed the idea of the "territorial trap" to account for the ways in which the hegemonic view of the nation-state, as the most common international geographical construct, has historically denied the use of other spatial scales in political economic analysis. The nation-state remains the central organizational principle of the world system, yet the emergence of discussion around scale has arisen in response to new spatialities engendered by processes of globalization. In revisiting the "territorial trap" thesis, Agnew (1999, p. 190) concluded that contemporary economic processes make analysis based on a nation-state territoriality less compelling: "today . . . development is increasingly a process determined by the relative ability of localities and regions within states to organize access to global networks." In this reassessment, the significance of diverse scales emerges around regional formations, and the role of places and regions in articulating globalizing processes in the context of nation-state territoriality. To contextualize these issues, the first part of this section focuses on administrative scale as the result of state-making strategies.[15] The second part of the discussion concerns theoretical perspectives on non-administrative scale relations in order to establish a basis for framing complex processes of *mobility* in the transboundary cultural economy.

In *The Production of Space*, Lefebvre also raised questions about dynamic qualities of state formation, and how the state stabilizes and continually adjusts political-territorial scale – global, national, regional, and local – as an accumulation strategy (Brenner, 1997, 2000). These ideas inform an understanding of China's long-term territorial coherence. The scaled administrative system of imperial China after the Ming dynasty – empire, province, city, county, and town – worked to stabilize the regions and knit the empire into a coherent whole. Stability, though, was not a static geography of bounded territories but an active and highly managed basis for the empire's massive accumulation strategies, especially through taxation, in cash and in kind, as officials funneled grain tribute from administrative territories to the capital. This spatial process engendered the commitment of officials in part because it mirrored their own desires: the imperial system of official examinations was also a system of scaled opportunities, a tiered set of examinations whose sequential passage earned appointments at correspondingly higher levels of scale in the urban hierarchy (Ho, 1962). Passage of the highest level examination, the metropolitan degree, won appointment to imperial bureaucracies in the capital. The millennial longevity of the Chinese empire is arguably the result of brilliant scale strategies, in the replication and reenactment of imperial power at all levels of administrative scale.

Changes in administrative scale strategies regularly accompany changes in political regimes. In the Maoist era, the state set up the *hukou* (permanent household registration) system to maintain societal control and prevent rural to urban migration by defining citizenship on the basis of either rural or urban residence. By separating rural and urban spheres, the *hukou* system created a spatial hierarchy of settlement status in a pyramidal order, from villages at the bottom to major cities at the top. Each higher level of settlement represented greater opportunities for state benefits, including school entry, job opportunities, and more (Mallee, 1996, pp. 4–5). This surveillance regime maintained control over the movement of individual lifepaths by tying people to their place of registration, and disallowing people from moving from rural to urban areas and up the urban hierarchy (see Cheng and Selden, 1994). It stabilized agricultural productivity and minimized demands on urban services. In China under reform the most stringent elements of the *hukou* system have given way, as the state effectively allowed millions of people to leave farm work, establish non-farm enterprises, and migrate for jobs. But the state did not ultimately yield control over administrative scale in loosening up the *hukou* system; it made administrative scale more porous and used it instead to propel surplus rural labor into new economic activities.

In many ways the spatiality of reform has been about the state loosening up control over administrative scale. Decentralization of power to the provinces under reform, and especially fiscal decentralization, whereby most provinces have been allowed to keep a greater portion of their own revenue, has challenged central government powers.[16] Yet arguably what is different is not that the center has lost power *per se*, but that the state has allowed greater flexibility between levels of administrative scale, and simultaneously opened up the national scale to allow greater intersections at all levels with global economic activity.[17] One reform era policy, *chexian gaishi* (abolishing counties and establishing cities), "scaled up" counties by making them higher level cities, which directly facilitated economic growth because cities have independent power to contract larger foreign investment projects and decide land use transformations (Zhang and Zhao, 1998). What has not changed is that these policies are also accumulation strategies, new articulations of scale and administrative geography adjusted by the state to increase production and promote economic growth in China under reform.

Globalizing processes have arguably pried loose the fixity of administrative scales, which in turn, and as reinterpreted by the state, have become more permeable and, in some cases, resilient. Globalization's appearances in new spatialities have also made us more aware of the importance of non-administrative scale configurations. The complexities of such new scale articulations do not necessarily correspond to bounded territorial regions and so cannot be disciplined by the map. Before we explore some theoretical issues, consider that one important aspect of non-administrative scale complexity is human *perspective* on relations of scale. From some distance, at a large scale, mapping coastal south China might seem a reasonable project. But the reality is that on the ground within this non-administrative region, the lie of a regional boundary line cannot be fixed. This reality about scale, that at larger scales, distinctions of human phenomena in space are more easily categorized, patterned and delineated – at the expense of difference – lets us understand how regions are constructed, and that regions are not objective, clearly defined spaces (until apparently we put a boundary on a map – another construction!), but rather represent belief in the homogeneity of a particular region. The lesson to be drawn is that insisting on beginning from any fixed scale is deeply antithetical to understanding dynamic processes of human realities. From a theoretical perspective, scaling social processes must allow any number of scales, different and changing scales – even simultaneously existing different types of scale – and flexible conceptualization defined by the social processes at stake. Because scaled activities and cultural imaginaries evolve as a result of diverse human processes over time, a scale framework should be viewed not as concretely fixed in a transhistorical sense but as changing to reflect transformations in social processes and human-environmental conditions. Yet thinking about scales alone yields a spatial metaphor of a tiered layer cake, or, in another common object metaphor, a set of nested Russian dolls. More realistically, levels of scale, as spatial imaginary, are discontinuous and dip and merge at discrete and changing points of contact, connected by activities of diverse agents, as people and institutions. Social processes articulate and move through scale relations. As Erik Swyngedouw (1997a, p. 169) has explained, "the theoretical and political priority, therefore, never resides in a particular geographical scale, but rather in the process through which particular scales become (re)constituted."

Scaling the region necessarily implicates dynamic interscalar relations, and grasping how social processes work through, transcend, and recreate scale. This issue is close to the conceptual tension about theorizing between the forces of globalization and the production of local and regional difference. What this tension speaks to is the real absence of linear causality characterizing social processes and instead reflexive and recursive aspects of human activity in spatial relations. The philosophical approach of dialectics has been useful for conceptualizing mutually constitutive scale relations and explaining dynamic interrelations among scaled processes (Howitt, 1993; Merrifield, 1993). David Harvey's (1996) treatment of the dialectical approach emphasizes its ability to question relations between subjects and objects, causes and effects, and understand relationships in constant and reflexive process, "entailing multiple changes of scale, perspective [and] orientation" (Harvey, 1996, p. 58). Processes in dialectical relations contain ripe sites for change, as transformation arises out of the tensions characteristic of complex interrelationships among multiple and intertwined activities, institutions, people, and events, but, importantly, often in different places and implicating different scales. In the logic of scale configurations, the transboundary cultural economy is a region of multiple and convergent trans-scale processes in dialectical relations. The transboundary cultural economy finds its historic embeddedness in diasporic journeys, through material processes of travel, migration, and capital relocation, which have linked places in south China across the world. The spatial relations of diaspora have been the context for regional and subethnic identity formation, in which social processes simultaneously transcend scales and converge in others, tying the threads of common experience among people in mobile and cosmopolitan communities with their overseas counterparts. Formed in processes of high mobility and ties to diverse places, people of south China and the Chinese overseas may have multiple and hybrid identities which are suited to particular places and events experienced along the diasporic lifepath (Wang, 1988; White and Cheng, 1993). In the context of identity formation, dialectical scale relations may be thought of as a set of "translocal" (Clifford, 1997) processes for the ways in which the translocal are those multiple places of attachment experienced by highly mobile people.[18]

The significance of place in the Chinese cultural imagination is an important transhistorical force of geographical orientation and individual and group identity formation. As Mingming Wang (1995, p. 33) has written, ideas about "place (*difang*) are intrinsic to the Chinese formation of social space and . . . ways of being in society." In China the pivot of identity formation has historically evolved in cultural conceptions of place and region, and also in scaled relations, from the village or town or city to the province and nation. Historian Bryna Goodman's study of native place sentiment and *huiguan* (native place associations) in Shanghai sets forth the located context of identity formation:

> For migrants who sought a living in Shanghai, native-place identity expressed both spiritual linkage to the place where their ancestors were buried and living ties to family members and community. These ties were most frequently economic as well as sentimental, for local communities assisted and sponsored individual sojourners, viewing them as economic investments for the community. (Goodman, 1995b, p. 5)

In this passage, place identity appears as a geographical imagination, a sentiment reenacted by diverse travelers. Place here is also scaled, from a home site of ancestral ties to community, county, city, and province, since *huiguan* commonly represented provincial, prefectural, or county-level associations. The second sentence of the quotation, in its emphasis on a cultural economy of long distance ties among members sharing a community of origin, suggests how place in Han Chinese society has been understood as actual located places, in villages, counties, and cities, and how it also works as a concept of social relations in spatial terms. This understanding of place exemplifies contemporary geographical conceptualizations, especially in Massey's idea of place as the spatial reach of social relations. Massey's (1995, p. 61) theorization of place sees "space as social relations stretched out," which makes place "the location of particular sets of intersecting social relations, intersecting activity spaces, both local ones and those that stretch more widely, even internationally." This concept of place subsumes scale relations in trans-scalar activity spaces, which are all of the diverse activities and their emplaced contexts to which people are connected in their daily lives. Malpas's understanding of places as nested, how "Places also open out to sets of other places through being nested, along with those places, within a larger spatial structure or framework of activity," also invokes scale relations. This nested character of places is a dimensionality or scaled quality that "makes possible a particular form of differentiated unity – a unity that would seem to play a particularly important role in the organization of memory" (Malpas, 1999, p. 105), here the place sentiments of migrants in Shanghai.

Still, how do various social processes actually work through discrete and changing sites of spatial activity? Culture and economy do not move, people move, and people significantly propel economic activity. How social processes work though scale must lie in human and corporate agency (after Giddens, 1984), as diverse agents negotiate or propel trans-scale activities. People on the move are agents of culture and economy, as individuals, corporate or state bodies, and they act in the interest of particular scales in order to realize certain goals. Varied types of high mobility – concerning people, goods, ideas, and forms of capital – characterize regional processes, and it is the movement of people, in their embodied individual and group contexts, that sends and delivers goods, carries and spreads ideas, and makes capital flow. Thus, the individual, through embodied practices, becomes the mobile agent of trans-scale activity and occupies and moves through the places of interscale negotiation. Understanding the body and bodily practices as socially constructed, mobile and variable – that bodies have histories and geographies – opens doors to diverse cultural milieux. In Casey's (1996, p. 34) terms, "To be located, culture also has to be embodied."

Historically, the Confucian concept and imperial institution *li* (ritual, ceremony) was the most important value that symbolically tied individuals, in the bodily performance of ritual activities, to the empire and values of the imperium. *Li* also generally defined the understanding of appropriate forms of societal behavior and conduct; it was the essence of self-rule in imperial society, as opposed to rule by regulations or law (Dutton, 1988). This is not the Western notion of ritual as activity of traditional religious belief. This is ritual as the embodied experience of symbolic culture in situated lifepath events, including seasonal and annual holidays, family and firm gatherings, and major life passages at times of births, weddings, and funerals. The significance of *li* itself in Chinese society was, critically, hierarchical and scaled. Angela Zito (1997) has conceptualized how *li* operated as a discourse and form of correct embodiment that tied the emperor's ritualized activities, in performing annual imperial sacrifices and producing calligraphic texts, into a state discursive formation of extraordinary power. The emperor served as the supreme embodiment of *li*, and local officials reenacted the imperial rituals and textual practices at all administrative levels down to the county, so they were well known at all scales. Performing *li* was utterly culturally symbolic and a transhistorical experience, in which Chinese civilization was simultaneously individually embodied and collectively reproduced through successive emperors and dynasties. What endures is the concept of embodiment as a situated, social and cultural practice: how the state and elites control society through the body and the individual through symbolic bodily qualities, and the site of the body as the mobile agent of trans-scale activity. An understanding of embodied ritual practices informs how individuals, families, and communities folded into a historic Chinese body politic and understood cultural Chineseness, and how individual and group identities formed through embodied life practices.

Historically and in the contemporary era the *jia* (the extended family as an economic unit or household), the hometown or native place, in the terms, *laojia* (informal spoken usage), the *jiaxiang* (common general usage), and the *guxiang* (literary usage), and the *qiaoxiang* (village of Chinese overseas) have existed as ideas and realities about individual and family located origin in China (Naquin and Rawski, 1987, pp. 33–54; Ebrey and Watson, 1986, pp. 1–13). In traditional China and up through the end of the dynastic period, the *jia* existed as much more than an economic unit; it was "a metaphor for the state and the foundation of correct – and hierarchical – relationships" (Naquin and Rawski, 1987, p. 34). The foundations of correct social relationships included embodied ritual practices though which Han society reproduced the norms of the imperial state in symbolically scaled relations, from the scale of millions of individual households to the one imperial household in the capital.

Lineage organization and lineage systems in China have varied regionally, with broad scale differences between north and south China. While village settlements prevailed across Han China, it was in southern China, especially in parts of Guangdong and Fujian, that the single lineage village prevailed. In work on regional distinctions in lineage organization in Guangdong and Fujian, Maurice Freedman (1958, 1966) found that village society operated on three levels: the local lineage at the village level, a higher order lineage of several villages, and at a larger level of the clan, which existed not as a territorial area but as a belief or imagined symbolic geography. Based on contemporary fieldwork in the Zhujiang delta, David Faure found a more abstract and symbolic place-oriented character of lineage organization. Faure (1986, p. 10) interpreted that rites at the ancestral hall did not necessarily focus on ancestors, but rather on local and regional deities, which "represents, in religious terms, the villagers' mental map of the community and its vicinity, a map which does not coincide with the lineage." Most places also had a local *tudi gong* (earth god) and other locality gods. Their cult practices were distinguished by procession festivals that made a "tour of the boundaries," which ritually secured the territory under their purview. The medium for these processions was incense; and a ritual officer, the master of the incense burner, followed the sedan chairs that bore the statues of the gods in procession festivals. In some communities, the procession stopped at every household, where the master exchanged burning incense sticks with the same from each household. In this way, "the burning tips and smoke of incense each year inscribe that territory and its thresholds, taken in from outside, re-consecrating each domestic altar as an installation of that territory, a place in a geography of places of origin" (Feuchtwang, 1992, pp. 23–24). The smoke of incense was the medium of communication with the gods, and the procession festival was the set of embodied spatial practices in village place-making. Because the local earth god acted as the intermediary between the stove god, in each household, and the city god, the relationship between locality gods represented a nexus of social relations that bound households, villages, neighborhoods, and cities into a scaled symbolic community (Naquin and Rawski, 1987, pp. 41–2).

In China under reform, returning to one's native place or home town remains in many senses a culturally symbolic if not ritualized journey that invokes these deep layers of historic social practices and forms of cultural symbolism on new terms. Interest in these journeys, especially when undertaken by prominent persons, is keen. In 1997 the Governor of the US state of Washington, Gary Locke, traveled to China on a trade mission and toured south to visit his *jiaxiang*, Jilong village, in Taishan county of Guangdong province. Surrounded by an entourage of local officials and distant relatives, reporters followed him and especially chronicled his movements, as he bowed before the family altar, in the house where his father was born, and kneeled to burn incense and offer a sacrificial pig at the site of his great-grandfather's grave (Zimmerman, 1997). Locke accepted these roles and so emplaced himself within a Chinese cultural ecumene. Taishan people claimed his origins, just as they proudly proclaimed his status as a government official on a big character poster: "Governor Gary Locke returns to his home town." This was more than a symbolic cultural event, as local officials were able to use the occasion of Locke's visit to secure funding for infrastructure improvements around the village. Understanding Jilong as Locke's *jiaxiang* represents the idea of translocal cultural identity, tied simultaneously and diachronically to functionally disparate communities. For Locke, born in Seattle, the collective lifepath represented by his male forebears laid claim to his symbolic identity, and its representational economic power, in Guangdong. Through processes of high mobility, (e)migration, and citizenship, ideas about place become processural and spatialized, and exist as trans-scale, translocal, and transhistorical aspects of human experience that continue to undergird transformation of the regional economy and remake regional meaning.

Contextual Geographies

The regional geographies presented in this book are contextual accounts, historic and contemporary, that provide ways of seeing the situated complexities of regional realities and transformations. Whereas a linear historical narrative would seek to explain or legitimize present geographies, a contextual approach opens up windows on the complicatedness of actual geographies, and so cannot feign to present a complete portrait of any one region. The material is organized to make important connections, between sometimes disparate fields of knowledge, with the concern to transmit complex understandings about places and the region in coastal south China. Two interrelated parts form the main structure of the book. The historical geographies are largely contained in chapters 1–4, followed by more contemporary perspectives in chapters 5–8. Yet each of the chapters also conjoins historical and contemporary material in order to consider a range of time–space questions about the relationship between past and present, in the problematics of scale, potentially transhistorical conditions and their transformations, juxtapositions of elements from cultural and economic spheres, and ethical considerations around the spatialities of difference.

The second chapter, "Region and Representation," compares changing regional representations of the south China coast and emphasizes the understanding of regions as both discursive, historically constructed entities, and sets of dynamic geographical realities, in order to recover both regional meaning and the realities of formative geographical processes. In representative images, from historic perceptions about the natural environment to concerns about bodily conduct and alternative life practices, the historic "south" in China emerges less as a defined place than as a *process* – of mobility, experience, and coming into difference. China's transhistorical concerns about regionalism have reemerged in the contemporary period and as new regional formations have contested the power of the capital. The idea of Greater China as a region is indicative of these complex regional positionings, especially as perspectives about its geographies shift from different vantages, in China, Taiwan, Hong Kong, Singapore, and beyond. Seeing different regional geographies in the idea of Greater China brings into focus how the spatialities of regions serve epistemological purposes for making sense of complex geographical processes.

The south China coast is a historic maritime cultural economy whose conditions in many ways challenged the orthodoxies of agrarian Han society. The third chapter, "Maritime Frontier/Mercantile Region," recovers the significance of historic social formations as a basis for understanding regional trading economies, mercantile practices, coastal urban development, and Chinese settlement in Southeast Asia. Notions of frontier marginality about south China owed to northern worldviews, whereas in local terms the south coast was a center of international maritime trade and tribute system ports. The normative causal explanation about the origins of the regional maritime economy in south China, in a low per capita arable land ratio, which supposedly compelled people to "make fields from the sea," may be reasonably reinterpreted in understandings of maritime coastal cultures which antedated Han in-migration, and later in uneven holdings of land wealth. The distinctive coastal origins of port city settlement between the Yangzi and Zhujiang river deltas challenge accepted models of urbanization for China, which have held that early settlement developed in productive agricultural regions and as a result of administrative settlement from the north.

The internationalization of the south China coast has a long history. The arrival of Western powers in the nineteenth century was one of the more recent phases of encounter, distinguished by being the most imperialist, and bound up in illegal trafficking of opium by Western mercantilists. Chapter 4, "Open Ports and the Treaty System," focuses on Ningbo, Fuzhou, and Xiamen, the smallest of the first five cities opened to foreign trade and residence in the nineteenth century. By comparison to Shanghai and Canton, the term "treaty port" has more often obscured the nature of these places, where foreign interests never gained a significant presence, and merchant groups and emergent civil society movements existed as the major forces of social and economic organization through the twentieth century. Geographical characteristics of merchant organizations, based in regionally specific dialect groups and *huiguan*, maintained control over local economies and long distance trading networks, and were a basis for social organization in the Nanyang. The transboundary cultural economy substantially emerged in this era as people from the south coast increased frequency of travel between south China and especially the British colonies of Malaya and Singapore.

The economy of the south China coast convulsed and stalled after 1949 as Maoist directives sought to rearrange the national space economy. Chapter 5, "Revolution and Diaspora," examines how the first phases of the revolution evolved in south China and found support in *huaqiao* (Chinese sojourners) or overseas Chinese communities. As the socialist revolution unfolded in China, Maoist directives marginalized the overseas Chinese and people with overseas connections. The simultaneous process of decolonization in Southeast Asia resulted in new nation-states and nationalisms, and *huaqiao* diminished ties with China, claimed local citizenship, and remade local communities. In the Southeast Asian region, Malaysia is an especially interesting country in which to assess issues around Chinese identity formation and postcolonial nationalism. Malaysia has the largest Chinese minority population on a world scale, and the significance of the history, size, and settlement of the Chinese population has significantly intersected with the state-making project in the postcolonial period. This chapter culminates in a landscape analysis of the conservation of a historic Chinese cultural site in Melaka, Malaysia, which affords a perspective on how Chinese overseas have retained homeland identities while also establishing themselves within postcolonial orders.

For readers more concerned with the conditions of south China under reform, chapter 6, "Gendered Industrialization," presents a challenge to existing literatures of the reform experience. By contrast to the bulk of analyses, this discussion does not explain regional restructuring in terms of the measured successes of economic growth. Instead, the chapter reviews reform from the perspective of the gendered space economy of development, and as the result of the intersections between five major reforms: the one-child policy, the household responsibility system, the loosening up of the *hukou* system, the open door policy and its dependence on low wage manufacturing, and the dismantling of the state-owned enterprises. The discussion foregrounds the gendered conditions of south China under reform because these issues, however typically marginalized, are central to China's successful human development. Women's labor has been the main source of surplus value in the export-oriented sector in south China, and the migration of women to regional manufacturing centers has resulted in new divisions of labor and new patterns of household formation. The analysis argues that the patriarchal state and normative economic policy have transfered "male bias in the development process," leading to uneven conditions for women, including diminished status in education and employment.

In the 1980s, export-oriented manufacturing was at the center of rapid growth in coastal south China. By the 1990s, industries associated with land development emerged as high growth sectors. Chapter 7, "Zone Fever," examines how SEZ development – originally a "disarticulated" element of reform in special areas set aside from general land use practices – became the basis of an uncontrolled development regime by the 1990s. SEZs were the located sites of policy experimentation with industrial land development, which Hong Kong property developers and elite state interests identified a basis for earning super profits. "Real estate fever" threatened the stability of the reform process by 1997, and the state began to issue policies to protect arable land. The impacts of development on China's arable land resources became the focus of debate, which served to elide problems of the rapid growth imperative at the basis of reform. The discussion examines the transboundary regional economy through implementation of the China–Singapore Suzhou Industrial Park, a flagship industrial park, which underscores how the central government promoted large-scale special zone development through the middle of the 1990s. At the end of the decade, the collapse of provincial trust and investment corporations associated with real estate development in Guangdong and Shanghai ultimately signaled the problems of the regime of rapid development and an end to real estate fever.

Shanghai and Hong Kong are the two leading cities of the south China coast and China's most internationalized urban areas. Chapter 8, "Urban Triumphant," assesses the Hong Kong Special Administrative Region (SAR) and the resurgence of Shanghai in the 1990s against the backdrop of wider regional interests in a Chinese cosmopolitanism. By contrast to the anti-urban ideologies of the Maoist era, under reform the cities of the China coast have reemerged as leading centers of transnational economic activity and cultural production. In the 1990s the primary destination of FDI shifted from Guangdong to Shanghai and the Yangzi delta, and Shanghai elites engaged the city's history of internationalism in attempts to establish Shanghai as China's leading world city. Debate over the 1997 Hong Kong handover has faded as the SAR has maintained its economic position among Asian capitals and joined a favored position in China's urban hierarchy. Hong Kong and Shanghai are China's centers of human and capital mobility, and the spatial processes at work in these cities are both the agents and subjects of globalization and its alternative forms.

Processes of regional formation in south China, contemporary regional conceptualizations, and the emergence of discussion around regionalism, as a set of responses to globalization, form the subjects of the epilogue. Focus on regional formations restores a measure of reality to would-be global tendencies of spatial expression by exposing material processes of globalizing capitalism, and the unevenness of development. In their material and representational forms, regional formations and their symbolic meanings serve as alternative bases of subnational and supranational power. These scale positions alternative to the national both question and confirm the significance of regionality in an era of globalist thinking. The apparent slippages in regional meaning suggest the conceiving of geographies that are less encumbered by paradigmatic characteristics of the nation-state and globalization rhetorics, and instead form a basis for unmapping economic development and the diverse world views it pretends to entertain.

NOTES

1 For discussions of the nature of economic reform see especially Dorothy Solinger (1993), Barry Naughton (1995), and George Lin (1997).

2 The treaty port era is the semi-colonial period in Chinese history from 1842 to the start of the Second World War; see chapter 3.

3 Susumu Yabuki (1995, p. 117) charts changes in provincial rankings by economic growth.

4 Major cities in the interior also received special privileges in the early 1980s, including Chongqing in 1983, followed by Wuhan, Xi'an, Shenyang, and

Harbin. The state gave these major industrial cities, along with Guangzhou and Dalian, independent economic management power equivalent to the level of provinces (see Solinger, 1993, p. 160). In 1992, five cities along the Yangzi and 18 provincial capitals in the interior were granted the same privileges as the open coastal cities. The state also allowed border provinces to set up border open cities. Still, coastal provinces regularly received about 90 percent of the total foreign investment; see Dali Yang (1997, pp. 33, 55). State decision-making about the selection, location, and size of the SEZs was a highly politicized process; see, for example, Jude Howell (1993) about the process of establishing the SEZ at Xiamen.

5 Deng Xiaoping had to first confront alternative reform proposals before his own prevailed; see Susan Shirk (1993) and Li Chengrui and Zhang Zhuoyuan (1984). Market-oriented reforms prevailed in part because real wages declined through the Maoist era. Yabuki (1995, pp. 61–80) has calculated that real wages, indexed to cost of living, finally recovered to the equivalent of the 1957 Maoist era peak in the early 1980s.

6 The World Bank (1993) was the leading institutional agent promoting the notion of "miracle" development. Jonathan Rigg (1997), has summarized the problems of the miracle debates. The economist Paul Krugman (1994) has debunked the miracle on different grounds. Krugman argued that growth in Asia will be short-lived based on the example of the former Soviet Union, where growth was based on rapid growth of capital inputs, but not on corresponding increases to factor productivity. Dwight Perkins, an economic historian and area specialist, has argued in a 1991 essay that China had already experienced gains in efficiency of production by the 1980s.

7 Coastal south China has been touted as the fifth member of this would-be group of economic "beasts." For an example of this perspective, see Yunwing Sung et al. (1995). Considerable and not unproblematic vocabulary is associated with the dragon and tiger metaphors for describing the NIEs, especially a range of adjectives including "tame," "paper," "on the prowl," "toothless," and much more. Some analysts have pointed out that discourses surrounding the NIEs sound like media coverage of sports teams rather than complex societies.

8 For an account of the demise of the Geography Department at Harvard see Smith (1987).

9 While the "impact-response" approach was dated within twenty years among rersearchers, numerous textbooks on China and East Asia adopted and repeated the paradigm long after scholars had rejected it. See the discussion by Cohen (1984, pp. 10–11).

10 Cumings was writing in partial response to debates over the influence of rational choice theory in the academy, and its privileging of game theoretical analysis in international political economy at the expense of area studies expertise.

11 This discussion is an abbreviated version of my 2002 essay, "Origins and evolution of a geographical idea: the macroregion in China." This book is a partial response to the problems of regional analysis and in China area studies.

12 Skinner conducted fieldwork there in 1949–50, but as Graeme Johnson 1998), has noted, Skinner was forced out by the war; Philip Huang (1985, p. 24) pointed out that the duration of the field work in Kao-tien-tzu (Gaodianzi) was three months.
13 In this regard, the marketing systems model also appeared to represent a partial answer to the challenge left by Max Weber, regarding the limitations of the Chinese urban system as a base for economic expansion.
14 In a more strident criticism, Frederick Mote (1995, p. 66) wrote, "We can speak of a Skinnerian age in studies of Chinese urban history in the last two or three decades, under which many historians believe that his system provides the only reasonable basis for analyzing all urban, in fact most social phenomena. It is fair to note that in the seventeen years since Skinner first adumbrated his brilliantly suggestive hypotheses, they have in many minds (perhaps also in Skinner's) hardened into something resembling an iron-clad law of Chinese history. The Skinner system's elevation to the status of doctrine has had the effect of tending to divide the field into those who accept it as established truth, and those who find it in part erroneous and therefore reject it *in toto*." The situation Mote described has resulted in either the use and acceptance of the Skinner models, or avoidance by omission. Mote also notes how no comprehensive focused reviews of the models have been published and neither have any organized retrospectives on Skinner's body of work taken place.
15 For contemporary geographical literature on scale see work by Richard Howitt (1993, 1998), who applies scale to emancipatory geographies and cautions what scale is not; Erik Swyngedouw (1997a, b), who best theorizes the social construction of scale; and Neil Brenner (1997, 1998, 1999, 2000), who discusses debates about the relationship between the nation-state and globalization.
16 For especially regional discussions of fiscal decentralization see Solinger (1993), Yang (1994, 1997), and Yabuki (1995, ch. 11). In 1980 Guangdong and Fujian provinces received a special set of fiscal policy relations with the central government that allowed them to retain a greater percentage of revenue. Guangdong's arrangement was to remit to the center one billion yuan annually, while Fujian received a 150 million yuan annual subsidy; these amounts were fixed for five years; see Shirk (1993, p. 157).
17 In research on new roles of large cities in China under reform, for example, such as the fourteen open port cities and other large cities given independent decision-making powers, Solinger (1993, pp. 205–22) has described this scale shift as one "from hierarchy to network," in which the state has expected major cities to organize economic activity with other cities and counties. Such urban networks are also scaled.
18 James Clifford (1997, p. 7) has used the term translocal "to articulate local and global processes," but my intent is to argue for a complex and material – as opposed to metaphorical – scale framework.

2

Region and Representation

> [E]very place or region "arrives" at the present moment trailing long histories: histories of economics and politics, of gender, class and ethnicity; and histories too, of the many different stories which have been told about all of these.
> (John Allen, Doreen Massey, and Allan Cochrane, 1998, p. 9)

Ideas about regionalism, in the rise of regional economic power, power relations between regions and the capital, and distinctive regional identities, have been enduring subjects in China, and they have resurged under reform.[1] The rise of the south China coast has yielded new perspectives on the role and meaning of regions in the national order, and rewriting the significance of southern China has become a new scholarly enterprise, both inside and outside the country. Notions about the rising influence of south China have proliferated, from popular sayings (Shenzhen is Hong Kongized; Guangdong is Shenzhenized, and the whole country is Guangdongized) to reinterpretations of canonical history. With the rise of the regions, scholars have been revisiting nothing less than China's basic evolutionary history to argue that, rather than northern culture simply diffusing south, Chinese civilization has evolved from diverse sites and mutually influential early kingdoms.

Regionalism, as a set of ideas constituted through the processes of regional formation, must be a contested geographical enterprise because it presupposes no particular territorial boundary definitions and imputes spatialities alternative to the state. But this does not mean that regionalism is necessarily antithetical to the state. Regionalism, especially in a large state, may form the basis for territorial coherence as well as dynamic processural change. As Diana Lary (1974, p. 8) has argued, to understand relations between diverse and contending regions in China it is important "to see regionalism as the sinews which held China together." Regional power increased in China as dynasties fell into decline, but enhanced regional power was not ultimately threatening to the center, and instead substituted for it as the center restructured and reconstituted the regime. Regionalism, then, in historic China "provided a continuity between the order that had died and the new one which would emerge" (Lary, p. 8). Lary found China's regions not just in the provinces but spanning multiple provinces, within provinces, and in transboundary and border zones. These aspects of regionalism and interregional relations have contributed to China's territorial coherence and to local and regional differences. In a reflection on the implications of imperial era regionalism for the contemporary period, Tu Wei-ming (1991, p. 12) has noted that "Although the phenomenon of Chinese culture disintegrating at the center and later being revived from the periphery is a recurring theme in Chinese history, it is unprecedented for the geopolitical center to remain entrenched while the periphery presents such powerful and persistent economic and cultural challenges." Tu's remarks also challenge us to consider the contemporary cultural, economic, and political forces in south China that together have produced a powerful regional formation in "Greater China."

If we are interested in the production of cosmopolitan geographical knowledges, it is important to see China how it sees itself – as a country of regions – and to ask questions about processes that create regional meaning and stitch China together in a coherent whole. The rise of the south China coast in the late twentieth century is a regional transformation, and not an unprecedented one, based on knowledge of significant and influential historic kingdoms in the Yangzi delta and in Fujian and Guangdong provinces. How should we recover such knowledges? Since the kinds of regional formations we are interested in are dynamic geographical entities, in "south China," "the southeast coast," "Greater China," and so forth, we can examine these regional formations through processes of transformation, as the empire expanded and sent its emissaries south, as the maritime cultural economy rose and fell, and as the regional economy has restructured. The history, multiplicity, and overlap of regional formations, in their sedimentation, erosion, and shifts over time, has inscribed and reinscribed their dynamic and transhistorical significance. The diversity of regional formations, and their representations, serves epistemological purposes for spatial processes and human imaginations about them, and, in their distinctions, as indications of cultural worldviews and contexts for the evolution of regional identities.

Geographical Representations

How is regional meaning produced? People produce geographical meaning through actions and experiences, and especially through the production of texts, and use of textual representations in constituting identity positions. For Chinese emperors and officials, the production of texts was a ritually embodied discursive act (Zito, 1997). The repetition and circulation of textual material, by and through individuals and communities, contributes to the construction of situated meaning, and carries forward ideas about geographical realities. We can recover regional representations and create comparative knowledges about them through a hermeneutic approach, by culling from diverse texts to produce new comparative accounts. [2] This analytical mode is a study of discourse, a cultural history of ideas and their contested uses. Yet even as we may recover regional meaning, that meaning must remain partial and interpretive, since the interest of this project is not to suggest a mimetic representation of any particular region of south China, but rather to underscore the significance of regional formations as dynamic spatialities that form and shift as expressions of human spatial processes.

The first section of the chapter explores how meanings of the south and south coast geographically shifted over time. The selected texts, from travelers' accounts, imperial records, poetry, and the fine arts, serve as the empirical basis for comparing ideas about south China, and how those meanings have been created and transformed.[3] Most are known subjects of historiographical and sinological analysis, and preserve knowledge as the memories and worldviews of the Chinese literati, their observers, and their critics. They leave a record of attempts to reconcile difference in south China by judging southern geographies in northern terms – as representatives of the north encountered and "othered" the south – and of experiencing incremental encounter in south China as nothing less than a civilizing process. In the process, the Han ecumene expanded, and ideas about locating south China shifted increasingly south. The Yangzi delta settlements, the geographical paragons of southern culture for northern elites, were high culture refuges during periods of dynastic restructuring. Northern ideas about the south as fundamentally other to northern society ultimately coalesced further south on the maritime frontier, where lifeways and maritime orientations contrasted sharply with the landed traditions of empire. Against this long sweep of imperial history, the rise of the south coast in China under reform has represented a kind of fundamental regional reversal.

The second section of the chapter explores this reversal through reinterpretations of the meaning of the south, and contemporary regional formations in the idea of Greater China. The use of the term Greater China has been popular in the West and places of Chinese transnationalism, but less so in China and Taiwan where different terms, in Chinese language iterations, have represented the transboundary regional economy and its alternative geographies. Especially, Western analyses of Greater China, in their limited linguistic positions, entangle the region's possible territorial configurations, and typically by comparison to the nation-state ideal. But rather than positioned against the state, the more complex way of seeing Greater China is as a regional intervention in the processes of globalization, and the ways in which globalization gives rise to regionalisms and regionalist thinking. I have selected topics and materials in this chapter with the recognition that representations of regional life cannot be separated from questions about the relationships of those representations to the imperial and national orders (after Williams, 1973). In the way that Prasenjit Duara (1995, p. 7), in *Rescuing History from the Nation*, unshackled the idea that "nationalism represents a unitary consciousness or identity," in favor of identity negotiation "within a network of changing and often conflicting representations," representational analysis allows us to assess the constituent elements of regional identity and how people may deploy representations of that identity to establish position within and challenge national and international orders.[4] Following from Duara, in this chapter I am interested in identifying changing and conflicting representations as a basis for rescuing the region from the empire, nation-state – and international political economy.

Northern vantage

For greater than two millennia, the vantage from northern capitals saw the lands of the south beyond the Huai River, which served as the first real and idealized boundary of the Han ecumene. [5] North of the Huai travel traditionally depended on horses and donkeys, and the staple crops were *kaoliang*, millet, and wheat. South of the Huai wetter lands materialized and overland travel became impractical. The land turned marshy, streams proliferated, and rainfall increased. Critically, rice could be grown south of the Huai.[6] In the coastal zone, low-lying marsh extended south of the Huai the length of Jiangsu province to the Yangzi delta, which limited coastal settlement and denied natural harbors.[7] The early twentieth century traveler Harry Franck described it this way:

> in modern China, there are distinct differences between the two sections of the country, shading into one another about midway between the Hoang Ho and the Yang Tze Kiang. Foreigners are prone to consider the latter the dividing-line, and there have been political tendencies of that kind in China itself; but unless we make a third division and call it Central China, as many do, the Yang Tze region has much more in common with the south than with the north. About the time he crosses the thirty-fourth parallel of latitude the southbound traveler in China will note an almost sudden change: camels, donkeys, Peking carts and the grassless, treeless, dust-blown north give way to water-buffaloes, traveling-chairs, and narrow flagstone roads meandering among endless flooded paddy-fields, a greener if not a cleaner land of many water-ways but without wheeled

> vehicles, except here and there creaking wheelbarrows competing with perpetual trains of coolie carriers. (Franck, 1925, p. vii)

Such assessments about south China do not so much map an accepted area, but reveal south China in more subtle terms as a process of travel and encounter: the perspective view is north China, the natural environment of south China begins at the Huai, the urban south emerges in the Yangzi delta, and the country becomes ever more southern with each decreasing degree of latitude.

With the rise of the Yangzi delta settlements during the Tang dynasty, the boundary of cultural and social orthodoxy extended to the Jiangnan, the fecund river valley region of the Yangzi in southern Jiangsu province that includes Nanjing and Shanghai. The Jiangnan has been the cultural transition zone on the coast between the north and south, but it is also, in a regional geography of cardinal directions, reliably the south. Its fundamental material importance to the north, even a millennium ago, was economic. As far back as the tenth century, when the site of Shanghai was an incipient fishing community, the larger deltaic region effectively financed the empire at large by regularly remitting to the capital more than its share of taxation revenue, thereby ensconcing itself in the imperial order as the leading regional economy (Marmé, 1993). The fiscal resources of the Yangzi delta region have indeed continued to disproportionately support the finances of the national government through the contemporary period (see White, 1989).

Coastal imaginaries

Realities of mapping south China have varied historically, from the largest scale distinction of the south, in the half of the country south of the Changjiang, to the most local scale regionalization of the far south coast in a circumscribed area around the Zhujiang delta of Guangdong province. In the discussion thus far, I have alternatively refered to south China and the south China *coast*, to distinguish the coastal zone from the larger extent of landed empire and to signal that the subjects of concern are about a set of regional processes that are more characteristic of the coastal zone than the coastal provinces at large. What should we understand about representations of a coastal region in China? The coast was not a landscape of desirability in traditional Chinese imagination. As Alain Corbin (1994) has argued, the *idea* of the coast as a destination of choice is a fully constructed concept which, in the West, did not emerge from earlier fearful images of the sea shore until the middle of the eighteenth century. Fear of the sea also characterized the classical Chinese landscape imagination, and even in contemporary China, by contrast to coastal destinations in Western and global tourist imaginations, the beach, as a leisure destination, does not really exist on similar terms.[8] The absence of a maritime and coastal orientation in Chinese landscape history is explained by the emphasis on riverine environments as hearths of civilization, and the fact that China's maritime tradition is regional, and based in the port cities of the south China coast.

From texts of the scholar-poet tradition of the Tang dynasty, Edward Schafer (1967, p. 138) evaluated how "Hua men" (men of China) saw the South China Sea as a "remote and frightening ocean," the "Swollen Sea." Richard Smith's (1996, pp. 85–6) work on Chinese cartography led him to the general observation that "The sea, an object of fear to most Chinese, was often depicted with a threatening aspect, such as violent wave undulations and fierce splashes of surf." In *Shore of Pearls* Schafer characterized the classical view of Hainan island, which could only be reached by crossing the sea:

> Through the ages Hainan has been regarded as the ultimate place of exile. It was horrible enough to be sent to live out one's days in misery in the fearful jungles of Annam beyond the Gate of Ghosts, but the passage of the strait that separated the island from the continent symbolized a divorce that was even more to be dreaded – the transit marked a sort of spiritual death. Even in modern times Hainan has been seen chiefly as a sink of desperadoes and outlaws – a "refugium peccatorum" as one writer has put it. So worthless was it for the purposes of decent men that Sun Yat-sen (according to one report) wished to sell it to a foreign power for fourteen million dollars. (Schafer, 1969, pp. 85–6)

In this passage images of the south combined with fear of the sea to construct Hainan as an especially loathsome place. Schafer lets us know that the renown of Hainan's special worthlessness remained as a transhistorical value, but its problem largely owes to the sea. In the Chinese classical imagination, pleasing landscape images of water have always been rivers and lakes of the landed environment. The term *shanshui* (mountains and water) denotes the preferred Chinese classical landscape imagery, in which water, *shui*, is also understood as a synonym for river – but never for ocean or sea. *Shanshui* also means landscape and scenery, and landscape painting, in which mountains and water form the typical subject matter (see Cahill, 1962). Unlike fine arts traditions of world maritime powers, especially Great Britain, the Netherlands, and the USA, and despite China's significant maritime history, Chinese painting entirely lacks a seascape tradition.

The idea of China as a river empire was fundamentally underscored as recently as 1988 when the state aired on national television *Heshang* (River

Elegy), a controversial six-part series that called into question traditional Chinese worldviews and Communist practices as the basis for the country's low status in the late twentieth-century world economy.[9] The appearance of *Heshang*, during a period of high inflation after the first decade of reform, was clearly meant to quell fears about China's path of economic restructuring. The political polemic of the series was obviously in favor of Zhao Ziyang, who was then general secretary of the Communist Party and had vigorously promoted reform. The series's film text attacked traditional cultural ideas and past practices through the critical evaluation of symbolic landscapes. The dragon, symbol of sacred mountains, the emperor, and Han Chinese ancestry, was shown to represent tyrannical rule. The loess plateau, cradle of Chinese agrarian society, was portrayed as a landlocked and earthbound culture region; enclosed by the Great Wall, symbol of the territorialization of Han China, *Heshang* depicted the Wall and the northern heartland as representations of parochialism, close-minded conservatism, and incompetent defenses. The series's most devastating critique barely coded condemnation of the socialist revolution, by portraying the peasantry as ignorant and the rural environment as backward. Throughout, the series focused on symbolisms of the Yellow River, traditional source of life of Chinese civilization, in the metaphorical qualities of a diseased artery, a chaotic river so laden and clogged with heavy yellow silt that it had prevented China from following a more dynamic evolutionary path.[10] For the Yellow River, and the Chinese nation, the way out of the problems of tradition lay in embracing the sea, the film's code for modernity and internationalization. The voice-over narrative tied the future strength of China, in the symbolism of the Yellow River, to the medium of the sea: "The Yellow River must rid itself of its fear of the ocean" (Xie and Yuan, 1991/2, p. 90). The sea emerged to symbolize processes of industrialization, and globalization, and the rise of the south China coast under reform; the new imperatives of *lanhai wenhua* (blue ocean culture) challenged the traditions of *huanghe wenhua* (Yellow River culture).

The Chinese intelligentsia and the state found common ground in critical reactions to the *Heshang* series: in its landscape and ocean-river symbolism, *Heshang* effectively advocated Westernization as China's path to modernization. As Jing Wang (1996, p. 118) has written, "*Heshang*'s scriptwriters were criticized for propagating a review of politics and history that advanced historical fatalism, geographical determinism, the 'fallacious backward ideology' of grand unification, Eurocentrism, total westernization, elite culturalism, the postulate of a nonsocialist new epoch, and the theory of the ultrastability of Chinese feudal society." The series engendered so much interest and caused so much controversy "that, for the first time, traditionalists in favor of the five thousand years of

Chinese cultural excellence in Taiwan and stalwart revolutionaries in the Communist Party shared the same ground as well as a common rhetoric" (Barmé, 1999, p. 23). No popular media form had ever before popularized central intellectual debates for a mass audience. After the Tiananmen incident, four of the series's most prominent writers emigrated, and Zhao Ziyang, widely characterized in terms of his regional identity as a southerner from Guangzhou, continues to live under house arrest in Beijing.

By contrast to the northern worldview, ideas about sea have been bound up in more complex symbolisms in south China, where the maritime tradition and popular religion mediated fear of the sea. People of the south China coast attributed special powers over the sea to the regional deities, Guanyin and Mazu, the two most important female deities in the Chinese pantheon. Guanyin, the Buddhist Goddess of Mercy, is worshipped throughout Asia, whereas Mazu worship is unique to the south China coast and places of diaspora. The cults of these goddesses have also symbolized defiance of the norms of filial piety and imperial society: both goddesses, as young women, refused to marry and met early deaths under patriarchal social pressure. Guanyin temples proliferated especially in south China, and in Taiwan, where temple practices have not been disrupted, Guanyin and Mazu temples are the second and third most numerous temples after Daoist sites (Qiu, 1979, pp. 103, 214). China's foremost Guanyin temple complex lies on Putuoshan island, in the Zhoushan archipelago off the Zhejiang coast, and has been an important pilgrimage site since the Tang dynasty. The first local history dedicated to Putuoshan, from the fourteenth century, recorded that early Chinese emperors tried in vain to locate the mythic islands of the immortals, but instead located Putuoshan, and advised that "although . . . situated amidst terrifying waves and frightening tides . . . one can reach it within a few days" (Yü, 1992, pp. 205, 6). Mazu's origin myth is based on the life of a living person in the late tenth century, a woman from a seafaring family in Meizhou, Fujian, to whom local people attributed special powers to protect seafarers from storms. Sailors of the south coast worshipped at Mazu temples before setting out to sea, including China's most famous maritime explorer, Zheng He (see chapter 3). Especially in the cult of Mazu, people of the south China coast forged a spiritual alliance with the sea, and, in relation to a larger scale of empire, through a deity whose mythic powers symbolized distinctively independent and anti-authoritarian sentiments. Over time, her cult spread to Taiwan, coastal Guangdong, and Hong Kong, and has been subject to such diverse interpretations that it has commanded a substantial following from people of diverse class backgrounds and identity positions (Watson, 1985). Mazu remains the most popular deity in many parts of south China, and new Mazu temples have become popular elements of the revitalized coastal landscape in China under reform.

Coming into difference

Images of historic regional character recovered from classical texts, as much as they represent the south, evince the authorial positions of pre-existing worldviews and notions of the importance of the Han ecumene centered in north China. Yet if we invoke Schafer's approaches to literary interpretations, in which he always *combines* understanding the material *reality* of the natural environment *with ideas* about the south, then we begin to see how meaning about the south as a region has evolved and endured in China. "When a true Chinese – a Hua man – of the Yellow River valley and its vicinity journeyed southward, he was conscious of striking changes in his physical surroundings. He expected them, since he was familiar with a long tradition of southern strangeness" (Schafer, 1967, p. 135). Strangeness about south China has been a type of otherness, differences that reminded imperial rulers and northern Han Chinese of the extent of the ordered world and the need to secure that world on its margins.

During the Tang and Song dynasties, when the empire expanded along its southern fringe, the increasing need to post officials to the southern mercantile cities, especially Guangzhou, left a genre of travel writing in journeys of exile. The earliest extant travel narrative is *Lai nan lu* (*Diary of My Coming to the South*), written by Tang dynasty official Li Ao (772–836) (Strassberg, 1994). Li Ao traveled from Chang'an by boat to Guangzhou, but not by sailing down the China coast. Li's party traveled over 2,500 miles through eastern China via rivers and canals, which avoided the real and perceived dangers of maritime voyaging (map 2.1). Li Ao's prose, sparing and formal, reads like a descriptive gazetteer, and yet his title, "coming to the south," confides the nature of those journeys. Li Ao's diary demonstrates the attention traveling officials paid to entering new realms, in crossing borders and boundaries, whether rivers or prefectural boundaries, and in arriving at cities, special temples, and other important local sites. His party lingered especially at Suzhou and Hangzhou, which were already famous cities during the Tang. After passing through the Yangzi River delta, Li's route shifted inland to the difficult and somewhat circuitous route via Poyang Lake and more inland rivers. For Li Ao the south also emerged in the journey, and the region is a set of arrivals and destinations that emerge through the experiences of mobility. Casey (1996, p. 24) has written about a region in this way as "an area concatenated by peregrinations between the places it connects."

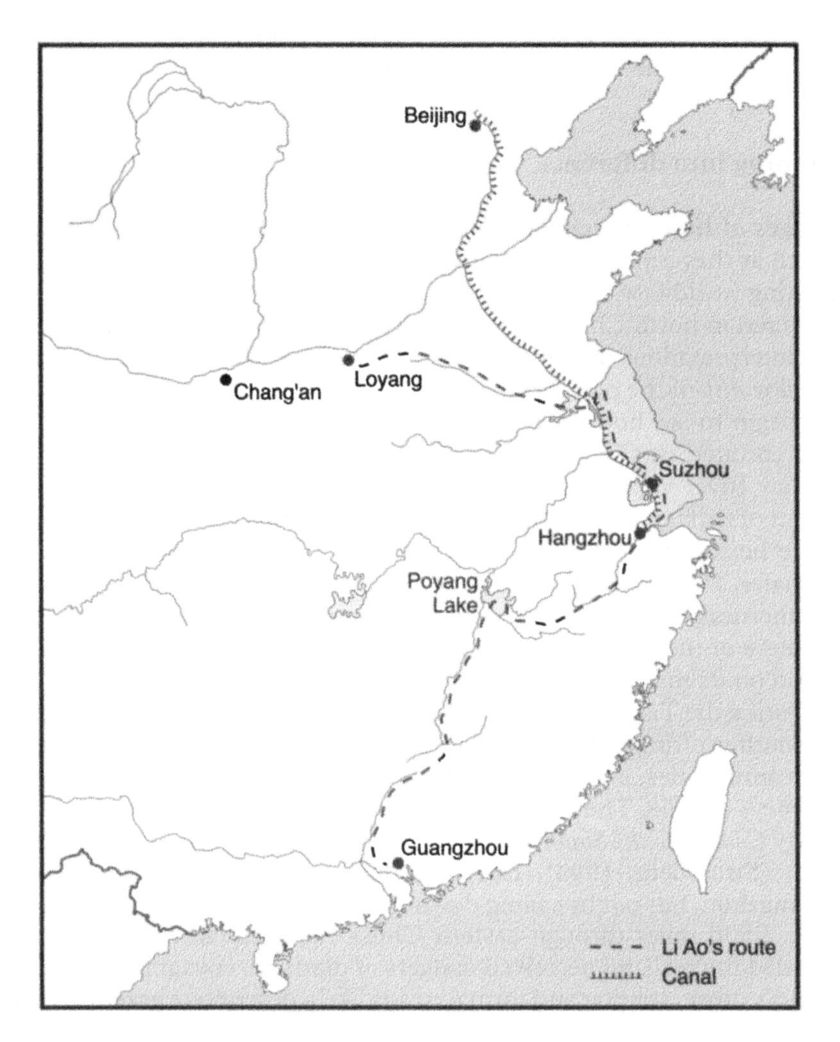

Map 2.1 Route of official journeys to south China.
Source: Strassberg (1994); line work by Jane Sinclair.

Richard Davis's (1996) study of the last years of the empire of the Southern Song dynasty (1127-1279) has problematized the south more than most. In Davis's analysis, the south is not transhistorically disdained, but initially othered as northerners encountered and ordered the south into their known world. Driven by military incursions from peoples on the north Asian frontier, migrants from the north "elevated the south initially to a place of political sanctuary, subsequently to a place of cultural conservation and rejuvenation" (Davis, 1996, pp. 141–2). Davis records that in pre-Song literary convention the south was often a synonym for the state of Chu. This assumption became less common with the advance of the Southern Song period as ideas about the location of the south shifted – increasingly south. With increased migration, "the Wu/Yue region overtook Chu as the symbol of the 'other' [and] the south emerged as a serious rival to the cultural hegemony of the north" (Davis, 1996, p, 142).[11]

> In the centuries preceding the unity of Qin and Han, the "south" did not extend much beyond the domain of Chu, a region politically and culturally quite autonomous from the "central states" directly north. Having emerged along the Yangzi River, the south was a land of perilous yet pristine wilderness: a land where hunting and gathering probably surpassed agriculture; a land populated with native shamans and aboriginals, immigrant recluses and exiles; a land possessing a familiar literacy yet an unfamiliar rhetoric, similar institutions but dissimilar customs; a land endowed with an abundance of lakes and rivers but also dreaded disease and pestilence – the antithesis of everything northern. It was the home of Daoism as well. This libertarian ideology, rejecting the social impositions of northern traditions like Confucianism, would advance a radical alternative agenda of individual expression and bequeath to the country, arguably, the most human of its many philosophies. But by the time of Christ, only two centuries after the imposition of unified rule, the south generally conjures up fewer images of hostile wilderness or Daoist reclusion, northern peoples and their ways having transformed the landscape. Yet from the perspective of those same northerners, the region remained irredeemably alien.(Davis, 1996, p. 141)

The realities of environmental differences of heat, moisture, color, scent, even the feel of the earth, contrasted substantially with what was otherwise familiar in the north. The human differences, of size, stature, skin color, spoken dialect, foods, arts, belief systems, and ideas about moral and bodily conduct constructed the south as a region of encountering others. Yet despite of this othering of southern differences, northerners embraced many southern sensibilities and enfolded especially the Yangzi delta settlements into the Han realm. As a prologue to an examination of events of the Qing dynasty and the Manchu concern to gain control over the south, Silas Wu has offered a similar interpretation on the south: "The Chinese South had developed a distinct culture as early as the last quarter of the sixth century. The Chinese who had migrated to this region from the North . . . had created a life-style that was more leisurely, elegant, and sensual than that of the North. The rise and fall of subsequent dynasties, Chinese or alien, had not altered the distinctive features of this culture" (Wu, 1979, p. 5). For Wu, in this passage, south

China is not a located area but a recommendation for a southern life-style, which he subsequently mapped in the leading settlements of the Yangzi delta.

Davis also pursued the qualities northerners placed on southern regional distinctions in sensuous and gendered terms. Where the northerners were characterized by "virile discipline," southerners were more "indolent and frivolous," their "dialects tended to be soft," and their "food delicate." Even southern modes of transport, junks and sedan chairs, were "sleepy" and "rather passive." Southern literature displayed "unmanly qualities," which also were presumed to signify that "all too frequently, southern men felt free to celebrate affections for other men, the fair countenance or delicate attire of some young boy" (Davis, 1996, p. 13). The Chinese word for south, *nan* (南), is a homonym of the word for male, *nan* (男), and in the language of sexualities *nanfang* (southern mode) and *nanfeng* (southern custom) have signified homosexuality.[12] Seventeenth-century playwright Li Yu complained that the "southern mode" was "prevalent in all parts of the country, but especially in Fujian. From Jianning and Shaowu onward, every prefecture and county is worse than the one before" (Hanan, 1990, p. 101). In Manila, where the Chinese community was derived largely from southern Fujian, the Spanish posted notices threatening fatal penalties for homosexual acts (Chan, 1978, p. 70). Homosexuality was also a known practice among pirate groups on the south China coast (Murray, 1987). It is not conclusive that homosexuality was more prevalent in Fujian and Guangdong than in other provinces, and literary sources record that homosexuality was present in major cities across China by Ming times, but the practice remained coded as utterly southern (Hinsch, 1990, pp. 120–4). These differences in perspective also reflected a north–south flesh trade that exposed not only southern predilections but northern desires:

> Nam-Viet was notorious as a supplier of slaves, especially females: "Slave girls of Viet, sleek of buttery flesh," wrote an appreciative Yüan Chen. Most of these unfortunates were aborigines, sold to Chinese and sent to the great cities of the north to tend the wants of the aristocracy. Neighboring Fukien and Kweichow were also sources of human flesh, and in the ninth century Fukien had the additional distinction of being the chief supplier of young eunuchs to the capital. (Schafer, 1967, p. 56)

Schafer's Nam-Viet is the region of the Nan Yue, the most southerly of the Yue groups, which encompassed an area that included Guangdong, part of Guangxi, and northern Vietnam. Here, regional characteristics of the south also emerged in ideas about alternative sexes and sexualities, in ideas about people whose positions lay outside the bounds of normative society. In this context, well known ideas about northern disdain for the south are more accurately revealed through the contradictions of northern desires.

Lesbianism also emerged in regional contexts and especially in southern China in the Golden Orchid Associations found in the Zhujiang delta of Guangdong province (Topley, 1975, p. 76; Hinsch, 1990, pp. 173–4). These groups were complex social forms, and lesbianism was one of several possible sexual identity positions their members might assume. They were a subset of a regional phenomenon of delayed marriage transfer in the sericulture areas of the delta, in which women delayed for up to three years moving to their husbands' households (Stockard, 1989). Marjorie Topley (1975, p. 67) contextualized the social practice: "For approximately one hundred years, from the early nineteenth century to the early twentieth century, numbers of women in a rural area of the Canton delta either refused to marry or, having married, refused to live with their husbands. . . . Typically they organized themselves into sisterhoods. The women remaining spinsters took vows before a deity, in front of witnesses, never to wed. Their vows were preceded by a hairdressing ritual resembling the one traditionally performed before marriage to signal a girl's arrival at social maturity." These sisterhoods, entry into which was marked by distinctive emplaced and embodied practices, were concentrated especially in parts of Nanhai, Panyu, Zhongshan, Sanshui, Heshan, Dongguan, and Shunde counties. They have also exited in Hui'an county, in Fujian outside Quanzhou (Jiang, 1989). Sworn spinsters worshipped Guanyin "because she is a woman and remained unwed" (Stockard, 1989, p. 80). Some were also literate and were socially perceived as more powerful, which positively influenced their opportunities for leadership in local Buddhist sects. Arguably in at least partial response to the intensive labor demands of the silk industry, women in the area did not have bound feet and female infanticide was uncommon (Topley, 1975, p. 70; Stockard, 1989, pp. 174–5n). These alternative marriage practices broke down in the early twentieth century with exposure of the regional sericulture industry to the world economy, as women lost livelihood security. The economic factors are important, but Helen Siu (1990b) has further argued that these alternative marriage arrangements were really part of a larger cultural complex that formed in the delta, even related to an older indigenous culture with some matriarchal elements that influenced the Han migrants. In these complex ways, embodied and emplaced social practices, in forms of marriage, household formation, gender orientation, and sexual difference, had evolved into diverse forms and further distinguished the region in south China. Although we cannot know definitively why these alternative practices evolved here, it is interesting that they did and that to some degree they were tolerated.

Wu's contextual history of the Qing dynasty examined differences of the south through politics of the imperial family that marked the southern tours of the great Kangxi emperor, who reigned from 1661–1772. In Wu's analysis too, the south is a process – of travel, mobility, and encounter – defined in part by the ritualized nature of the journeys, and by the differences of southern destinations. The imperial southern tour always took the same route, from Beijing down the Grand Canal to the Yangzi delta region (map 2.1). The Kangxi emperor made six southern tours, for the ostensible reasons of inspecting river conservation works and to learn firsthand about local conditions, but also to promote political stability and ties with southern literati, many of whom harbored antagonisms toward the Manchu victory and the new Qing dynasty that brought down the Ming and ended Han rule. The Yangzi delta region was powerful and it had to be managed. The Kangxi emperor also liked southern culture, especially southern style theater and food. The emperor brought his son and designated successor, Yinreng, on the tours, to socialize him for appropriate imperial duty. But Yinreng proved a questionable heir by disregarding norms of filial piety: he was blatantly disloyal and made gallingly clear his interest in his father's early abdication. One year after the final tour in 1707, the Kangxi emperor finally deposed Yinreng. He had committed a litany of crimes, but none angered his father as much as those that took place on the southern tours. It was imperially sanctioned affections for southern theater that pointed to Yinreng's scandalous conduct.

Southern theater had become an art form and a center of elite socialization. Drama in the southern style had its roots in the Southern Song and became the dominant theatrical school by the Ming dynasty, which opened its reign from Nanjing. The southern play, by contrast to the northern style, was longer and livelier, distinguished by more characters on stage, choral as opposed to individual singing, and a more languid tempo. The love affair was central to the plot of every southern play. Dramatists regularly hold out *The Peach Blossom Fan* as perhaps the most famous in the genre (Birch, 1976). Its story is the downfall of the Ming dynasty, and its imagery draws on the faded glories of Nanjing, the great southern capital. The first scene of the play opens with several key southern motifs: verdant landscapes, travel and encountering difference, and the possibilities of exotic love interests:

On Grieve-Not Lake beside the Poet's Tower,[13]
The weeping willows burgeon once again.
The sun is setting: hill and river blend

In perfect beauty, and the traveler is tempted
To drink, recalling beauties long ago,
Painted and powdered in the southern courts.
Sad thoughts come with twilight, while the swallows
Frolic regardless of the fall of kings.[14]

Southern theater was a center of elite socialization, but Yinreng maintained a world of interest beyond patronage of the arts. The Kangxi emperor was loathe to digress on Yinreng's conduct, but managed one charge: "I never allowed women from outside to frequent my palace; nor did I ever allow fair-looking boys to serve me at my side. I have thus kept my body absolutely clean, without blemish. . . . Now the Heir Apparent had done all these!" (Wu, 1979, p.92). The southern mode was one thing, but Yinreng's pedophilic predilections were apparently out of control. Yinreng sanctioned a trade in children, bought by agents protected by his imperial charge, and while selling children into service was not uncommon among actors, whose low social class was rigidly defined, "the human trade in the South had ballooned into a scandal that damaged the image of the throne" (Wu, 1979, p. 96). For these and other atrocities, Yinreng's deeds were ultimately deemed mad and he was removed as unfit to rule.

During the Southern Song dynasty, as Nanjing, Suzhou, and other major cities became refuges for literati from the north, Han elite culture became increasingly infused with elements of southern style. In the process, representations of the south became more complex and contradictory for northerners, who began to view southern culture as more desirable and different in the sense of remaining intriguingly exotic without the elements of southern strangeness – at least not until one ventured further south to the unruly shores of Fujian. Silas Wu's treatment of the Kangxi emperor's attempts to integrate into the realm of southern culture portrays the south as a region of seduction that ruined lesser men. For the Confucian official, sensualized ideas about the south, infused with images of sexual alterity, combined with the general notion of southern strangeness, could yield perceptions of a southern region as alternatively a zone of seduction and a land of cultural and environmental fear and moral lawlessness. One model of imperial rectitude, in the emperor's own journeys and southern tours, offered exiled officials some spiritual refuge in coming to the south. But the conduct of the heir apparent ultimately complicated and contradicted that model. Compared to the conduct of the father, the son led a life of moral debasement that only mocked the Confucian order. For northern elites, the south represented both a high culture refuge and a region where things could go wrong.

Rewriting South China

In the imperial period, historians followed the patterns of traditional textual practice and reproduced images of the south that upheld the superiority of northern culture, which undergirded Confucianism and its role at the basis of state ideology and legitimacy. Practices of statecraft transformed in the twentieth century, but the Communist regime had its own nationalisms to build and intellectuals were compelled to write in service to the state. The Maoist national project commandeered the historic discourses of the northern vantage and promoted ideas about a northern origin of Chinese civilization and its superior influences on the rest of the country. Certainly it mattered that the civil war base of the Communists was at Yan'an, deep in the loess plateau of Shaanxi province, coursed by the Yellow River. Edward Friedman (1994, p. 70) has argued how "Mao's anti-imperialist revolution was the culmination of Chinese national history as an ascent from Peking man through an expansionist, amalgamating, and unifying Han culture to the founding of the People's Republic." Maoist era nationalism explicitly condemned mercantilism and internationalism, and focused on isolating China from what Communist ideology dictated as exploitative foreign influences. When Friedman (1994, p. 81) writes that Maoist ideology "stigmatized the south as the enemy of the nation," this "south" is the south China coast of treaty ports, Chinese who went overseas, and the compradore class of Chinese merchants who developed international trade relations.

Significant promotion of nationalist ideology and state propaganda continues in China under reform, yet the reform era also made into policy Deng Xiaoping's maxim, "seek truth from facts." Recent interpretations of China's past have abandoned the unilinear "north to south" model, and the long-cherished ideal that the only true origins of Chinese civilization, as evolved from the north China Neolithic period, lay in the Yellow and Wei river valleys. New interpretations have favored understanding China's archeological past as a rich historical geography of diverse centers of formation, so that "the late Neolithic and Bronze Age of China were a rich amalgam of influences from many areas, including those outside the Yellow River valley itself" (Olsen, 1992, p. 4). Reinterpretations of the geographical origins of Chinese civilization have undermined the Maoist era dictums that maintained the northern vantage, so that China's contemporary problems, as in "River Elegy," may now be attributed to a historic lack of openness to the world. With the rise of power in the provinces, China's scholars have been rewriting regional and national history.

Bai Yue

Knowledge about the historic southern kingdoms has proliferated in China under reform. From the Zhou dynasty in the second millennium BCE to the Tang, several independent and sometimes loosely confederated kingdoms ruled over the south China coast, from Zhejiang to Vietnam and inland as far as Hunan and Guangxi. These kingdoms are known in Chinese history as the Bai Yue or Hundred Yue peoples, but only recently has their history become a common project. In the first decade of reform, Chinese scholars published over 1,000 papers on the Bai Yue (Peng, 1990, p. 3).[15] Three of the more prominent Bai Yue groups are the Wu Yue, which geographically correlates with the Jiangnan area, the Min Yue, of the Fuzhou area, and the Nan Yue, of the Lingnan area in Guangdong and Guangxi provinces. The kingdom of Nan Yue had been incorporated into the empire during the Qin dynasty (221–207 BCE) and again during the Han (206 BCE–220 CE), but was an independent kingdom for nearly a hundred years after the fall of the Qin.

Data for contemporary chronicling of China's historical geography leans heavily on the country's archeological potential, and in China under reform the frenzied pace of new construction has led to some major finds. In 1983, construction excavation in Guangzhou uncovered the intact and richly endowed tomb of the second king of Nan Yue. The discovery was important in itself, but the fanfare accompanying the find was, for the purposes of reconstructing southern regional identity, perhaps more significant (Lary, 1996). Guangdong provincial authorities sponsored an imposing museum to house the new Nan Yue artifacts, on the site of the north gate of the former city wall. The collection traveled for display to regional museums, especially in Hong Kong. Collection descriptions affirmed the unique designs of the artifacts, their superb quality, and their autonomous origins in historic Lingnan. Before the tomb discovery only smaller Nan Yue sites in more remote areas had been uncovered, and Nan Yue existed more as a mythological landscape than a material reality. The Guangzhou discovery and its promotion reconstituted Nan Yue and confirmed its existence: the Nan Yue artifacts, two millennia old, were celebrated as evidence of contemporary regional identity. In 1990, in the style of southern drama, a Guangzhou dance ensemble presented "King of Nan Yue," a performance that told the story of the first king of Nan Yue, a northerner from Hebei, who fell in love with a Nan Yue woman and became converted "from harsh, joyless northernness to lush southernness" (Lary, 1996, p. 5). In this imaginary, the identity of a leader from the north was reconstituted on southern terms and by a woman, and the south emerges in contemporary images of the most ancient symbolisms, a new desirable *yin* to the north's old harsh *yang*.[16]

Kingdom of Chu

The most fundamental rewriting of regional significance is the traditional account of the northern hearth of Han Chinese culture and its historical diffusion to the south, enveloping cultures along its advance into a Han ecumene (see p. 102, n. 4 below; Friedman, 1994, pp. 80–3). This canonical view has held that the cultural and political complexes of the northern kingdoms, especially the Zhou and Shang, in the Wei and Yellow river valleys, diffused south to infuse Chu with significance and richness. The kingdom of Chu (Spring and Autumn 722–481 BCE and Warring States 475–221 BCE periods) ranged from modern Hebei province to Hunan, Anhui, and southern Henan, and existed for over 800 years until it was conquered by the state of Qin in 223 BCE (map 2.2). Now, in Chinese scholarly literature as well as in the popular press, the interpretation of the historical significance of Chu has transformed from a cultural receiving ground to a center of indigenous and influential cultural and political leadership. Much of the debate over defining Chu has focused on the reassessment of great archeological finds from Henan province in the 1920s. Originally interpreted as representative more of northern culture, some of the Chu finds have been reinterpreted to be emblematic of the "new style formed in the state of Chu and then spread northward" (Li, 1991, p. 12). Further, "The Chu bronze culture was therefore among the most advanced of the bronze cultures that flourished in the other states at the time. In fact, contemporary arguments hold that during the Warring States period, the state of Chu had already united the southern half of China, laying the foundation for subsequent unification during the Qin and Han dynasties" (Li, 1991, p. 21–2). Such perspectives suggest a fundamental spatial reinterpretation of Chinese historical geography.

Can the idea of "southern strangeness" also be unraveled by new understandings of Chu? Robert Campany (1996) has argued that "anomaly accounts," or writings about strange phenomena, formed a regional genre in Chinese literary tradition during the seven century period from Han to Tang times.[17] Anomaly accounts were not part of the Confucian canon, and most authors of these texts were officials in the lower Yangzi region. On the margins of literary tradition, anomaly accounts typically took the form of descriptive lists and were concerned with strange landscape elements of distant regions, including odd peoples, marvels, and beasts. The oldest surviving anomaly account, the

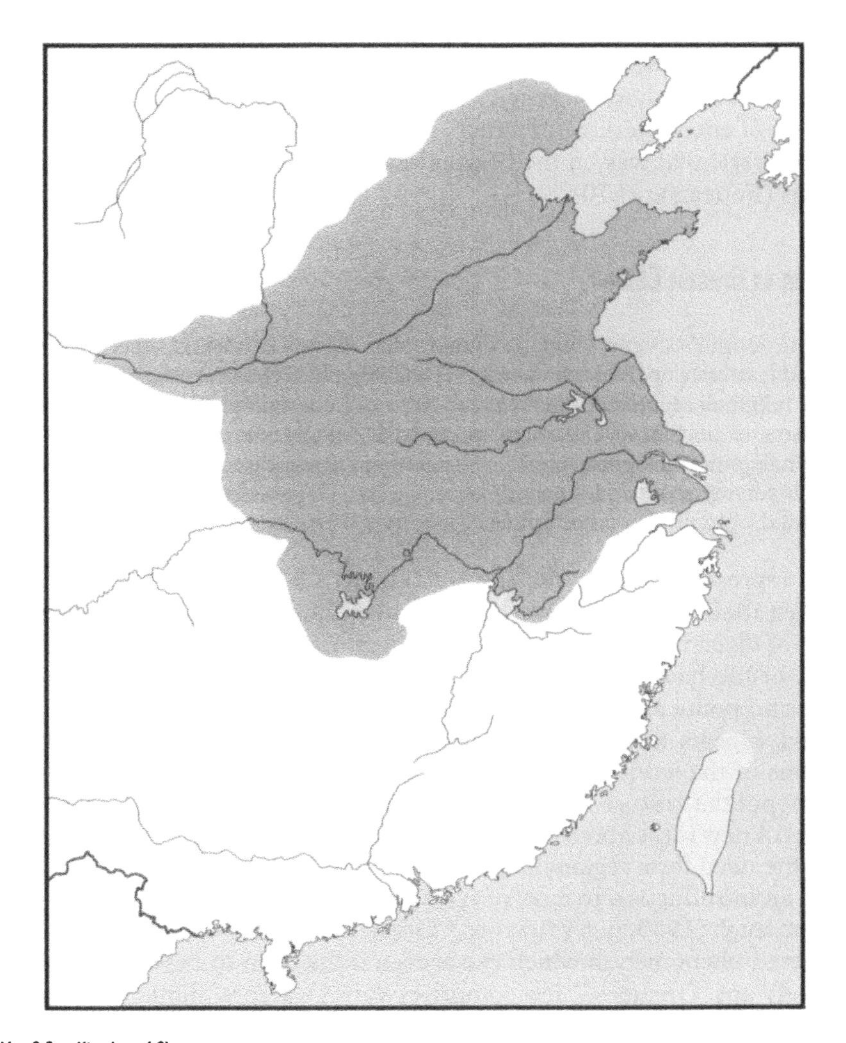

Map 2.2 Kingdom of Chu.
Source: Herrman (1966); line work by Jane Sinclair.

Shan hai jing (*Record of the Mountains and the Seas*), has southern origins and is likely a product of Chu. The contents of the *Shan hai jing* bear a spatial organization broadly corresponding to regions of the known world in sections named "classics of mountains," "classics of seas," and "classics of the great wastes," the last "beyond the pale of all civilizing influence" (Campany, 1996, p. 35). In the context of historic migrations, the record of anomaly accounts suggests that literati, as arbiters of empire, were discursively encompassing and organizing ideas about difference in the south, a practice which would eventually trade perceptions about the wild and barbaric for increasingly known and civilized landscapes. Southern strangeness had become not only a subject but a project of encountering difference, and representing that difference in a literary style that was, in the Foucauldian sense, an ultimate ordering of things (Foucault, 1970).

Region as Greater China

> The south's Greater China, a China that crosses borders, is successfully and fearlessly open to the world. It is a China that celebrates a multiplicity of religious sects and world views, one not constrained by the anachronisms of northern Confucianism, with its denigration of the young, the female, and the commercial – and one not manipulated and glorified by self-serving, parasitic northern bureaucrats. The southern project is not afraid of multiple communities of identity. (Friedman, 1993, p. 271)

This representation of "the south's Greater China" contravenes many images about China at large. It sets south China apart as a transnational zone of diversity and difference, where cultural, political, and economic ways of life have transformed beyond the stolidities of the patriarchal state and political directives of the northern capital region. This Greater China appears as China's contemporary counterpart to other leading regions of the world where dynamic economies and global ideas about cosmopolitan culture distinguish the area from its nation-state. The realities of new ideas about the importance of south in China under reform fuel the need for a regionalization like Greater China.

In an introduction to a set of essays devoted to Greater China, David Shambaugh (1993, p. 653) wrote, "Greater China is a complex and multifaceted phenomenon which exists even if the term to describe it is not entirely apt. Greater China comprises various actors, dimensions, and processes. Together they pose a potential challenge to the regional and international order." The term Greater China arose to designate new strong economic ties between Hong Kong, Macao, Taiwan, and especially coastal south China. The idea of Greater China has subsequently evolved beyond the recognition of transboundary economic relations to go as far as encompassing the cultural, political, and economic processes in China at large and between China and Chinese overseas on a world scale. Greater China, in its diverse forms, has been referred to as a phenomenon, a region, a possible political-economic entity, and a form of globalization. But it lacks entirely institutional representation and corresponds to no recognized territory. Debate ensues over what it means and where it lies, but still few deny its importance. Like globalizaion, Greater

China is a communicational device whose slippages in meaning reflect diverse positions and goals.

In this analysis I treat Greater China (hereafter greater China) in the first instance as a transboundary cultural economy, in recognition of the causal forces that have given rise to the concept, and, in its representational capacity, as a term that signifies a range of complex globalizing processes associated with a transnational China, and people who identify with Chinese culture.[18] In these ways greater China works as a figurative region and a geographical imagination through which people and institutions engage in multiple and shifting issues about relations between China, Taiwan, Hong Kong, and the Chinese overseas, and the cultural and economic processes generated in and between these places and groups. As Aihwa Ong and Donald Nonini (1997, p. 9) have written, "imaginaries – of the nation-state, of Greater China, and of Asian values – are produced from different sites, and they are in varying degrees of tension with each other; they also ultimately have points of application 'at home'." This critical understanding of greater China, as a discursive trope and signifier of diverse ideas, clashes with some of the social science efforts to pin down greater China as a defined political economic entity. In attempts to make greater China appear more "scientific," political scientists and sociologists especially have used transboundary regional designations, including "natural economic territory" and "growth triangle," in place of greater China to designate a political economic region. These attempts to come to grips with a rapidly evolving transboundary region signal both a need to conceptualize the regional formation and the reality that greater China is subject to multiple regional representations.

In one of the more succinct discussions of globalization, Fredric Jameson (1998a, p. xi) prefaced his remarks with a caution "about the inevitable recursiveness of definitions," and the recognition that globalization as a concept "knows its own internal slippages among various zones of reference."[19] This consideration of greater China similarly has no interest in providing a singular definition for greater China, but instead in showing how the geographical term has worked to convey particular processes, ideas, and goals. Like Jameson's estimation about globalization's self-reflexive character, ideas about greater China deployed in Asia tend to know slippages in meaning, whereas Western debates around the idea of greater China do not reliably reveal their contradictory elements. The problems of misunderstanding the possibilities of greater China conflate economic activity with sovereign political power, and an unbounded transboundary economic region with Chinese territorial sovereignty, ultimately supporting the idea of a "China threat" in the geopolitical world order. This line of logic also suggests how ideas about greater China may be seen as alternatives to forces of hegemonic Western capitalism in its globalizing capacities. Is it too predictable that Western analyses about this possibility have uniformly interpreted greater China as the hegemonic force?

The contemporary term greater China – and that of greater Hong Kong – came into use in the 1980s to express the development of new transboundary economic realities in south China. The idea of greater Hong Kong evolved from the perspective of the pre-1997 colonial territory in recognition of increasing ties between Hong Kong, Macao, and Guangdong province. Greater Hong Kong was a challenging, if ironic, counterpart to greater China, by signifying all the ways in which Hong Kong's economy and cultural trends had begun to gain influence in China, from real estate investment to television programming. Greater China, by contrast, originated in Hong Kong and Taiwan and spread much more widely into the international arena to designate a southern China transboundary region extending from Hong Kong and Macao to Guangdong, and between Taiwan and Fujian.

Chinese language expressions of the greater China idea originated in Hong Kong and Taiwan at the turn of the 1970s. As Harry Harding (1993, 1994) has explained, the direct translation of greater China, *da Zhongguo*, was not among the first terms in use. Instead, phrases like *Zhongguoren jingji gongtong shichang* (Chinese common market), *Zhongguoren gongtongti* (Chinese community), and *Zhongguoren jingji jituan* (Chinese economic group) were among the first to appear, which indicates the reality of the different positionalities at stake in the evolving transoundary region, where political–territorial congruence, especially in the case of Taiwan, would be an elusive goal. This is why some Taiwan-based discussions of the greater China idea prefer renditions like Chinese common market, which implies equal political standing among members, after the pattern of the European Union (Cheng, 1993–4).[20] Some Taiwan-based interpretations of greater China have also favored the inclusion of Singapore, in translations such as "Asian Chinese common market," for implied treatment of territorial members as states. This interpretation of greater China also invokes ideas about a larger-scale region based on Chinese cultural values in Confucian thought. PRC scholars, on the other hand, have generally disfavored interpretations of the transboundary economic region by any means that would suggest state-to-state political terms. Harding (1994, p. 235) has recorded the term's widest possible set of renditions, both Chinese and English, in "no fewer than 41 variations on this particular theme, ranging (in alphabetical order) from the 'Asian Chinese common market', through the 'Chinese economic community', and the 'Greater China economic sphere', to the 'South-east China free trade area'. These formulations differ along two principal dimensions: the scope of the territory regarded as being included, and the degree of formal integration being proposed." The shared ground among these terms, according to Harding (1993, p. 660; 1994, p. 235), is that they all refer to some element of a common phenomenon: "the construction or revival of economic, political, and cultural ties among dispersed Chinese communities around the world as the political barriers to their interaction fall."

Let us consider the actual geographies implicated by this discussion. The different terms for a greater China suggest variability in territorial extent, and this slippage in territorial meaning has different consequences in different contexts. For political analysts concerned with the fundamentals of the nation-state, implications of the discussion of a potential territorial component of greater China connote an expansionary territory. From a pragmatic international perspective, the concept of greater China has existed to designate the reality that policy formulation toward China can no longer proceed without accounting for the cultural and economic connections across the Taiwan Strait and the role of Hong Kong in the Chinese economy. Michael Yahuda (1993, pp. 689–90) has suggested that extensive use of "Greater China" is a particularly American phenomenon, tied to discussions in US government policy circles. By emphasizing the idea of revival of ties among "Chinese communities around the world as the political barriers to their interaction fall," Harding's discussion constructs a boundless globality for greater China. But such assumptions raise questions about the differences between realities and possibilities in an era of globalization rhetorics. The end of martial law in Taiwan in 1987 and the lifting of the ban on travel to the mainland witnessed the fall of an extraordinary barrier, but what other political barriers have given way? The opening of China in general after 1978 allowed Chinese overseas, like everyone else, the opportunity to travel to the PRC. But how does this lead to strengthened ties among "dispersed Chinese communities around the world?" Does this perspective suggest a particular kind of view about ethnic association and uniformity (also common to traditional regional geography), and an essentialist priority of race-based ties? Here, the slippages are between new possibilities for communications, with the internet or in "world meetings" of Chinese business leaders or clan associations (see H. Liu, 1998), and problematic interpretations of the new globalizing geographies of information and mobility through the lens of traditional geopolitics.

Historic counterparts of the greater China term promote such notions about imperialist and expansionary geographies. "Greater China" is related to the designations for imperial Han China or China "proper" and its "frontier dependencies," which have historically included the areas of Manchuria, Mongolia, and Tibet (Uhalley, 1994). The terms "inner" and "outer" China have also been used to distinguish these realms. The symbolic properties of inner and outer China worked to indicate the relative strength of the Chinese empire through the vagaries of dynastic cycles: when the dynasty was strong, the imperium held sway over "outer" China; a weak dynasty contracted to the area of eastern or "inner" China. "Greater China" has also been used to designate China, including "outer China" on Second World War-era maps. Thus, in the language of mid-twentieth century geopolitics, "Greater China" was an imperialist power at its most expansive territorial extent. Recollections and repetitions of this idea give rise to contemporary notions about a greater China on a much larger scale than the transboundary economic region in south China, and as a potential imperialist power.

Alternative cultural interpretations of greater China in a "new Confucianism" have intersected with political economic representations to script a regional modernization narrative for parts of East and Southeast Asia. Singapore has played a central role in promoting the regional reinvigoration of Confucianism, in what Arif Dirlik (1995) has called a "reinvention of Confucianism" from the "borderlands." In 1982, against claims that young people were becoming too Western and individualistic, Singapore initiated a program of courses in ethics and religion, including Confucianism (Wong, 1986).[21] Then Prime Minister Lee Kwan Yew legitimated the program by inviting eight of the world's leading scholars of Confucianism to confer on the role of Confucianism in Singaporean society (see Tu, 1984).[22] Official support for renewed interest in Confucianism appeared in China in 1985 when the Chinese Confucius Society was established as "the first non-governmental national research organization" (*Xinhua*, 1989). The state sanctioned Confucian revivalism through the early 1990s, in the intellectual vortex after the Tiananmen incident and Deng Xiaoping's banning for "at least three years" evaluations of the intensified pace of reform sanctioned by his southern tour in 1992 (Dirlik 1994; H. Liu, 1998). In China in 1994 Jiang Zemin turned out to greet the participants at the meetings of the International Society of Confucianism, which installed Senior Minister Lee of Singapore as its honorary president. For Chinese political leaders, this new Confucianism appeared as a discursive detour around problems of state authoritarianism, and as a rhetorical reply to problems of uneven development and increasing income polarization (see J. Wang, 1996; Barmé, 2000). Unburdened of early twentieth-century Weberian ideas about Confucianism's impediments to modernity, the new Confucianism emerged as a would-be humanist condition at the basis of Asian regional development, and an explanation for rapid accumulation as an apparently "natural" regional phenomenon. As a cultural economic position, the new Confucianism served as an ideological challenge to Western sociological theories of capitalist organization, and an alternative regional identity formation – another communicational device – challenging Western globalization narratives.

Myth of the "ungrounded empire"

The problems of understanding greater China in the West are often based on assumptions about greater China as a container of national interest. Such misunderstandings are apparent in media coverage of political events about transnational processes between the USA and Asia. The example of the US "Donorgate" debacle, in which the Democratic National Committee (DNC) received foreign campaign contributions during the 1996 federal election, suggests the nature of the problem. In 1998, the US Senate report on Donorgate sought to substantiate that illegal donations to the DNC originated from "bank accounts in the Greater China area." On this basis, the report assumed that Beijing had attempted to influence US foreign policy and that particular Chinese-American citizens had been acting illegally as intermediaries for foreign-sourced campaign funding (Sanger and Van Natta, 1998). Without coursing through the details of the debate, it is arguable that, both at the political institutional level and in the popular press, many US analysts have viewed greater China as a networked region of Chinese ethnicity, which has conflated Chinese ethnicity with Chinese citizenship and sovereign representation.[23] This "networked Chinese" thesis finds inspiration in diverse and politically unrelated sources.

A spate of popular tracts attempting to account for economic growth in Asia has promoted race-based globalization narratives and contributed to problems of misunderstanding the transboundary cultural economy in south China. One needs only a glance at the title of Sterling Seagrave's *Lords of the Rim: The Invisible Empire of the Overseas Chinese* (1995) to apprehend implications of a global Chinese nation. John Naisbitt (1996, pp. 20, 26), in the first chapter of *Megatrends Asia*, entitled "From nation-states to networks," highlighted in bold print that "The economy of the borderless Overseas Chinese is the third largest in the world," and that "The Cold War is over and the Chinese won." Murray Weidenbaum and Samuel Hughes (1996, p. 23) chose perhaps the most unfortunate title, *The Bamboo Network: How Expatriate Chinese are Creating a New Economic Superpower in Asia*, in their attempt to explain how "the overseas Chinese have developed a bamboo network that transcends national boundaries." These books, apparently designed for popular markets, have treated issues of race unproblematically, and have assumed shared ties and values among Chinese individuals without evidence to support such ties.

Critical scholarship on greater China also entangles its regional geographies. In *Ungrounded Empires: The Cultural Politics of Modern Chinese Transnationalism*, Aihwa Ong (1997) has identified different tropes that would claim to define greater China as "an overseas Chinese capitalist zone" (p. 173) and an "open-ended space of Chinese capitalism" (p. 179). However apparently deterritorialized, such a zone as a financial entity, combining the foreign currency reserves of Taiwan, Hong Kong, Singapore, and China "would place the Chinese bloc way ahead of Japan as Asia's first-rank economic giant" (pp. 175–6). Such deterritorialized perspectives find theoretical basis in the work of Gilles Deleuze and Félix Guattari (1987) among others, but the open spatiality assumed in the discussion, and apparent assumptions of a deterritorialized/reterritorialized frame, privilege the nation-state and its boundary-making project, and so do not read the slippages in regional/global formations. Scholars in China reject the greater China concept, Ong continues, because "any ideological recognition of a Chinese transnational capitalist zone will undermine China as a territorially based political entity" (p. 176). Ong (p. 175) uses the *da Zhonghua* translation for greater China, "a term coined by overseas economists to describe the increasing economic integration among China, Hong Kong, and Taiwan produced by globalization," but does not mention any of the many different Chinese translations of greater China in use. The character for *hua* (華) in the *Zhonghua* rendition of China is a historic translation of the country name, compared to the modern *Zhongguo*, in which the character for *guo* (國) indicates the character for a nation-state. Using the *da Zhonghua* translation recalls a historic China larger than its present boundaries ("outer China"), which carries the connotations of an imperialist state. These different interpretations of greater China reflect the complex conditions of the regional formation, but simplifications about them can have real policy consequences.

The extensive scholarly literature on Chinese business networks (e.g. Hamilton, 1991; Hsing, 1998; Yeung and Olds, 2000) has found in business networks the "culture" of the regional cultural economy. Cultural factors of Chinese business network formation typically focus on the traditions of familial and social life based in ancestor worship and filial piety, kin-based connections, and trust-based relations and *guanxi* practices (see M. Yang, 1994). But between the scholarly literature and the popular sphere, the complexities of diasporic cultural economy are often lost, especially from the perspective of the Western gaze that sees first the ethnicity or race of economic agents. The reality is that contemporary business networks are internally differentiated and distinctive, especially based on differences of citizenship, class, and dialect, or ethnic sub-group. As Brenda Yeoh and Katie Willis (1999, p. 357) have critically observed, "while reductionist perspectives on 'Chinese business networks' can lead to views about the 'Chinese' as a racially bound population group, the social, cultural, economic, and national differences among them are equally crucial and provide an intricate grid on which the politics of sameness and difference are constantly played out." Here I add to this the concern that part of what is at stake in the reductionist "networked Chinese" problematic is the "network" metaphor. "Network" may be the metaphor of our times, of globalization and the Internet, but there are at least two concepts of network that we need to understand.[24] One spatial concept of network would encompass an entire field, connecting all elements within it (see also Latour, 1993). Images of the Internet and information systems are presented in this way – but the reality is that the Internet is unevenly connected and can only be perceived discretely in parts. You can never get a fix on the whole of the network, and thus networking represents itself in a particular way that is not accurate to the ways in which networks actually work. If we instead think of network through place, we realize that even as places are interconnected, it is not possible to determine those places all at once, or to assume that all places are equally interconnected. The same holds true for business networks and suggests why understanding cultural economic practices of diaspora through axes of difference, place, and scale may allow an ethical intervention in those perspectives on Chinese business networks whose underlying logics otherwise transmit inappropriate notions about connections among all members, as if hinged to the production of a race-based globality.

Transboundary economies and the territorial trap

The social science attempts to conceptualize greater China raise some of the same problems, also common to traditional regional geography and the problem of the territorial trap, in which the regional boundary is understood to circumscribe an area of some internal homogeneity or uniformity. These conceptual problems are found in the common approaches used to frame transboundary economic regions in Asia during the 1980s and 1990s: the growth triangle (GT) (Lee, 1991; Parsonage, 1992) and the natural economic territory (NET) (Scalapino, 1991; Jordan and Khanna, 1995) (see figures 2.1 and 2.2). In general, the GT and NET approaches, each of which has been used to frame analyses of greater China, emphasize contemporary regional processes and acknowledge geographical and historical characteristics of rapidly developing regions, but treat minimally the spatial realities of transboundary economic development. In that way they are nominally geographic, and so

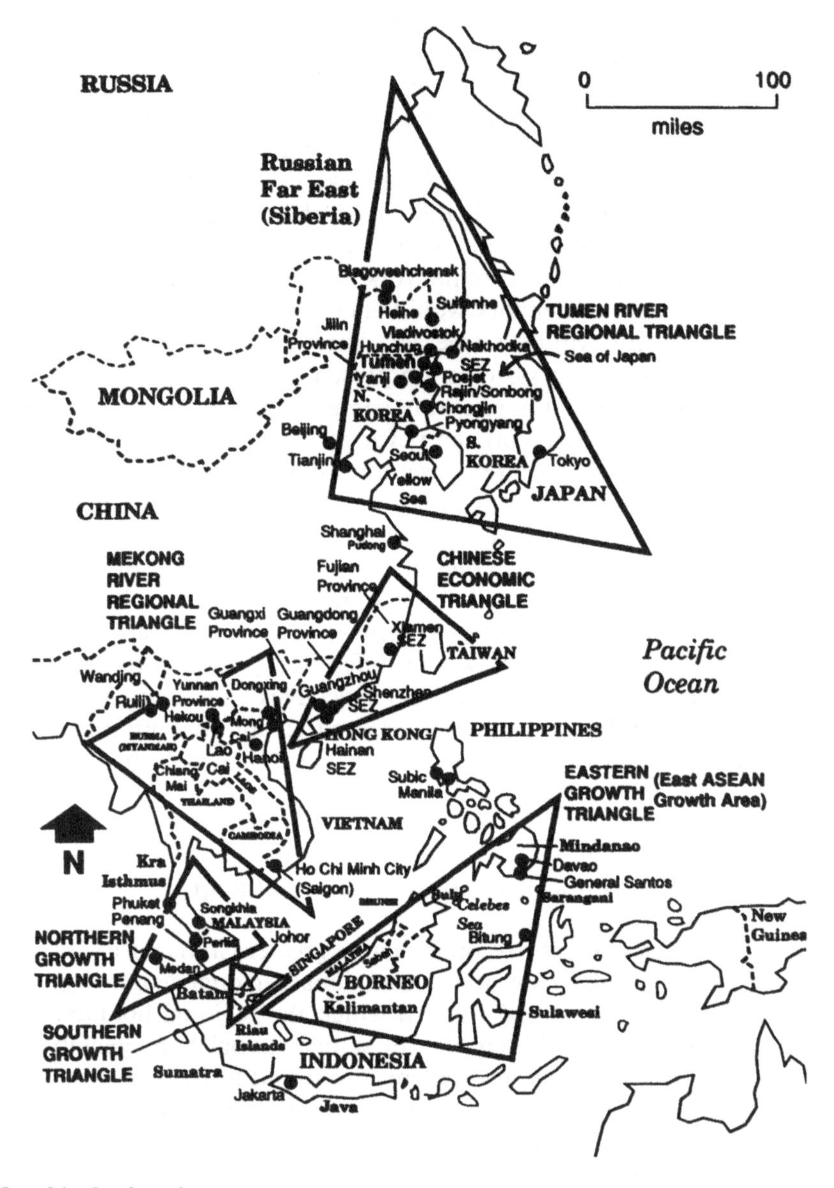

Figure 2.1 Growth triangles.
Source: Xiangming Chen (1995); originally printed in the Christian Science Monitor, December 1, 1993.

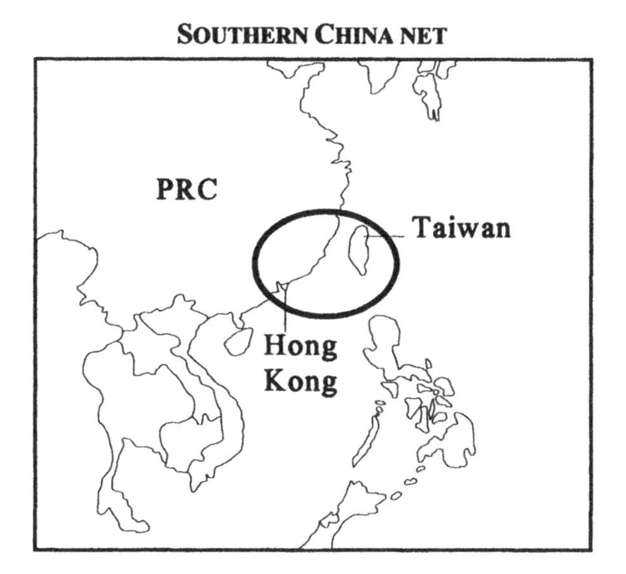

Figure 2.2 Southern China natural economic territory.
Source: Jordan and Khanna (1995).

also serve as tropes of regional analysis. These schemes exhibit properties of the "territorial trap" in which the act of naming and fixing a boundary – however crudely in the case of triangles and circles – idealizes set representations of space and region irrespective of historic and political context. Once transboundary economic regions are lent characteristics of the state, they may be treated as if they share similar institutional characteristics, and as if they exist prior to the formation of a localized society and as a "container" of that society. Such significations of state-like characteristics infer ideas about shared identity among the constituent population groups akin to national identity, and shared political economic practices, as in the idea of "networked" Chinese people in the greater China region. The conceptual problem in these political economic treatments, both scholarly and popular, is the conflation between the transboundary cultural economy and bounded territorial regions. The idea of a regional formation called greater China accommodates overlapping and divergent world views because, in part, it represents globalizing processes which cannot be circumscribed within politically bounded state territory.

History of the North and the South

> The most important contradiction: Beijing people go to Guangdong to learn the essence of dealing with the central government. (Qin and Ni, 1993, p. 10)

In the early 1990s, ideas about the economic strength of the greater China transboundary cultural economy led to ideas about new formations of political power and the notion that regional tensions could lead to the collapse of the Chinese state. In the wake of the collapse of the Soviet Union, the publication of *Nanbei chunqiu: Zhongguo hui bu hui fenlie?* (*The History of the North and the South: Will China Disintegrate?*) in 1993 appeared to suggest a break-up of the country, and various scholars have portrayed the book in such a way. But, written as a collaborative project of the state Historical Materialism Institute, the book really engages a different subject: the traditional leading region of the north has fallen behind, and northern ways have been the root of the problem. Further, China is not about to collapse; rather, China's historic interregional relations are the basis for national-scale coherence. After hundreds of years of northern influence on the south, in China under reform the north must learn from the south. In lengthy, comparative regional analysis, the book assesses changes in the provinces under reform, particularly successful examples of reform leadership practiced by cities and provinces in south China, and which events in two millennia of Chinese history have laid the base for the late twentieth-century transformation that turned the tables on the northern heartland. The authors launch the account by defining south China in different terms: "the south lies below the lower reaches of the Yellow River and below middle to lower reaches of the Changjiang, and the north is all the area outside the south" (Qin and Ni, 1993, p. 46).[25] By this definition the south is Hainan, Guangdong, Jiangsu, Zhejiang, Fujian, Shanghai – and Shandong. The inclusion of Shandong underscores the issues at stake. Shandong has enjoyed a rich reform experience in high rates of foreign investment and economic growth, and so must be characterized as one of the leading provinces of reform. Although in the early historic period there is some precedence for the south emerging below the Yellow River, in this account the south becomes a regional trope for the successful reform experience, and a discontinuous region, while the north represents all regions that have lagged behind and insufficiently adopted reform-minded ways.

In *Nanbei*, Southern success under reform is generally attributed to institutional and economic innovation and unbridled entrepreneurialism. Numerous examples in *Nanbei* laud reform policies innovated in the south, while simultaneously characterizing the limits of northern ways. The southern standard holds that "as long as policy is not prohibitive, we will do it," while the comparative northern perspective is "if there is no policy, we cannot do it." (Qin and Ni, 1993, p. 77). Even if Shenzhen securities brokers were selling shares on the street without state permission, *Nanbei* makes use of the incident by highlighting how Shenzhen people would wait in line several nights in a row just for a form to buy shares, while the Beijing government was still arguing about whether to set up a share market. In these ways *Nanbei* constructs and concretizes contemporary stereotypic representations about the two regions, and through paired comparisons which depict the north and south as the two essential geographical regions of the country. The contemporary south has served as the model for the northern transformation, the politically savvy region where northerners can be relieved of their fears of reform and liberate their thoughts.

Regional Thinking

Using greater China as a fulcrum for analysis about regional formation raises the idea of regions encompassing parts and wholes of nation-states and other territories, or their representations, which suggests opening up a new language of regional formation independent of the state and territorial boundary inscription. Human geographies of greater China, characterized by diaspora and high mobility, require theorizing a concept of region in ways that encompass both material places and concepts of place and embodied spatial practice. Economic geographies at the foundation of the idea of greater China demand an understanding of capital organization in ways that compel a cultural economic geography, one that recognizes the cultural and social contexts of emplaced economic practices. Historicizing greater China is critically important, because in its historical geography lie the elements of knowledge that would ideally free people from the scripts of greater China created by business and governments (and scholars). Greater China as a term and an idea has a long history. Greater China as a transboundary region in south China has a much longer one.

These historic and contemporary geographies have been presented and retold, and, in this text, continue to construct ideas about south China. From perceptions about the natural environment to concerns about individual and bodily conduct and the changes wrought by economic restructuring, the south in China has emerged as a set of dynamic regional formations, their human imaginations, and changing representations. The region from the northern vantage, the one in motion, of overland travel to the lands of southern strangeness that Li Ao endured, to the one of seduction whose lifeways and cultures relieved certain northern sensibilities – the south has been a process of incremental encounter. Apparently disparaging representations of the south were contradictory motifs which ultimately revealed the perspectives of northern wayfarers who had yet to acculturate themselves to southern style alternatives. In times of change, northerners have embraced the south and absorbed its distinctions as a basis for northern transformation.

Once the Huai River defined the limits of the habitable world for an early Chinese empire. Later, the most proximate version of the south, the Jiangnan region, became by the Song dynasty an exotically attractive high culture refuge. Yet Schafer (1967, p. 263) described how the Yangzi basin "gave only partial preparation for the deeper south." The more problematic notions of the south increasingly shifted south, to Fujian and Hainan. There too, challenges to northern orthodoxies have proliferated and sustained. Like the changing significance of the Jiangnan region, the regional role of the longer south China coast has transformed over time, challenged the national order, and, in the activities of its leaders, incrementally rewritten regional meaning on its own terms. Despite repeated notions derived from the northern vantage, of the south China coast as some realm beyond the limits of acceptable civilization, the position of the south has shifted and currently ascended in the national–regional order. From cultural forms in fine arts to real economic power, the population centers of south China have commanded regional significance and leadership in national and international realms. In China under reform, the south China coast is a region of internationalization – the leading region of transboundary and transnational connection and transformation – and the northern vantage has been revealed as a pure social construction of the imperial era and Maoist political economy. Even from this preliminary exploration, we can see how the significance of regions emerges not in transhistorically bounded territories, but in reflections of transformative geographical processes, their representative spatialities in dynamic and shifting regional formations, and the need for spatializations of such processes to take form in regional imaginations.

ACKNOWLEDGMENT

The extract from Kung Shang-jen, "The Peach Blossom Fan," edited/translated by Chen Shih-hsiang and Harold Acton (Berkeley: University of California Press, 1976) is reproduced by permission. Copyright © 1976 The Regents of the University of California.

NOTES

1 Compare especially the work in David S. G. Goodman (1989, 1997), David S. G. Goodman and Gerald Segal (1994), Solinger (1993), and Yang (1997).
2 See James Duncan and David Ley's (1993) introduction to *Place/Culture/ Representation* for explanations and use of this approach in geography.
3 Assessments of social construction approaches have been politicized, and arguments against their merits usually fail to acknowledge that social construction concerns ideas about empirical realities as opposed to a lack of attention to those realities. In environmental studies see William Cronon's (1996) discussion of the problems in the preface to the second edition of *Uncommon Ground.* In China area studies, see Jeffrey Wasserstrom (1998, p. B5).
4 See also the discussions of identity formation in Allen et al. (1998).
5 Exceptions about the located heart of empire in China's long history are Lin'an, the progenitor of contemporary Hangzhou, as capital of the Southern Song and Yuan dynasties, Fuzhou, as the capital of the "Southern Ming," 1644–62, an interregnum between the Ming and the Qing, and Nanjing, which served as capital for the early Ming dynasty and in the twentieth century for the brief period of Chiang Kai-shek's mainland rule during the "Nanjing decade." Nanjing means "southern capital" just as Beijing means "northern capital."
6 The significance of this latitudinal division is similar to the significance of the hundredth meridian in the USA, which has served as the traditional division between the more well watered landscapes of the Middle West and the arid lands of the West.
7 Changes in the stream channel of the Huai also exacerbated perceptions of an unruly environment in the south. In 1324 the Yellow River shifted course and deposited so much sediment in the valley of the Huai that the Huai no longer flowed to the sea but ended in a series of lakes (see Zheng Zhaojing, 1993, pp. 32, 136–161). The flood and drought regime in the Huai basin – more than 900 documented floods and as many droughts during the two millennia period from 246 BCE to 1948 CE – gave the Huai a notorious reputation, and shoring up the river became a major hydrologic project during the early communist period.
8 In the end pages of their popular history, *The Beach,* Lena Lencek and Gideon Bosker (1998) include a substantial list of "world class" beach destinations which features a list of beach resorts for every populated continent except Africa. The reasonably substantial section on Asia includes no listings for China. China's two major beaches are located at the shores of cities that served as foreign settlements during the treaty port era, on Gulangyu island in Xiamen, the special economic zone, and in Qingdao, a former German-dominated settlement, on the Shandong peninsula, and one of the fourteen open port cities.
9 Reactions to "River Elegy" provoked sustained response in China and Taiwan, in the form of several book-length replies, conferences, and scores of

articles (see also Rosen and Zou, 1991/2, 1992a, b; Wakeman, 1989).

10 It is poignant and a serious natural resource problem that, as a result of rapid industrialization in China under reform, in the 1990s the Yellow River has run dry every year, with the dry section extending 600 km from its mouth to the upper reaches in Henan province. The 1997 record flow cut-off lasted for 130 days; see *China Daily* (1998a).

11 The Wu/Yue region is effectively the south China coast. Wu is a historic name corresponding to the Jiangnan region and is the contemporary name for the dialect in the Shanghai area. See the penultimate section of this chapter for discussion of the Yue peoples and corresponding geographical area.

12 I am indebted to Maram Epstein for these insights, as well as for discussions about the importance of a southern literary genre.

13 These features were located on the west side of Nanjing city outside the wall.

14 From the translation by Shih-hsiang Chen and Harold Acton (1976).

15 At Nanjing University the new journal *Dongnan wenhua* (*Southeast Culture*) is devoted to the scholarship of regional issues and early kingdoms.

16 The symbolic gender and directional correlates of *yin* and *yang* are, respectively, female and male, and north and south.

17 Sinologists have also proposed the existence of two separate literary cultures in pre-Han China, one a northern tradition based in the Yellow River valley, and the second a southern tradition based in the State of Chu – hundreds of years before the evolution of a northern and southern tradition in Chinese drama; see David Hawkes, 1985.

18 This analysis uses the lower case "g" in greater China in order to signal that the phrase does not represent an accepted geographical entity. In a basic toponymic sense, "Greater" is used to indicate a social and economic region that extends beyond the political boundaries of the city itself, such as Greater London, Greater Los Angeles, etc. Applied to a country, of course, "greater" has different implications.

19 Fredric Jameson gave a series of lectures on the postmodern condition at Beijing University in 1985, and thus this element of the analysis is not a case of deploying Western social theory for its own sake but in recognition of the context that Jameson's work has become part of contemporary intellectual debate in China. Jameson's lectures were published in Chinese in 1987 and in a subsequent 1997 edition.

20 Proposals for a Chinese common market from Taiwan have included direct trade between Taiwan and China, the elimination of all trade barriers, guarantees on capital flows and repatriation of profits, stabilized or fixed currency exchange rates between Taiwan and Hong Kong dollars and the Chinese yuan, and the formation of transnational joint venture corporations for heavy and high technology industries.

21 In 1980, then Prime Minister Lew Kwan Yew attributed his People's Action Party's (PAP) political dominance to the inheritance of Confucian values, which allowed the PAP to attempt to reassert party loyalty with the Chinese population and without making direct comparisons with other eth-

nic groups (see Kamm, 1980). Based on 1998 statistics, ethnic Chinese form 77 percent of the population in Singapore, and Malays and Indians make up the largest minorities.

22 The Singaporean government simultaneously planned alternative courses in Buddhism, Christiantiy, Islam, and Hinduism. When the courses were finally offered, more than twice as many students chose to enroll in Buddhism (44.4 percent) as in Confucian ethics (17.8 percent); see John Wong (1986, p. 289).

23 One of the people implicated in funneling the contributions was John Huang. Huang was born in Fujian province, China in 1945, and in 1949 his family migrated to Taiwan. In June 1997, in the midst of intensive interest over his role in the campaign finance scandal, Huang was giving a talk to veterans of a Taiwan military academy. This is an unlikely profile for a would-be Beijing agent, but news accounts have omitted these facts. See http://www.american-politics.com/061797JohnHuangSpeaks.html

24 I am indebted to Jeff Malpas for discussing these ideas about the network metaphor.

25 Wuhan is the break point between the middle and lower reaches of the Changjiang, and so this description indicates the area of the Changjiang south from Wuhan to Shanghai.

3

Maritime Frontier/Mercantile Region

> The chief motive for Chinese expansion towards the south was an economic one. The value of the deltas of the West River around modern Canton and of the Red River around Hanoi had long been recognized. In the earliest surviving account of the campaigns against the Yüehs, the 2nd century work of *Huai-nan Zu*, it is said that Shih Huangti sent five armies totalling some 500,000 men because of "the expected gains from the lands of the Yüeh with their rhinoceros horns, elephants tusks, kingfisher feathers and pearls."
> (Wang Gungwu, 1958, p. 8)

Sometime during the era of the Qin Dynasty (221–207 BCE) a settlement of imperial record rose in the area of Fuzhou, contemporary capital of Fujian province. Its site, upstream from the mouth of the Min River, appeared remote; no other settlements of administrative notice existed higher up in the Min watershed or nearby on the coast. The existence and location of Fuzhou contradicts the logic of standard accounts about settlement evolution in China, which have explained urbanization as a result of systematic imperial colonization along the southern frontier, following routes of access along navigable river valleys. Sen-dou Chang (1963) named the Han dynasty "river empire" because of the close association between urbanization and river transportation. Fuzhou's hinterland did not become systematically occupied with towns for several hundred years until the Tang era, with the increase of commercial tea planting in the mountains. When we look at the settlement record for the south China coast, the earliest settlements antedate the land-based, systematic southern expansion of the Chinese empire (map 3.1). The nascent urban settlement at Fuzhou must have been settled from the sea (see Bielenstein, 1947, 1959). Fuzhou, as the capital of Min Yue, is now better understood as the center of an important early regional settlement and a sometime independent center of social transformation. The epigraph by Wang Gungwu, from his study of the Nanhai or South China Sea trade, uncaps a seemingly pre-Confucian worldview of mercantile imperialism, a Qin dynasty calculating the worth of regional exotica, and a lively trade across the South China Sea basin. The Qin dynasty

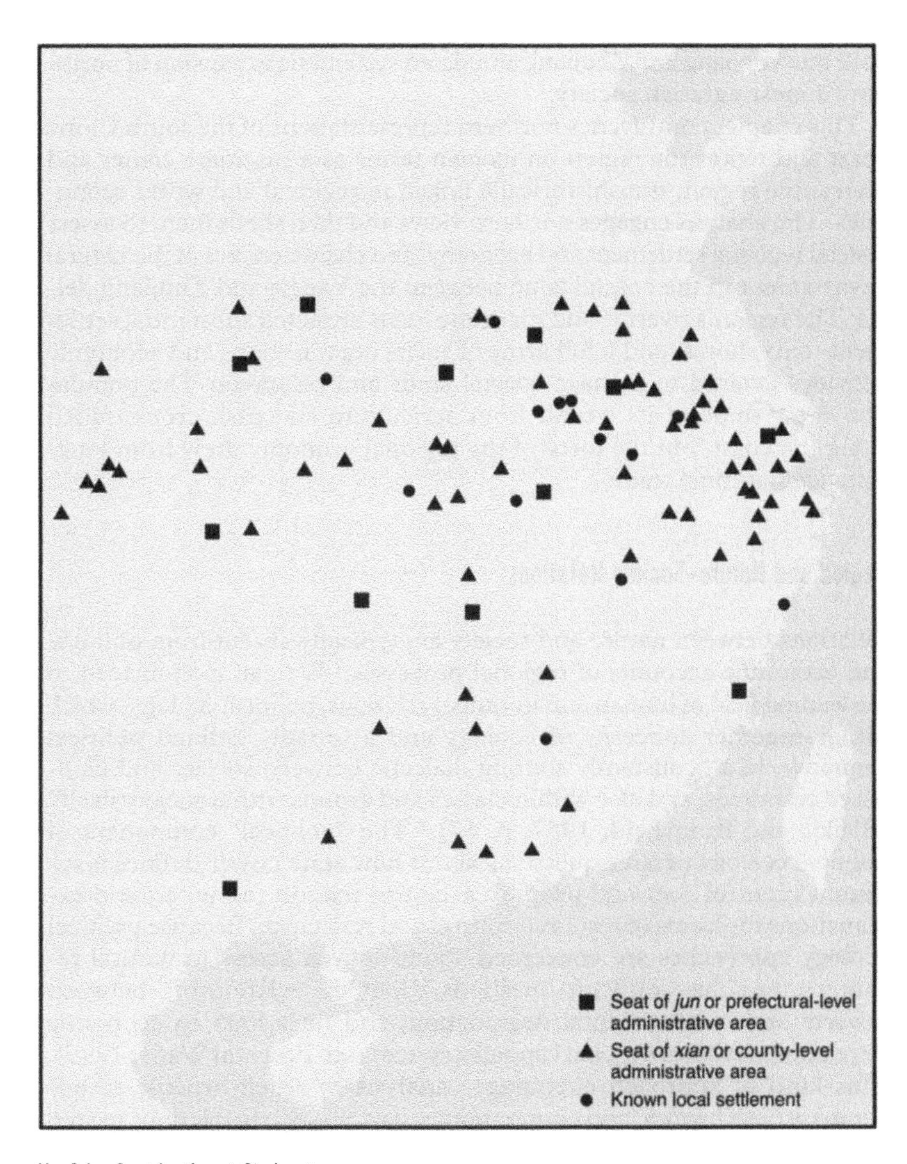

Map 3.1 Coastal settlement, Qin dynasty.
Source: Zhongguo lishi dituji, volume 2 (1982); line work by Jane Sinclair.

unification forced the Bai Yue people into the empire and sent ethnic Han administration to Fuzhou and Guangzhou, in the area of contemporary Panyu in the Zhujiang delta, for the first time. In the third century BCE the empire brought two coastal southern ports within its orbit – and underscored how the origins of settlement in the deltas of both the Minjiang and Zhujiang antedated systematic expansion of northern Chinese agrarian society.[1]

This chapter contravenes northern representations of the south China coast and writes the region on its own terms as a maritime center and mercantile region, transhistorically linked to regional and world economies. The analysis engages northern views and then sheds them to assess instead regional settlement and economy, and characteristics of the natural environment in the coastal zone between the Yangzi and Zhujiang deltas. The region's riverine and maritime focus attracted substantial settlement to its shores, and a full array of social organizations and economic activities evolved to manage coastal lands and resources. The population drew some of its wealth from agriculture and cash crops traded along the coast, but the force of the regional economy drew from long-distance maritime trade.

Region and Nature-Society Relations

Relations between nature and society are typically absent from political and economic accounts of regional processes.[2] We can look instead to the leading field of human–environment relations, political ecology, which brings together concerns of ecology and a broadly defined political economy, in a "constantly shifting dialectic between society and land-based resources, and also within classes and groups within society itself" (Blaikie and Brookfield, 1987, p. 17).[3] The "political" component of political ecology presses questions about how state power defines institutional control over and people's access to natural resources, and explanations for forms of resource control and resistance. Because political ecology approaches are concerned about uneven access to natural resources they raise difficult questions about the relationship between poverty and environmental degradation, and their links to economic development within a global capitalist system (see Peet and Watts, 1996). This kind of approach encourages analysis to reach broadly to encompass both historic and contemporary nature–society relations in step with larger sets of social processes that constitute regions and regional meaning.

Yet if political ecology has concerned issues of poverty and uneven resource distribution, how is it suited to regional conceptualization in coastal south China? The "miracle economy" discourse imposed on the south China coast has practically masked the fact that poverty has also been embedded in the coastal region, especially in the rural sector in parts of Fujian and Guangdong (see Delman, et al., 1990). Regional poverty, combined with opportunities generated in the maritime mercantile economy, gave rise to the Chinese diaspora. The less careful analyses of regional human–environment relations, however, simply point to the low per capita arable land ratio as the major cause of the maritime tradition and the diaspora, and ignore questions about uneven access to land and other resources. But any environmentally determinist account ignores the existence of local coastal cultural ecologies – maritime ways of life – that antedated the influx of Han people. The more complex analyses of regional transformation have also assessed the social relations of production and people's access to resources in the coastal zone. Regional historian Fu Yiling (1961) identified uneven control over land resources and increasing tenancy in the agrarian economy as the critical cause of regional poverty and large-scale emigration (see also Fu and Chen, 1987). Evelyn Rawski (1972) argued that rather than rural poverty, rising peasant expectations and increased commercialization brought by the influx of Western trade impelled maritime trade on the Fujian coast. Ng Chin-keong (1983) reasoned that complex combinations of these factors and others drove the regional economy, and argued for a situated account of the nexus between rural and maritime economies to demonstrate how both the local state and rural society supported maritime expansion. This chapter reinterprets these debates through accounts of historic nature–society relations, and considers specifically the growth of the regional port cities, uneven accumulation of land wealth through coastal and riverine land reclamation, and forms of social organization geared to long distance trade. The evolution of the regional economy, especially in southern Fujian, made Xiamen the center of the "Amoy network" and the leading hearth of diaspora, whose emigrants controlled trading bases around the South China Sea and beyond at the time of the European advance on the Asian trade.

Fringe of Empire

From the perspective of the capital, the south coast below the Jiangnan was one of China's frontiers, a zone of cultural, political, and environmental marginality.[4] In the realm of literati culture, an intellectual frontier also lay in a narrow range of northern Zhejiang, in the Shaoxing area to the Ningbo plain. Zhexi, in northwestern Zhejiang on the southern border of Jiangsu, "with its flat country of swamps and lakes, had customs that were 'extravagant and refined,' much like that of neighboring Suzhou and Changzhou, while Zhedong (northeastern Zhejiang) shares the 'simple and straightforward' customs of Fujian. . . . Zhexi, together with Jiangnan, was the center of philological studies of the Confucian classics . . . Zhedong, in contrast, was home to a scholarly tradition that stressed historical writings and criticisms opposing the autocratic nature of the imperial system" (Yeh, 1996, p. 17). Like the symbolism in "River Elegy," these landscape images signify intellectual differences; in this case, regional differences of scholars south of the Jiangnan who produced worldviews alternative to the Confucian order.

Realities of the maritime frontier also emerged in the physical landscape south of the Jiangnan. The area of northern Zhejiang province around Hangzhou Bay is a flat deltaic plain where the topography rarely rises more than twenty meters above sea level. Although it accounts for only one-quarter of the total area of the province, its two most important cities, Hangzhou, the provincial capital, and Ningbo, are located there. From Ningbo south, the coast narrows into less generous landscapes of smaller port cities and less productive hinterlands, strips of coastal plain, bounded by rocky shores and granitic mountain interiors, until it meets the southernmost of China's major river systems, the Zhujiang, in Guangdong. Most of the coast south of the Yangzi River is ria coastline, a coastal plain of drowned river valleys met by a relatively steep coastal platform offshore. These geomorphological conditions create the conditions for fine natural harbors and dramatic viewscapes around the world: Hong Kong, San Francisco, and Rio de Janeiro share this predicament. Inland, a series of northeast to southwest trending mountain ranges rises up in the west to effectively form a barrier that shields central and southern Zhejiang from central China. The Zhoushan archipelago, a major fishing zone, is the seaward extension of Zhejiang's mainland topography. The mountain systems limited historic access to the interior and isolated Fujian from adequate communications systems and more substantial overland migration. The coast road was a minor artery until rapid development in China under reform pressed needs for new transportation infrastructure, and despite the construction of new rail lines, there is still no coastal railway. Overland rail lines to Xiamen and Fuzhou were constructed in the Maoist era and considered monuments of the socialist revolution. Topographically and economically, the region has embraced the sea.

In contrast to the economically rich agricultural deltas in the lower Yangzi and Zhujiang, along much of the intervening southeast coast arable land was in short supply and local people combined agriculture, fishing, and mercantile economies to guarantee means beyond subsistence. The imagery of a prolific wet rice economy in southern China belies the conditions of the natural environment in the coastal zone of the southeast where, except in limited strips of coastal plain and the deltaic areas of a few small river systems, soils were poor. Local perspective on these conditions yielded a well recognized general observation on human–environment relations in the region: *renduo dishao* (large population and little arable land) or a low per capita arable land ratio. The introduction to a nineteenth-century Xiamen local history contextualized this condition, and its solution, in more poetic terms: "Fields are few but the sea is vast, so people have made fields from the sea."[5] This observation encompasses both the literal making of fields from the sea, in extensive land reclamation practices on the coast, and the maritime mercantile economy. What livelihood could not be earned by land was recouped at sea, in fishing and trading economies that extended beyond the region to other Asian empires. Fujian depended on imported rice from Vietnam to supplement inadequate local production as early as the tenth century, and by the late twelfth century Fujian relied on rice from Guangdong as a matter of course (see Rawski, 1972; Ng, 1983). By the second half of the sixteenth century Guangdong's surplus had turned into a deficit, and Guangdong became dependent on importing rice from neighboring Guangxi province (Lin, 1997, p. 39). Fujian later imported rice from domestic surplus areas in China and also from Taiwan and especially Thailand (Ng, 1983).

The southeast coast was an economic frontier where overseas trade played a larger role than in any other area of China. Although merchants occupied a low position in the idealized Confucian social order, and numerous modern texts have repeated ideas about an anti-merchant ethos in China, this perspective has not prevailed transhistorically and uniformly across China.[6] Indeed, based on the work of philosopher Ying-shih Yü (1997), merchant culture in China has been completely underevaluated. Ideas about the *shidao* (market way) and merchant culture emerged early, in the third century BCE, and by "the sixteenth and seventeenth centuries a quiet but active social movement known as 'abandoning Confucian studies for commercial pursuits' (*qiru jiugu*) swept China" (Yü, 1997, p. 45). Wellington Chan (1977, p. 20) has found that "through a succession of dynasties the legal restrictions against merchants were gradually relaxed until there were hardly any real ones left by Ming and Qing times." Historian Thomas Metzger (1970) compared primary documents on domestic commerce over the centuries to conclude that official views were contradictory on commercial practices, especially where commerce and trade were necessary and thus tolerated activities. On the south China coast the region depended on long distance trade and merchants did not exist on the fringe of society. Eduard Vermeer (1990, p. 7) has described the situation in terms of the resource constraints of the region: "overseas migration and trade represented an escape from the social and economic limitations of rural Fujian. Merchants and laborers voted with their feet (and ships). In spite of the dutiful remarks to the contrary by Chinese literati-historians, it is doubtful whether either the landlord or the official was held in higher esteem than the merchant. In this society, money counted and the government was feared rather than respected." Fu Yiling (1956) argued that on the southeast coast relations between the rural gentry and merchants were close, and that merchants ultimately sought to convert their mercantile capital into landed wealth.

Rescuing the region from the empire on the south China coast is complicated by the fact that international trade, as it is understood in the modern world economy, was not a systematic category of imperial record. Unlike in other countries with great maritime traditions, in China the record is occluded by changing realities and imperial representations of foreign trade. In historic China international trade emerges alternatively as the junk trade, a material element of tributary relations, a sometime imperial policy, banned as a result of coastal disturbances, practiced privately during tribute missions, marginalized because eunuchs typically managed it, and sometimes indistinguishable from piracy.[7] None of these representations recognizes the basic existence, characteristics, and practices of transregional and international maritime trade over the *longue durée*. Historic interpretations of long distance trade have also emphasized perspectives about trade rise and decline, licit versus illicit trade, official ports versus clandestine smuggling sanctuaries, and the like.[8] Policy and law, though, is often symbolic: it was important to the constitution of the imperial world view that trade occurred as a result of tribute, or that it was made illegal when disorder flared. Trade was also valuable, and the imperial regime obtained not only exotic goods but duties and taxes on the movement of cargo. The tribute system was an appropriate umbrella for these contradictory realities. Changes in imperial trade policies also produced real effects, because international mercantile activity irregularly and abruptly shifted in location.[9] In Fujian from the ninth to the fifteenth centuries, Quanzhou was the main center of trade, but later in the 1540s, Yuegang in Zhangzhou prefecture emerged as the leading harbor.[10] These and many other realities attest to how we cannot in any linear way follow the historical geography of international maritime trade. We will instead recover the mercantile region in historic nature–society relations, and ways that economic processes intersected with local social formations to produce a maritime world in south China.

Early Coastal Settlement

It is difficult to imagine, from knowledge of contemporary landscapes of rapid development in south China, that the biogeography of the south coast was once characterized by mixed forests and forest megafauna, the

Asian elephant (*Elephas maximus*), the south China tiger (*Panther tigris amoyensis*), and, prehistorically, the panda (*Ailuropoda melanoleuca*).[11] Although elephants were common in Fujian and Guangdong and roamed as far north as the Yangzi basin in the tenth century, settlement and deforestation of the southern uplands fundamentally altered the habitat of these animals by the time of the Yuan dynasty (Marks, 1997). Local histories chronicled tigers in terms of tiger attacks on villages, no doubt as a result of habitat encroachment, through the nineteenth century.[12] Despite considerable deforestation, in historic and contemporary times, the mountains of Fujian have served as an important source of commercial timber, which has been floated down the Min River and wholesaled through Fuzhou.[13] Timber was historically Fuzhou's number two commercial commodity, behind tea, and related industries in paper production and bookmaking were two of the oldest industries in the province. Robert Gardella (1990, p. 317) has made the important point that commercial primary products of the Fujian highlands, which included tea, medicinal herbs, indigo, ramie, tung oil, and camphor, in addition to lumber, were not consumed locally but more dependably traded inter provincially into coastwise and international markets.

Early regional population groups, before Han in-migration, had settled in both the mountains and the coastal plains. The most significant of these groups were the Yue, of the Bai Yue, who cultivated wet rice and dominated the coastal region from the shore to the fertile valleys of the interior mountains. Peoples of the interior and uplands in south China often practiced swidden or "slash and burn" agriculture, especially the Yao (Eberhard, 1968).[14] From the Yangzi river south, Han migrants settled among the Yue in river valleys, while the coastal seafaring Yue distinguished the littoral.[15] The contemporary "boat people," known as the Dan, are a recognized ethnic group, especially on the Guangdong coast and in the Zhujiang delta, and may have descended from the Yue and the Yao groups. Knowledge about early population groups is relatively scarce. William Meacham's (1983) analysis of the prehistory of the Yue concludes that a regional culture was under formation from 6000–4000 BCE, which coincides with the end of a major sea level rise that submerged at least a 130–km wide strip of the coast.

Given the general pattern of Han in-migrants settling in coastal lowlands, it is not surprising that historic reports on diminished arable land per capita begin to appear by the Song dynasty. The Song was a period of considerable southward migration and increased long distance trade, which led many Han officials to conclude that the shortage of land gave rise to the regional maritime economy. Of course the reality was that a maritime economy antedated Han arrival. Nevertheless, a twelfth-century official described the situation:

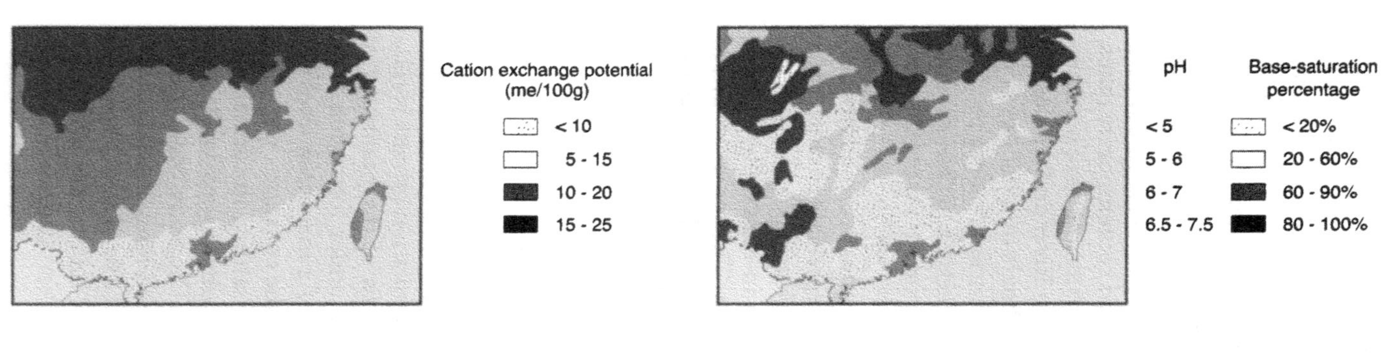

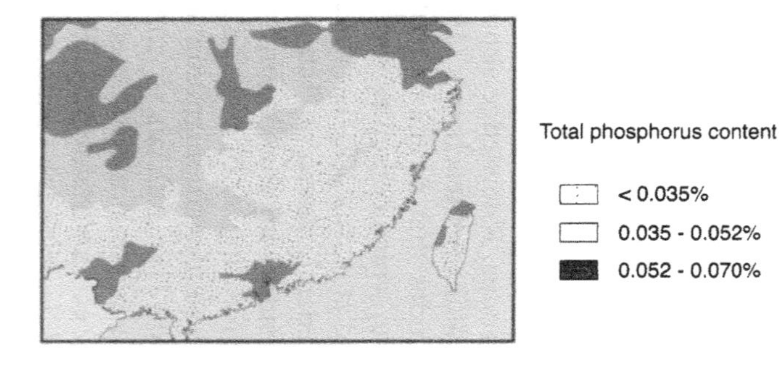

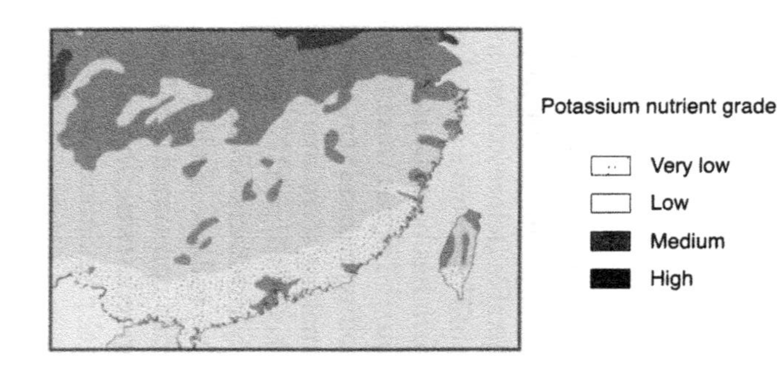

Map 3.2 Regional soil quality.
Source: Soils of China (1990); line work by Jane Sinclair.

> In today's world there are people from everywhere who turn from agriculture to become scholars, monks, and artisans, but nowhere is this as common as in Fujian. Land in Fujian is very limited; it is insufficient for food and clothing. Thus people scatter to the four corners. . . . Those who have scattered are daily more numerous, while those who stay are growing fewer all the time. If the people are few, than livelihood will grow easier, but if the people are many, then livelihood grows more difficult.[16]

The realities of the natural environment speak to elements of the problem: 90 percent of Fujian's territory is hilly or mountainous, and the four irrigible plains of the province, around Fuzhou, Xinghua, Quanzhou, and Zhangzhou, constitute just over 1.5 percent of its total area (Sun, 1962, p. 451). The quality of the soil along the coast, by comparison to the Yangzi and Zhujiang deltas, is poor. Low soil fertility answers why by the nineteenth century one of Fujian's most important and regular domestic imports from the north was "beancake" – pressed nitrogen-rich soybeans used as fertilizer.[17] Map 3.2 depicts soil fertility as a combined function of cation exchange capacity, which indicates the nutrient-retaining ability of a soil, the base saturation percentage and pH, and the content of phosphorus and potassium, basic plant nutrients.[18] Low levels in each of these qualities contribute to low soil fertility, and in Fujian, as the maps show, these factors reached some of their lowest levels across the region.

The limited productivity of the agricultural sector was fundamentally transformed by the arrival of new rice strains from Champa, and nearly half a millennium later, American staple crops. Characterized by drought resistance and a faster growing cycle, the Champa varieties reached maturity in 100 days or less, a third less time than indigenous varieties, which promoted double-cropping. They were in Fujian by the late tenth century and widespread by the early eleventh century (Ho, 1956). The Fujian coast experiences occasional summer drought, and so new rice varieties offered increased food security when the rains failed.[19] The peanut (*Arachis hypogaea*) was the first American crop plant in China, introduced through the south coast in the middle of the sixteenth century, likely by the Portuguese, and found its niche in sandy soils along riverine margins where nothing else of value could be farmed. Peanut oil, which stands a hotter temperature than most oils before burning, subsequently became a cooking staple across China. Sweet potato (*Ipomoea batatas*) and maize (*Zea mays*) soon followed, and were grown in the colder environments of the coastal uplands. Sweet potato became the third most important staple crop, after rice or wheat, and the leading staple where rice was unavailable or unaffordable (Ho, 1955, 1959, pp. 183–95). The Irish potato (*Solanum tuberosum*) appeared in the late 1600s, was grown in soils too poor for maize and sweet potato, and became the staple food

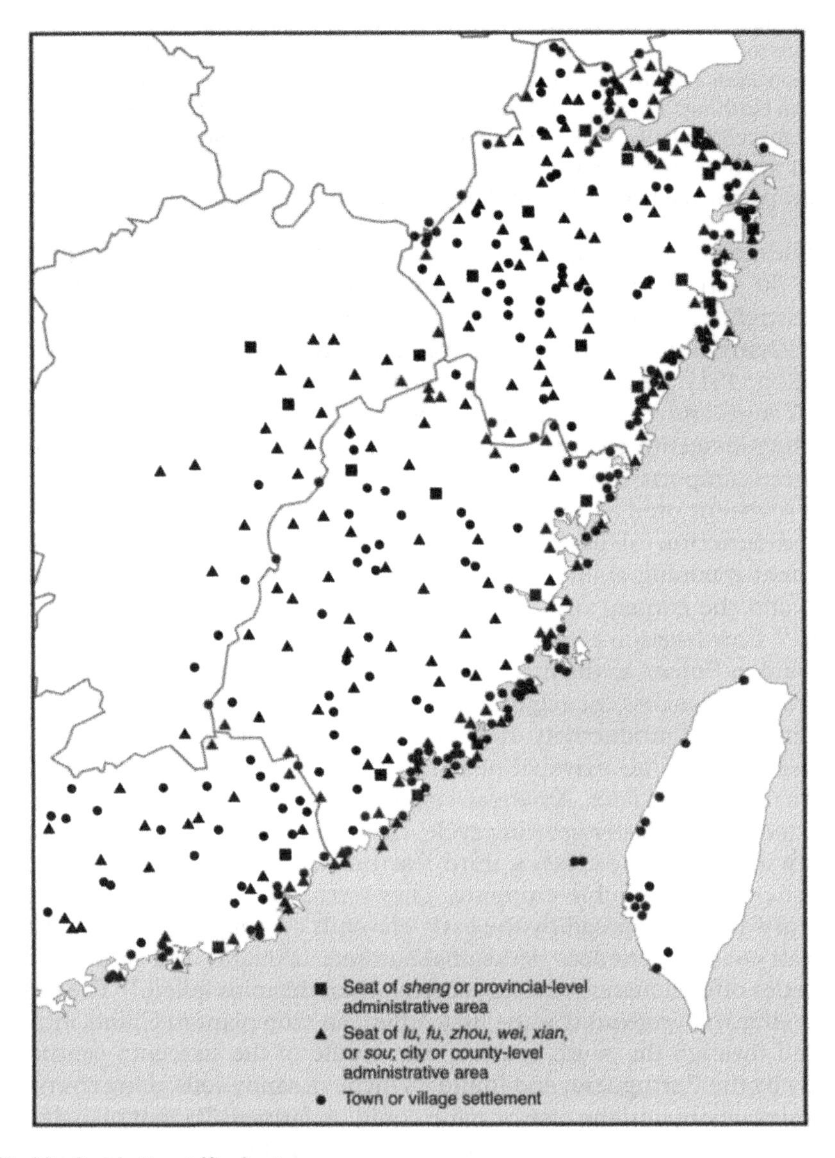

Map 3.3 Coastal settlement, Ming dynasty.
Source: Zhongguo lishi dituji, volume 7 (1982); line work by Jane Sinclair.

of people on the most marginal land. As a result, after the seventeenth century, landscapes of natural vegetation in accessible mountain areas gave way to fields of corn and potatoes, and in some areas, managed forests. Ho Ping-ti (1959) has made a substantial argument for the correlation between the arrival of these staple crops and a significant upsurge in China's population. The distribution of coastal settlement increased significantly during the Ming dynasty, which reflects both greater agricultural development and increased maritime trade (map 3.3).

Cities of Min

Coastal Fujian was sufficiently prosperous by the Tang dynasty that its leading settlement, called Min, at the site of Fuzhou, served as capital to a short-lived regional empire from 893 to the 940s. After the fall of the Tang dynasty, during the interregnum between the Tang and the Southern Song, the settlement at Min distinguished itself as a separate state by paying tribute to the ruling Chinese dynasty in the north (Schafer, 1954). Intra-provincial competition for leadership between the elites of Fuzhou, and a branch of the ruling family in Quanzhou, its economic rival in southern Fujian, kept the empire less unified than it might have been.

One significant arena in which Fujian elites contended was the practice of Buddhism (Clark, 1991). Buddhism flourished in Tang dynasty China, and Fuzhou emerged as a significant center of Buddhist practice. But elites at Quanzhou, always the greater port, poured mercantile wealth into building new Buddhist monasteries and, consequently, the local power base. Such activities had real effects on society and economy by creating extensive monastic landholdings in the suburbs and hinterlands of Fuzhou and Quanzhou. Landholding elites granted lands to monasteries, and landowners sometimes appropriated the status of monks because monastic lands were taxed at lower rates (Ng, 1972; Clark, 1991, pp.61–4; Gernet, 1995). Elite support of Buddhism effectively turned into a tax haven strategy, and resulted in monasteries holding a disproportionate amount of land, and some of the best land in Fujian. This concentration of landholding in monasteries lowered the regional per capita arable land ratio on secular lands and compelled more intensive cultivation. It also pushed Han settlement into the uplands and stimulated new rounds of land reclamation in riverine and coastal wetlands. In the Zhangzhou area, an important agricultural area inland from Xiamen on the Jiulong River, 70–85 percent of land was recorded under monastic control (Clark, 1991, pp. 144–5).[20] Taxes on monastic lands remained low through the start of the Ming dynasty, but the patterns of ownership concentration made landlord–tenant relations uneasy at best and tenant rebellions became common. The moral economy was also in disarray: monks were widely known to have abdicated vows of celibacy and maintained families (Ng, 1972, p. 201; T'ien, 1990, p, 83). At the peak of regional mercantile activity in the Song and Yuan, nuns and monks were also engaged in commerce (Yuan, 1992, p. 106). Imperial edicts sought to restrain the situation by the end of the Yuan, by compelling secularization of Buddhist properties and increasing taxation. Patronage of monasteries subsequently began to decline and some closed. In light of these conditions, the historic problems of "population pressure" in Fujian must be understood in the context of historic Han in-migration and complex social relations of production, in Buddhist and elite control over landed wealth, and uneven accumulation of land resources.

Quanzhou and Ningbo

The major settlements of the south China coast occupied sites in the lower reaches of the region's rivers, where the range of human activities, from agriculture and land reclamation to marine trading and fishing, could take place with relative ease and security. Ningbo, Wenzhou, Fuzhou, Quanzhou, Shantou, Zhangzhou, and Guangzhou are all sites of early riverine settlement. They were not established on the open coast, but several kilometers upriver, and supported by small towns downriver which served as garrisons and first lines of defense. Among the major coastal cities, only Xiamen was on the open coast. Xiamen was originally a garrison settlement on a large island at the mouth of the Jiulong River serving Zhangzhou and other cities in its delta (figure 3.1).

The Song dynasty was an era of previously unmatched economic expansion and urbanization, and especially in the Yangzi delta and on the south coast. Encouraged by imperial policies, China's international maritime trade grew to unprecedented proportions during the Song and became one of the great oceanic trades in world history. As Laurence Ma (1971, p. 29), in his study of Song urbanization, has contextualized its significance, "it was across the oceans that much of China's cultural influence was disseminated." The maritime trade on the south coast was also constituted regionally: Guangdong and southern Fujian specialized in the import trade from the Nanyang, while Ningbo and the Yangzi delta concentrated on exports to the northern trading markets in Korea and Japan. Quanzhou merchants acted as wholesalers for the empire by trading the Nanyang imports north through the major Yangzi delta ports and beyond. Quanzhou was the leading port of the era and in 1087 the Song dynasty formalized existing trade there by establishing an office of the *shibo si* (superintendency of trading ships), which was effectively a

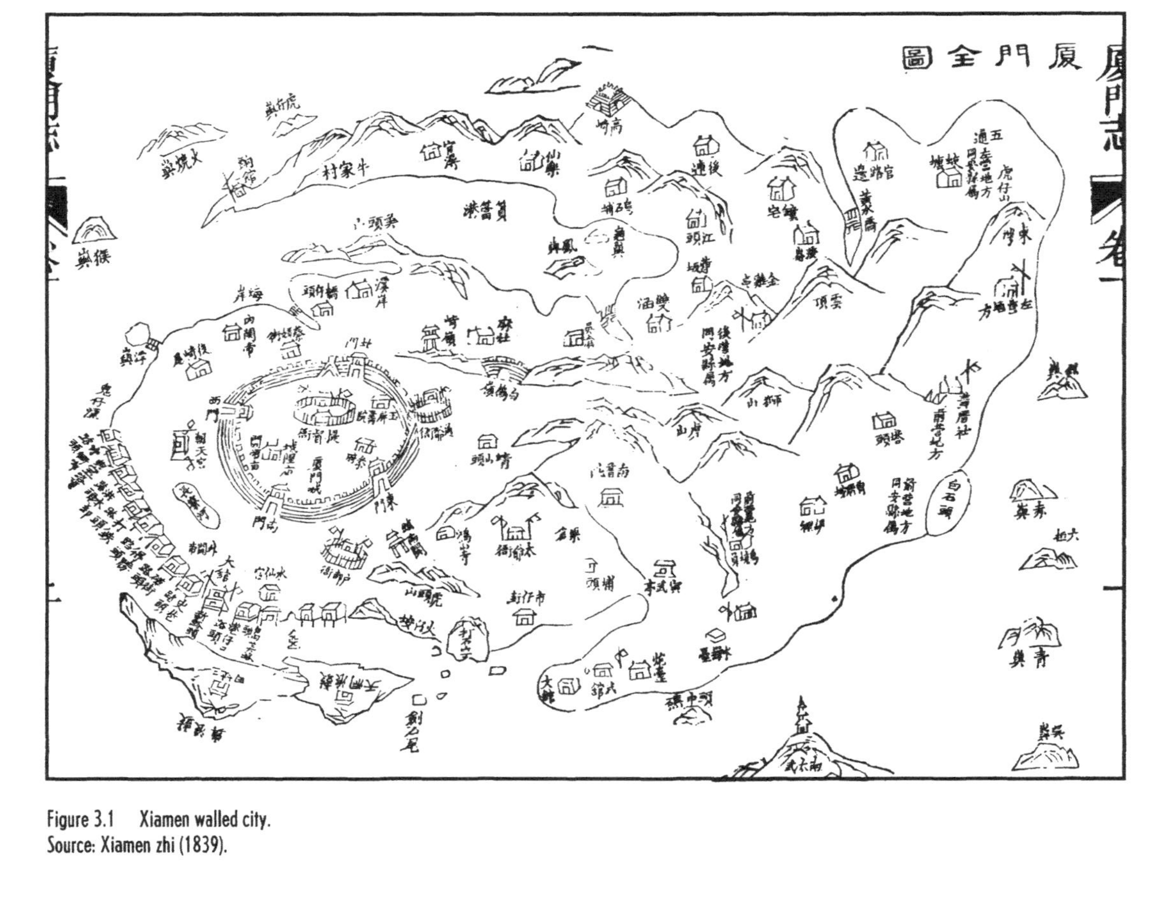

Figure 3.1 Xiamen walled city.
Source: Xiamen zhi (1839).

customs office (Wheatley, 1959; Ma, 1971, p. 35; Clark, 1991, pp. 128–32). During the Ming dynasty, *shibo si* were at Ningbo, Quanzhou, and Guangzhou, which confirmed the significance of these ports and the maritime trade.

During the Song, Quanzhou city, just inland from its eponymous bay, emerged as the leading center of China's long distance trade and eclipsed Guangzhou as the largest port in the empire. A great diversity of luxury goods passed through Quanzhou: aromatics, drugs and medicinals, spices, dried and preserved edibles, hardwoods and rattans, precious stones, metals, porcelain, rare textiles, and more (Chau, 1911; Wheatley, 1959). Mercantile wealth fueled the growth of the city and its hinterland. The city wall was rebuilt twice, from an original one mile long in the late eighth century, to seven miles in the tenth century, to about ten miles long in 1230 (Clark, 1991, pp. 138–9). The city wall enclosed an enlarged foreign merchant quarter, which was distinguished by merchants from the Islamic ecumene who had originally settled in Quanzhou and Guangzhou as early as the Tang dynasty. Settlement expansion required land, and wetlands were reclaimed between the original settlement site and the Jin River. A regular binge of bridge building tied Quanzhou to its hinterland and helped develop the agrarian economy in favor of cash crops, especially sugar cane, lichee, and cotton cloth (Clark, 1991, pp. 95–110).

The origins of another major settlement point to how cash crop agriculture and long distance trading economies fueled urbanization. Ningbo arrived from a more easterly site to its present location at the head of the Yong River in 738, when it was relocated "in response to growing commercial opportunities" (Shiba, 1977, p. 392). Yoshinobu Shiba described the area as "rather isolated; it also suffered from floods, droughts, and tidal waves, elements hardly conducive to the development of a large city" (p. 395) and "originally unfit for production owing to the sandiness of the soil" (pp. 391–2). The area was not brought under widespread cultivation until the Tang dynasty, when the local administration began to engineer massive public works projects, building dams, sluices, and floodgates across the Ningbo plain. Thus again, at Ningbo, settlement origins do not appear in the fruits of agriculture; indeed, in this case they appeared in the absence of it. Trade, commerce, and land reclamation substantially preceded agrarian society around Ningbo.[21]

For several reasons Ningbo's location in northeastern Zhejiang was the most convenient in the Yangzi delta region for an entrepot. Hangzhou Bay and the mouth of the Yangzi itself were subject to treacherous shallows at low tide. By contrast, the entrance to the Yong River downstream from Ningbo was easily passable by ocean-going junks. Water conservancy projects installed several canals that connected Ningbo to Hangzhou, the nominal terminus of the Grand Canal, so that in fact

Ningbo became its southern terminus and the main international port in the Jiangnan region. Both Japanese and Korean ships traded through Ningbo, and by the time of the Southern Song, Ningbo was a major port and the most significant entrepôt on the coast between Quanzhou to the south, and the Shandong peninsula to the north. Markets in luxury goods, a cosmopolitan population, and vigorous economic activity infused the city with prosperity. The city's merchants managed China's precious exports of silk and porcelain, both imperial monopolies, and like the pattern of merchants who dealt in the more important and valuable commodities on a transregional basis, many merchants doing business in Ningbo hailed from other key maritime centers. During the period 1015–1138 more than half of the Chinese voyages made from Ningbo to Korea were by merchants from other port cities, and Quanzhou merchants dominated the route (Shiba, 1977, p. 396). Ningbo's rise as an international trade port also positively influenced the growth of Putuoshan as a Guanyin pilgrimage site. In the Zhoushan archipelago, Putuoshan was an important marine anchorage on the sailing route from Ningbo to the Liuqius, Korea, and Japan. The Guanyin temple on Putuoshan first received official patronage in 1080, after an imperial envoy en route to Korea reported that Guanyin protected them from a fierce storm in the Zhoushan (Yü, 1992, p. 217). Imperial and mercantile patronage of the Guanyin temple on Putuoshan, in provision of land and grain for the monastery, and through the Ming and Qing dynasties, secured its position as China's pivotal center of Guanyin worship. In the way that historical practices are reconstituted in contemporary cultural and economic forms, Putuoshan has become a major tourism attraction in China under reform and remains a significant Buddhist pilgrimage destination.

The deltas and land reclamation

After substantial Han in-migration, settlement evolution depended on land reclamation along the coastal plains and estuarine wetlands, to the degree that "the draining of coastal salt marshes for crop land was one of the most important features of the great demographic and economic surge of south China from the eighth to the thirteenth centuries" (Clark, 1991, p. 141). The idea of China as a river empire is bound up in the role of the local state in organizing water management, irrigation and transport canals, and drainage and flood control projects (see Chi, 1963). The state's role in water management was the basis of the geographical–political contract between the imperium and society: until the Ming dynasty taxation was paid in kind, in grain tribute, and so the production, collection, and transportation of grain depended on a stable agricultural environment and its water supply. The most important north–south transport artery, the Grand Canal, was constructed to tie the productive Yangzi delta to the northern capital region. Dependence on southern grain had developed to such an extent that as early as the Tang dynasty the land tax of the Jiangnan region, which at that time included the areas of Jiangsu, Zhejiang, Jiangxi, and Anhui provinces, contributed 90 percent of the country-wide total (Chi, 1963, p. 125).

Fertile lake shores of the Yangzi River's great deltaic wetland were a convenient target for reclamation during the Southern Song. Wealthy families assembled laborers to *weihu* (reclaim lakeshore), often in the name of military colonization, but more realistically for private accumulation of land resources (Chi, 1963). Lake reclamation was apparently so extensive that it disrupted local drainage patterns and gave rise to serious flooding in surrounding areas. A Southern Song dynasty official explained the problems:

> During slight drought, the owners of the *wei*-land take possession of the upstream sections of the rivers and monopolize the advantages of irrigation, and the people's land is deprived of the use of water. When the water of the rivers and lakes overflows, the surplus water is sent downstream and the people's land is used as a watershed. Even if fortune favors the *wei*-land with a good harvest, and rent increases and taxes are double collected, the people's land is reduced to waste whenever there is flood or drought. The damage thus done to the normal tax income is indeed incalculable! (Wei Chin in Chi, 1963, pp. 136–7)

Widespread wetland reclamation undermined government ability to control water resources and created new social and economic problems in the agricultural sector.

Historic records of land reclamation in the Zhujiang delta date back to the Song, and by the early Ming dynasty the state had assigned military colonies to undertake reclamation and settlement (Siu, 1989, pp. 20–7). The focus of reclamation was riverine shores of recently deposited silt. As the sands appeared from the water they were reclaimed, and so the Zhujiang delta lay in relative stages of reclamation, agricultural development, and productivity. With increased settlement in the region wealthy lineages and *huiguan* petitioned government to reclaim the *shatian* (sand fields), and much new agricultural land became concentrated in the hands of the elite. Land reclamations of this sort were highly capitalized long-term investments, which the state encouraged by offering land tax exemptions for working the sands. Ambitious reclamation schemes emerged, such as the deposition of stones, layer by layer, to increase the rate of sediment accretion, followed by construction of dyke and polder systems. Class relations also formed on the sands, since actual reclamation and cultivation activities were not carried out by lineage members, but by tenant farmers who were commonly the Dan, the boat people. Land owners did not allow the Dan to own land or erect substantial housing, and refered to them as *shuiliu chai* (floating twigs) and *waimian ren* (outside people). They lacked the cultural complex of the *limian ren* (inside people), or the *bendi ren* (original people) who deployed their wealth and status to maintain socially constructed class relations and economic opportunities (Liu, 1995, p. 35). Wealthy lineages invested in land reclamation, avoided taxes by partial land registration, appropriated the labor surplus of the Dan, and monumentalized their wealth in ostentatious ancestral halls, temples, and graves. Together with associated cult festivals and ancestral rites, these visible landscape features displayed wealth in society, built up power of particular lineage groups, and served to construct social status (Siu, 1990a). The process indeed spurred the invention of fictive lineage accounts, in which families recorded false and pretentious descent relationships from early Han kingdoms (Faure, 1989). The degree to which some lineages gained control over the sands made land reclamation in the delta "a process of regional economic development as much as a social and cultural construction" (Liu, 1995, p. 21). In this social formation in Guangdong we especially see how temples and ancestral halls, place contexts of local identity formation, were tied ultimately to natural resource access and control.

Mercantile region

> There are few things in the world, the classics of Chinese medicine warn, more dangerous than wind.
>
> The imagination of winds reached beyond medicine and meteorology to encompass ideas of space and time, poetry and politics, geography and self. (Kuriyama, 1994, p. 23)

The sea was frightful enough in the classical Chinese imagination, but the winds were also problematic. Winds were fickle: they bore rains, changes of the seasons, and also sickness. Certainly the cultural subtleties of understanding the winds form their own subject of study, yet how difficult would it be for inland sensibilities to embrace the winds as practical elements of knowledge and regional life? The seasonal change of the winds, in the Asian monsoon system, drove long distance mobility on the south China coast. The monsoon also brought rain, but unreliably so, especially in the northern hemisphere summer when drought could take hold and threaten the rice crop, especially in Fujian.

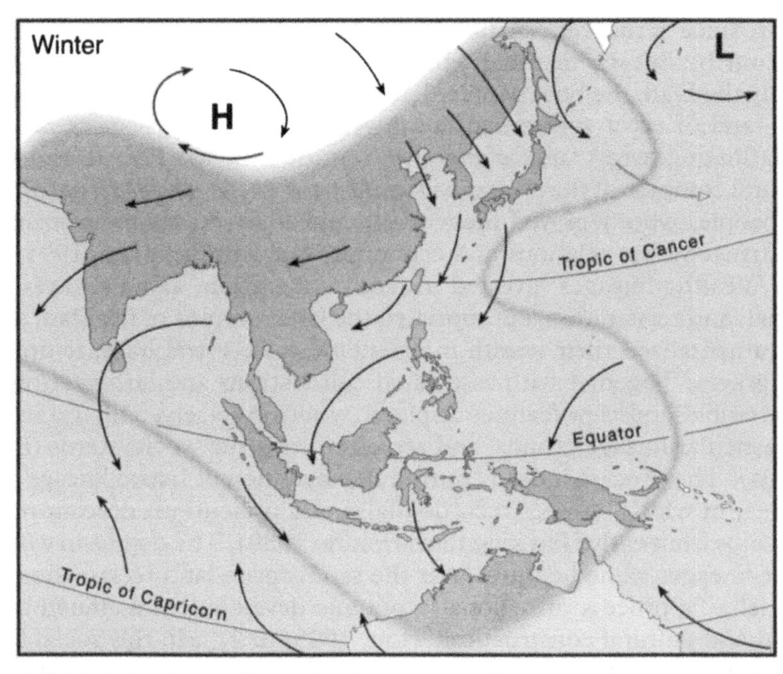

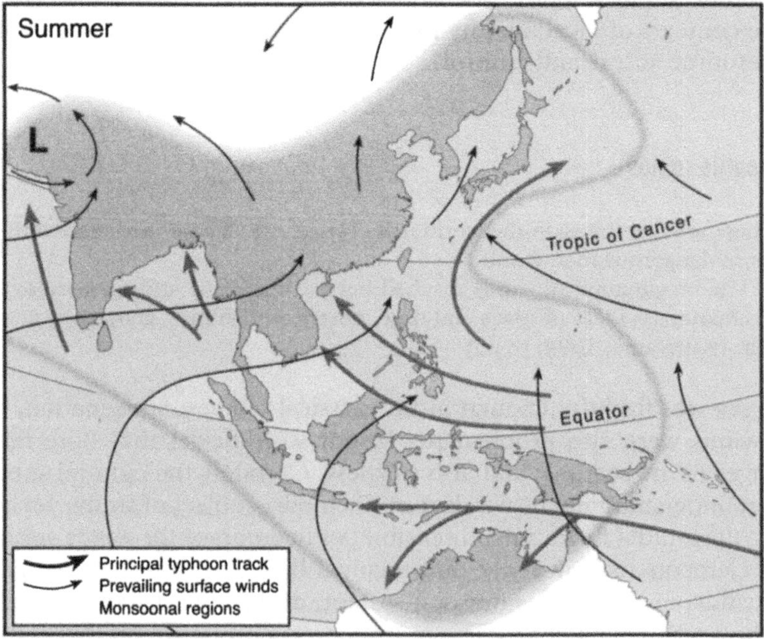

Map 3.4 The monsoon
Source: Robinson (1967), Henderson-Sellers and Robinson (1986), and Musk (1988); line work by Jane Sinclair.

The monsoon wind system has two major seasons, winter and summer, any many local variations throughout East and Southeast Asia. In winter, cold air over the Eurasian continent creates high pressure that moves air masses outward toward the warmer Pacific. This land to ocean seasonal air movement is the northeast monsoon over the China coast, which propelled sailing craft from north to south, from Quanzhou to the Nanyang. In summer, continental low pressure draws in moisture-laden air from the Pacific. This southwest monsoon powered sailing craft from the South China Sea north up the China coast, and brought late spring and early summer rain (map 3.4). Mariners sailing the coastal routes understood their journeys in terms of the reversal of the monsoons. In historic Southeast Asia, ports at the pivot of the monsoon, where localized South China Sea patterns give way to Indian Ocean patterns, were the places where sojourneying merchants established overseas communities and waited out the reversal of the wind system. Melaka, established ca.1402 on the west coast of the Malayan peninsula, served this entrepôt function, and later Singapore.

The Melaka sultanate, the first Islamic sultanate in the Southeast Asian region, was in place less than a decade when China learned of its significance and sent official envoys to make it a tributary state. In 1409, China's maritime explorer Zheng He bestowed upon Melaka's ruler Parameswara the designated tributary paraphernalia, especially the silver seals for signing tributary memorials, and the imperial clothing, hat, belt, and robes. Performance of these rituals made Melaka a kingdom tributary to China and under its diplomatic protection.[22] Annam, Burma, Cambodia, Japan, Korea, Laos, the Liuqiu kingdom, Melaka, Siam, the Sulu sultanate, Tibet, and tribal groups in Central Asia were Chinese tributary states during either the Ming (1368–1644) or Qing (1644–1911) dynasties, or both. Through the tribute system, the Chinese empire constructed a political–territorial interpretation of the Asian realm whereby tributary states symbolically acknowledged the superior position of the Chinese emperor in the regional world order (Fairbank, 1968). The presentation of tribute was also a form of imperial trade, which made tribute relations a mode of diplomacy and economic exchange.

The Zheng He voyages

By contrast to the policies of the Song and Yuan dynasties that promoted international maritime trade, the return of native Han rule under the Ming dynasty in 1368 resulted in a regime of international retrenchment and a series of attempted bans on private trade. But during a brief period at the beginning of the dynasty the empire also briefly promoted

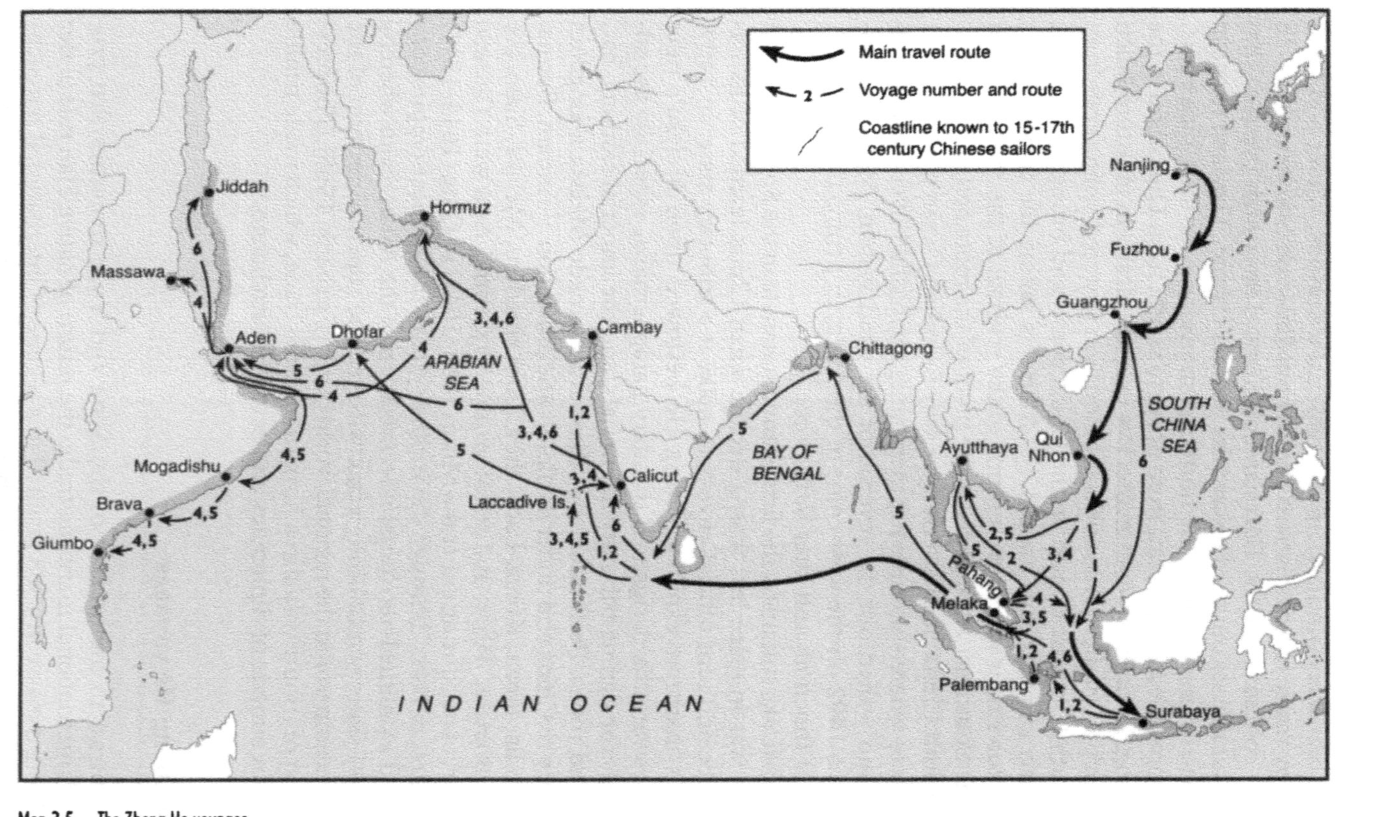

Map 3.5 The Zheng He voyages.
Source: Blunden and Elvin (1983) and Qian and Tan (1995); line work by Jane Sinclair.

maritime exploration and contacts with distant empires. Ming dynasty emperor Yongle (reign 1403–24) commissioned admiral Zheng He to lead voyages into the Nanyang realm of China's would-be tributary states and beyond, as far as the Western shore of the Indian Ocean in contemporary east Africa. Zheng He was a Muslim originally from Yunnan province and was suited to diplomatic engagements in the Arabian trading ecumene. The voyages called at all major settlements in Southeast Asia, around the Indian Ocean, and in the Arabian world.[23] These missions celebrated the return of Han rule after the Mongol era, returned with intelligence about the activities of distant kingdoms, collected exotic treasures for the imperial court, and reinvigorated the tribute system (map 3.5).

From 1405 to 1433, the Zheng He voyages were mounted on a scale unmatched in the world. The political and technological significance of the voyages is evident in the sheer size of the missions, which dramatically out-scaled the Columbian voyages nearly a century later (Yang, 1985). Each voyage amassed as many as 200 vessels, which were manned by a total crew of over 27,000 men. Zheng He's flagship was an enormous ship for its era – 1,500 tonnes – which outweighed the largest of the Columbian vessels by fifteen times and was nearly five times the length of Columbus's 30–meter *Santa Maria*. Yet Zheng He's historical legacy arguably does not match his contribution to maritime history and only in China and Southeast Asia are Zheng He's maritime achievements well known. These limits to history are best explained in a scholarly sense by the fact that Zheng He's logs, the primary records of the voyages, were destroyed, likely purposefully taken from the imperial Ming archive and burned. In the official Chinese bureaucratic tradition of careful treatment and storage of texts, this is a flagrant, extraordinary act. That it happened symbolizes shifting imperial perspectives about the meaning of trade and overseas exploration in historic China, and the role of eunuchs.

The political power of eunuchs in China was never greater than during the Ming dynasty (Hucker, 1978; Tsai, 1996). Castrati had long been necessary servants in the imperial household because a castrated service class posed no threat to the purity of reproduction of the imperial family line. But more eunuchs were employed during the Ming than ever before, especially after the Yongle emperor moved the capital from Nanjing to Beijing, and they were granted more powers. Originally poor men with marginal future prospects, imperial eunuchs gained an opportunity for education within the imperial city and official appointment. Eunuchs staffed 24 official agencies of the imperial household, from construction and repairs of the imperial palace, to procuring and preparing foodstuffs for the imperial kitchen, and conducting imperial trade. Ming emperors regularly commissioned eunuchs as purchasing agents for important and rare commodities, including medicinal herbs, carpets, silks, porcelain, and especially pearls from the south coast (Hucker, 1978, pp. 92–6). Commodity acquisition became especially associated with eunuchs, as they competed for imperial favor by endeavoring to present to the court ever more exotic goods. The famous arrivals of the first giraffes in China served this role, received as tribute in 1414, 1415, and 1433, from Bengal, Melinda, and Mecca, respectively, as a result of contacts made during the Zheng He voyages (Duyvendak, 1939, pp. 399–412). Through such practices, however, eunuchs began to exceed the bounds of their service classifications. Confucian officials disdained their activities and found reprehensible their rise to power. In Henry Tsai's (1996, p. 19) estimation, "the extent to which they influenced Ming diplomacy was both unprecedented and unfathomable." In response, as Duyvendak (1939, p. 398) has assessed, "a secret war was waged by the officials against the encroachments of their power by the eunuchs."

Zheng He was a eunuch, and eunuchs staffed the office of *shibo si*. Ningbo *shibo si* received ships from Japan, Quanzhou handled Liuqiu's envoys, and Guangzhou was the port of call for ships from Champa, Siam, and Western Ocean countries. In the fifteenth century, the Quanzhou superintendency was moved to Fuzhou, and Liuqiu missions began to trade directly with Fuzhou. The *shibo si* also sought to command the comings and goings of the private junk trade. The valuable trade in exotic luxury goods generated a considerable amount of revenue for the empire, and from the official worldview, eunuchs were in charge of handling it. During the Ming dynasty eunuchs came to represent exploration, trade, and the collection of exotic goods.

The Yongle emperor died in 1424, and subsequent court debates over the imperial succession and the Zheng He voyages implicated power struggles between the eunuchs and the literati:

> Zheng He was stationed at Nanjing in charge of the beautification of the capital. While his ships lay idle and gathered rust, as it were, there was an outbreak of conniving, feuding, and finagling between the literati and the castrati. The winners were usually those who could garner the strongest backing of the emperor. In the summer of 1430, the fifth Ming emperor, Xuande, decided that he wanted to revivify the Ming imperial prowess and ordered the seventh and final expedition. (Tsai, 1996, pp. 163–4)

Not long after his return from the final voyage, Zheng He died at the age of 65. The politics of power internal to the court continued to implicate struggles between titled officials and the eunuchs. During the reign of the Chenghua emperor (1465–87), eunuch Wang Zhi, Inspector of Frontiers, sought to retrieve the records of the Zheng He voyages from the imperial archives in anticipation of mounting a sea expedition to the

Vietnam coast – but the logs could not be found. A high official with access to the archives, bent on curtailing eunuch powers, likely destroyed the records (Duyvendak, 1939). The rationale driving the destruction of some of the most extraordinary records of exploration and discovery, on a world scale, is remembered in the following terms:

> The expeditions of the San-pao[24] to the Western Ocean wasted tens of myriads of money and grains, and moreover the people who met their deaths may be counted by the myriads. Although he returned with wonderful precious things, what benefit was it to the state? This was merely an action of bad government of which ministers should severely disapprove. Even if the old archives were still preserved they should be destroyed in order to suppress these things at the root. (Qin, 1937, pp. 3–4 cited in Duyvendak, 1939, p. 396)

This passage may also suggest that the records, in their context of production, challenged appropriate imperial modes of conduct. Understanding associations between correct imperial embodiment and the eunuch production of official documents raises new questions about the ideological problems represented by the Zheng He voyages and trade during the Ming dynasty. In a social order that maintained concern for appropriate and embodied ritual practice, eunuchs were a "third sex" (see Herdt, 1994). They could not adopt appropriate positions among the "five relationships" that characterized Confucian society or spiritual–ideological positions of *yin* and *yang* (Mitamura, 1963). Ideas about eunuch bodies made them other to the normative social order in which voluntary bodily deformity was reprehensible. Bodily characteristics of eunuchs, including loss of beard and especially higher-pitched voices, marked their differences. Lacking penises, urinary flow might leak, yielding orders unfit for social discourse. Eunuchs retained their testicles in a sealed container, so that "wholeness" might be achieved in the next life. Other to society, their management of trade made trade the object of the other. Thus trade, as eunuch office, could be seen as doubly othered practice. These socially constructed problems about eunuchs, eunuch bodies, and third sex embodiment must have contributed substantially if implicitly to vitriolic debates over imperial support for trade and exploration during the Ming dynasty.

Zheng He and Tianhou

Epigraphical materials about the Zheng He voyages have survived, inscribed on stone tablets erected at temples dedicated to Tianfei (Heavenly Lady), the Ming dynasty name of imperial goddess Tianhou, locally known as Mazu, patron goddess of the south China coast (Duyvendak, 1939). One of the temples was at Liujiagang, near Suzhou, where Zheng He's fleet assembled in the Nanjing capital region, and the other was at Changle in Fujian, where the expeditions awaited the northeastern monsoon. The Liujiagang tablet, erected in 1430, records the dates and basic itineraries of previous expeditions, and attributes the success of the missions to the powers of goddess Tianfei. The inscription at the Tianfei temple in Changle, likely erected in late 1431, recorded Zheng He's performance of ritual sacrifices and his improvements to the temple, which paralleled the practices of the office of his imperial appointment at Nanjing, in charge of palace construction. It also explained how powers of Tianfei quelled fears of the sea:

> The power of the goddess having indeed been manifested in previous times has been abundantly revealed in the present generation. In the midst of the rushing waters [and winds] suddenly there was a divine lantern shining in the mast,[25] and as soon as this light appeared the danger was appeased, so that even in the danger of capsizing one felt assured that there was no cause for fear. When we arrived in the distant countries we captured alive those of the native kings who were not respectful and exterminated those barbarians robbers who were engaged in piracy, so that consequently the searoute was cleansed and pacified and the natives put their trust in it. All this is due to the favours of the goddess. (Duyvendak, 1939, p. 350)

Zheng He's propitiation of Tianhou, as an imperially sanctioned cult, tied the empire and the south coast into a scaled, symbolic regional identity formation that bound the tribute system and international maritime trade into a transboundary cultural economy. The elimination of the imperial voyages could not fundamentally alter the conditions of this regional formation. Despite the bans against private trade, maritime trade and commercialism expanded through the sixteenth century and the Ming–Qing transition. The seventeenth-century Ming loyalist movement on the south China coast, led by the Zheng family in Fujian, symbolized both the wealth of the coastal economy and the realities of a maritime empire in coastal south China. The Zheng family routed the Dutch from Taiwan, and the Kangxi emperor honored Tianhou for protecting the sea expedition that allowed the Qing dynasty to incorporate Taiwan within the empire.[26] Subsequently, trade and migration across the Strait substantially increased.

The Amoy Network

In Beijing in 1987 Deng Xiaoping reminded US Nobel laureate Yuan-tseh Lee, "you are from Taiwan and your ancestors are said to be brought there by Zheng Chenggong, a great hero of the Chinese nation" (*Xinhua*, 1987). Two years earlier, high on a rocky promontory over Xiamen harbor, high-ranking provincial officials held a ceremony unveiling the statue of Zheng Chenggong, who liberated Taiwan from Dutch occupation in 1661. In 1979 the Zheng Chenggong Memorial Hall, on Gulangyu island just seaward from Xiamen, was reopened to the public after a complete renovation. It was first opened in 1962, the tricentennial anniversary of the restoration of Taiwan by Zheng Chenggong. The 15-meter high granite statue of Zheng, dressed in the robes of the scholarly official class, his right arm extended high and pointing in the direction of Taiwan, presents a different image than the Ming loyalist leader who ravaged Qing military forces for more than two decades. Whether in Mao's history or Deng's, Zheng Chenggong has served the post-1949 Communist agenda (Croizer, 1977). Zheng brought Taiwan under Chinese rule for the first time – no matter that the Zheng regime, backed by superior naval forces and thousands of followers, was at war with the Qing dynasty government at the time, or that the Qing actually engaged the services of the Dutch to eliminate Zheng Chenggong from the coast. In the early years of the Qing dynasty Zheng Chenggong led a Ming loyalist movement which drew its strength from dominance of the maritime economy of Fujian. The Zheng regime also transformed the island of Xiamen into the home base of the "Amoy network," and the leading port in Fujian (map 3.6).

After the Manchus conquered north China one of their major challenges to unification of the empire emerged on the south coast, where local forces remained loyal to the Ming dynasty and Han rule (Struve, 1984). Zheng Zhilong was a trader from Quanzhou prefecture whose family controlled the mercantile economy of the coast from the 1620s. He had vast experience in coastal and regional trade, from Macao to Japan. In 1627 he took Xiamen island from government troops and established it as a key entrepôt in southern Fujian. Zheng Zhilong's leadership qualities were legendary and tied to acute understandings of place identity and their intersections with larger-scale political and economic processes:

> it was his solid bond with the ancestral village, his qualities as a grass roots level organizer combined with an extraordinary knowledge and understanding of the local terrain, his rivals, Chinese pirates and administrators and Dutch company servants alike, which enabled him to emerge in the

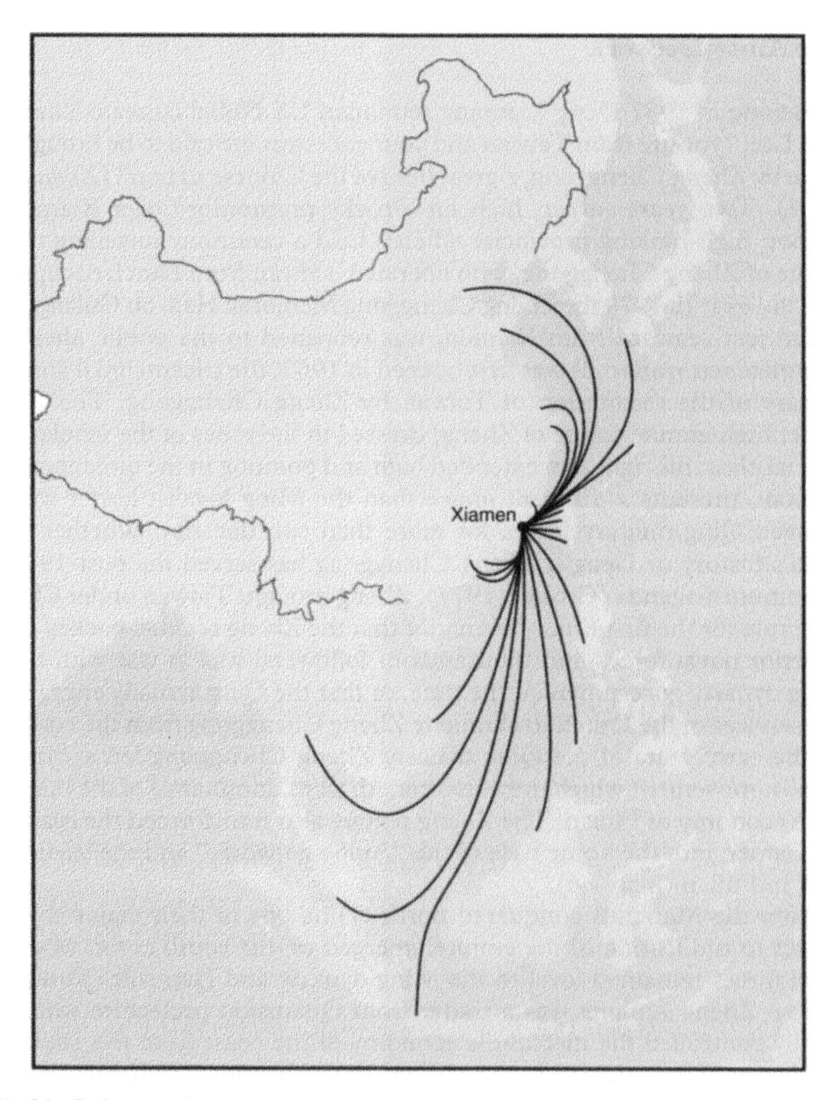

Map 3.6 The Amoy network.
Source: Ng Chin-keong (1972); line work by Jane Sinclair.

> 1630s as the undisputed powerbroker in the Zhangzhou and Quanzhou regions. (Blussé, 1990, p. 264)

He later accepted imperial titles in return for alliance with the Qing, but his son, Zheng Chenggong, born into the maritime world of his father's empire in Japan, and half Japanese, refused to capitulate. With the family mercantile empire behind him – "the gigantic merchant fleets returning from Japan and Southeast Asia" (Ng, 1983, p. 50) – Zheng Chenggong controlled most of China's trade with Japan and ruled the coast from Fujian to Guangdong and across the Taiwan Strait.[27] He early established a base on Xiamen island and financed the movement with trade and tax levies from areas under his control. A Qing edict of 1656 attempted to suspend maritime trade along the whole of the coast from Shandong to Guangdong, but this made the commodity trade only more lucrative and the trade largely continued as before. Zheng was even emboldened to lead an unsuccessful attack on Nanjing in 1659. Qing military forces attacked Xiamen and Zheng sought a more distant refuge on Taiwan, where he captured Fort Zeelandia, the Dutch trading base for the China and Japan trade, in 1661. He died the next year but his son Zheng Qing carried out local colonization on Taiwan, concentrated around Tainan, and reestablished a strong presence on Xiamen by reconstructing and enlarging the town (Ng, 1983, pp. 51–2). Subsequently, the Dutch assisted Qing forces in their attempt to eliminate the Zheng regime from Xiamen. In the ongoing battle over control of the coast, the Xiamen base was a fulcrum of resistance and focus of attack. Throughout, the Fujian loyalists were decidedly the superior seafarers and maintained their regime on both land and sea.

The Qing finally resorted to a more desperate scheme to bring down the Zheng regime, in crimes of war designed to eviscerate the rebels' lineage and regional economy: execute Zheng Zhilong, Zheng Chenggong's father, held prisoner in Beijing, desecrate the Zheng ancestral tombs outside Quanzhou, offer generous rewards to rebels who surrendered, and strictly enforce the trade ban and defense of the sea coast. Enforcement of the trade ban culminated in a policy to evacuate the coastal population. From 1660 to 1666, the *jianbi qingye* (fortify the walls and empty the fields) policy evacuated strategic areas of the Fujian coast and parts of the Guangdong and Zhejiang coast about 30 *li* (10 km) inland (Kessler 1976, pp. 44–5).[28] Imperial troops finally drove Zheng Qing out of Xiamen in 1663, but he managed to retake the island six years later, which engendered another round of coastal evacuation. The effects of the policy must have been limited, since the British East India Company approached the Zheng regime to establish a factory at Amoy in 1676 (Morse, 1926, pp. 44–5). In 1680 Zheng Qing withdrew to Taiwan for good, where he died in 1681. In 1683 the Qing finally led a successful invasion on Taiwan, and Zheng Qing's son, Keshuang, surrendered. Taiwan subsequently came under the administration of Fujian province, and the following year the Qing government designated Xiamen the imperial port for the Nanyang trade.

The activities of the Zheng regime are one of the great accounts of regional formation. In the name of empire, Han society, and the native Ming dynasty, the Zheng family maintained appropriate ideologies of imperial rule as well as trading relations to Japan and the Nanyang. They brought increased Han settlement to Taiwan, and transformed Xiamen into the great port of southern Fujian. Wealthy merchants were officials in the Zheng regime, and many of them owned their own ships and capitalized the maritime voyages. Several large mercantile firms operated under Zheng control, and some of them had a base in Hangzhou, in the Jiangnan, where they collected silks for transport to Xiamen (Nan, 1966). The Zheng regime also capitalized small-scale merchants who worked to provide shipbuilding and weaponry materials, especially iron tools, which had long been produced in the Zhangzhou area (Ng, 1983, p. 54). Local lineage practices also supported the maritime economy, especially the usual regional practice of adopting sons to send out on the dangerous sea voyages (Freedman, 1966, p. 7).[29] Across class and regional divides, the Zhengs tied mercantile Minnan into an extraordinary formation of regional power.

The junk trade

The Amoy Network might be rethought in the translocal terms of diaspora if we understand how the social formation of southern Fujian was reproduced in distant ports. Recall that the macroregion analysis for the southeast coast did not distinguish between the activities of Western mercantilists and the regionally important junk trade with Nanyang states. The junk trade was China's native trade and maintained coastal routes both north and south, and international routes, regularly to Japan and Siam. The junk trade demanded a boat-building industry, supported especially by the lumber products of Fujian, social relations of production geared to manage wholesaling operations, and forms of societal organization that supported long distance trade. Fujian held the reputation for constructing the highest quality large junks. The *haichuan* (sea-going boats) mostly traveled coastwise and to offshore islands; the *yangchuan* (ocean-going vessels) undertook voyages to the Nanyang. These junks were grand ships and also representations of native place, distinguished by painted designs and colors associated with particular ports of origin (see Audemard, 1960, 1970; Worcester, 1966). Fujian junks had a green bow and those from Guangdong were painted red (Cushman, 1993, p. 51).

What does it mean in the context of early Ming tribute relations, from 1385 to 1435, that China *gave* Liuqiu large ships and granted skillful seafarers for the maintenance of the tribute missions? Liuqiu might have been engaging in tribute trade, but their ships were Fujianese and so were the mariners at their helm (Chang, 1990).[30] Most historic texts do not point out that many of the merchants dealing with the Western trade at Canton were originally from Fujian families (Cordier, 1902). It is also the case that Fujianese managed Siam's tributary trade, and a Sino-Siamese community managed private and royal trade for the elite at Ayutthaya (Viraphol, 1977, p. 40).[31] Macao, regularly associated with earliest Portuguese settlement in East Asia, was originally a colony of Fujianese merchants (Blussé, 1990, p. 253). The Portuguese carracks sailing the Japan route from Macao often carried Chinese pilots from Fujian (Boxer, 1968, p. 18). Mariners of Fujian had been completing the circuit through Manila Bay for a few centuries when the Spanish arrived. The Spanish made Manila the chief city of the colony in 1571 and trade with China was by far its most important business; up to sixty junks a year were required to carry the demanded quantity of Chinese products to Luzon (Schurz, 1939, p. 71). The great Manila galleon trade, across the Pacific to Acapulco, depended on the Chinese sojourning population from Fujian to build the galleons and pack the goods. When the first of the horrendous massacres of Chinese took place in Manila, in 1603, an estimated 25,000 died, and some 80 percent of the dead were from Haicheng, the port seaward from Zhangzhou (Laufer, 1908, p. 272). By the eighteenth century, during the long reign (1736–95) of the Qianlong emperor, when trade on the south coast was, according to the macroregion analysis, supposed to have entered a "dark age," the imperial court had in fact established a reward system to encourage merchants to import rice from the Nanyang (Li J., 1990). Increased commercialization of the agricultural economy during the Ming dynasty had diminished rice land and rice prices were increasing. In the middle of the seventeenth century, diverse merchants from counties in southern Fujian – Denghai, Haicheng, Longxi, Nanhai, Sanshui, and Tong'an – all received imperial acknowledgement for importing rice. In a revisionist study of the Qing maritime customs, Huang Guosheng (1999) has demonstrated that between 1688 and 1830 the Qing government reorganized the management of customs in the southern coastal provinces to support the expansion of domestic and regional trade. What emerges from even this short list of distinctions is that southern Fujianese had early formed substantial mercantile relations and settlements throughout the ports of the region. Fujian's maritime economy was maintained through the Ming–Qing transition, and, supported by Qing policies, the arrival of Western mercantile powers to the south China coast. Ultimately, it formed the basis of Western trade, as early Western traders grafted on to it, plied the same routes, called at the same ports.

NOTES

1 This empirical foundation also denies compatibility with the macroregion model and central place theory as an explanatory model for settlement origin. Central place theory holds that urbanization results from increased diversification in local and domestic economies – not long-distance trade.
2 Environmental concerns were absent from the *Rethinking the Region* project, substantially overwritten by processes of urbanization, layers of investment, and rounds of economic restructuring.
3 See also Raymond Bryant and Sinéad Bailey (1997). Recent work in political ecology is based in diverse theoretical approaches and concern for human difference and socially constructed forms of knowledge about nature–society relations.
4 Harold Wiens (1954) and C. P. Fitzgerald (1972) have treated the southern frontier as a region of Han colonization. Fitzgerald's introduction to *The Southern Expansion of the Chinese People* adopted the common northern vantage: "Chinese influence, Chinese culture and Chinese power have always moved southward since the first age of which we have reliable historical evidence. When 'China' meant the ancient confederacy of states acknowledging the overlordship of the Chou dynasty, Son of Heaven, and covered only the basin of the Yellow River, her influence, and soon her culture, began to penetrate the then alien peoples of the Yangtze valley."
5 In *Xiamen zhi* (1839, chapter 15/2a); see also Jennifer Cushman (1993).
6 Widespread historical repetitions of the anti-trade and anti-foreign ideologies of the Qianlong emperor, reflected in the Qianlong emperor's famous edict to the King of England that China had no need for England's manufactures, during the MacCartney expedition in 1793, have arguably highlighted one worldview of China's foreign relations at the expense of understanding variations in trade policy that characterized the realities of coastwise and Asian regional maritime trade.
7 See Kwan-wai So (1975) and Dian Murray (1987), for discussions of regional piracy. Kwan-wai So especially shows how official representations of piracy blamed Japanese, while most "pirates" during trade bans were Chinese merchants and fisherman. Murray found the origin of piratical dynasties in Guangdong in fishing communities of the Dan, who lived their lives largely at sea and as nuclear families. Women of this subculture commonly lived aboard ship and sometimes engaged in pirate raids. One of south China's most awesome pirate leaders was a woman, known only as Cheng I's wife; see Murray (1981).
8 Eduard Vermeer's (1990) introduction to *Development and Decline of Fukien Province in the Seventeenth and Eighteenth Centuries* emphasizes economic decline in Fujian after the sixteenth century, yet Vermeer is also compelled by the evidence to allow that decline was partial, geographically specific, and relative to the Yuan dynasty trade through Quanzhou. Evelyn Rawski (1972, p. 67) described as illegal the continuing sixteenth-century Fujian trade through the Zhangzhou area ports in the Minnan region, followed by

an analysis of significant Zhangzhou-based trade with the Portuguese, the Spanish at Manila, and Japan.

9 For example, after new imperial trade bans in the 1530s and 1540s, a center of maritime trade emerged offshore on Shuangyu island in the western Zhoushan archipelago, and trade, in the eyes of empire, became piratical smuggling; see Kwan-wai So (1975) and Pin-t'sun Chang (1990).

10 The Qing dyansty later made Yuegang an administrative center in order to regain some control over the trade, and changed the name of the port to Haicheng; about Yuegang see Ng (1983) and Chang (1990).

11 Periodic climate change and cooler temperatures also likely affected the historic distribution of these animals, especially the panda. Pleistocene fossil records have placed the panda throughout Lingnan and northern Vietnam. One Han dynasty site of panda evidence is in the mountains of the Fujian–Jiangxi border; see George Schaller (1993, pp. 62–3). Forests of evergreen and deciduous broad-leaved species predominated at higher elevations above 1000 meters. The tropical and subtropical needle-leaved forest is mostly located on low mountains and valleys and has become common where primary forest has been removed. Along the coast, Fuding, near the Zhejiang province border, is the northernmost limit of mangrove (*Rhizophera* spp.) distribution. For the distribution of vegetation see Zhao Songqiao (1986).

12 The expatriate community in Fuzhou repeatedly reported man-eating tigers: "in one district, not more than thirty miles from Foochow, nearly one hundred people, according to popular report, have been carried away by tigers within one year. Man-eating tigers are old tigers which cannot capture the active wild creatures, and which find in man an easier prey" (Kellogg, 1925, p. 35). The south China tiger has survived in a sliver of its original habitat range in the mountain forests along the Guangdong–Guangxi border, and is endangered.

13 In hilly areas below 800 meters the common pine is *Pinus massoniana*, the red pine, which has been used for shipbuilding because it is denser and stronger than the other common commercial species, the fir, *Cunninghamia lanceolata*. *Cunninghamia* grows quickly to harvestable size in twenty years and has become a common contemporary afforestation species.

14 Yue and Yao are large-scale designators and include distinctive smaller culture groups. Eberhard (1968, p. 88), wrote about the Yue that "Chinese regarded them as non-Chinese who were 'civilized'." The Yao, by contrast, have an origin myth which explains their descendence from dogs, which the Han have used to construe the Yao as a relatively marginal cultural group.

15 Yue (越) as a general term for early non-Sinitic people of the south coast region should not be confused with Yue (粤), which is another name for Guangdong and the formal linguistic name of the region of Yue dialects spoken in over 100 counties in Guangdong and Guangxi provinces. Cantonese is the Yue dialect of the city of Guangzhou; see Jerry Norman (1988, pp. 214–21).

16 From a text by the official Zeng Feng, adapted from the translation by

Hugh Clark (1991, p. 141).

17 Fertilizer solutions were many and varied, including soybeans, crushed oyster shells, mixtures of pig bristles, animal bone or wood ashes, sesame cakes, bamboo or bracken leaves, and human waste, or "night soil", near cities; see, for example, *Zhangzhou fu zhi* (1573, chapter 11/43b–44b).

18 The interactive effects of particular soil conditions vary considerably, and also with respect to the demands of individual crop plants; see *Soils of China* (1990).

19 Local farmers selectively adapted new seed strains to meet a variety of local conditions, and by the Ming period over 150 different types of rice seed were in use; see Rawski (1972, p. 40). The use of a relatively salt resistant seed confirmed the push to reclaim coastal and estuarine shores (in *Quanzhou fu zhi*, 1612, chapter 3/38b-40b). The *Zhangzhou fu zhi* (1573 chapter 5/6b) records different types of reclaimed land, including low grade *haitian* (sea fields or land reclaimed from the sea).

20 Local officials began to protest the situation by the end of the twelfth century; see *Zhangzhou fu zhi* (1628, chapter 8). But T'ien Ju-K'ang (1990), cautions that landholding concentration varied substantially. Local records from Putian, Xianyou, and Zhangpu counties, all coastal counties, showed that Buddhist landholdings amounted to no more than 20 percent of total farmland; compare *Xinghua fu zhi* (1503, chapter 54) and *Zhangpu xian zhi* (1700, chapter 8). By comparison to the Zhangzhou case, it would appear that landholding concentrated in Buddhist temples in the vicinity of wealthy towns as a potential strategy to maximize accumulation from the agricultural sector. Zhangzhou lay in the heart of the province's largest plain and was an important center of sugar wholesaling by the Song dynasty.

21 Ningbo's origins, like those of other coastal settlements, are incompatible with central place theory and its modeling of settlement origins in locally productive agricultural economies.

22 From the Southeast Asian perspective, Melaka was only one of several regional ports in the early 1400s. China's notice of Melaka constructed its status as the leading regional port; see Anthony Reid (1988, pp. 205–6).

23 Zheng He did not accompany the second voyage. Duyvendak's (1939) careful account addresses problems of dating the voyages, widely repeated in the *Ming shi* (*History of the Ming*) and standard histories as numbering seven, by pointing out that the sixth voyage likely never took place.

24 San-pao (in *pinyin* Sanbao), which means "three treasures," is a sobriquet of Zheng He.

25 This "divine light" was likely an electric charge at the top of the mast, known as St Elmo's fire in the Western sailing tradition.

26 See discussion over the date of Tianhou's imperial designation in James Watson (1985, pp. 299–300, n 25).

27 Like the arrangements made with his father, the Qing government offered to pardon Zheng Chenggong and members of his regime, bestow on them official rank, and yield some control over the custom's revenue in southern Fujian, but Zheng Chenggong would have nothing of the Manchu overtures; see Lawrence Kessler (1976, pp. 39–40).

28 Debate over the question about which areas were affected suggests that some walled cities were also evacuated, but comparison of sources suggests the brunt of the removal affected the rural population and fishermen.
29 Son adoption is also referred to in the *Xiamen zhi* (1839, chapter 15/13a). Adopted sons who successfully returned would be more substantially incorporated into the lineage.
30 After 1435 ships were no longer granted, but still the Liuqiuans were permitted to build ships in Fujian (see Chang, 1990, pp. 66–7).
31 Junks were also built in Bangkok because of the availability of teak, one of the best hardwoods for boat building (see Viraphol, 1977, pp. 4–5; 40–1).

4

Open Ports and the Treaty System

> Forms of economic organization do not develop in a social vacuum: they are rooted in cultures and institutions. Each society tends to generate its own organizational arrangements. The more a society is historically distinct, the more it evolves in isolation from other societies, and the more its organizational forms are specific. However, when technology broadens the scope of economic activity, and when business systems interact on a global scale, organizational forms diffuse, borrow from each other, and create a mixture that responds to largely common patterns of production and competition, while adapting to the specific social environments in which they operate.
> (Manuel Castells, 1996, p. 172)

The Liuqiu missions arrived biannually at Fuzhou and sustained longer and with greater regularity than those of any other empire (map 4.1). They continued until 1851, and so in the final years of the tributary era the newly installed British consul watched with interest the arrival of tribute junks from Liuqiu – for possible introduction of British manufactures to the Japanese market, because Liuqiu, in the context of obligatory tribute missions to both countries, was effectively the commodities wholesaler between China and Japan (see Ch'en, 1968; Sakai, 1968).[1] The fleet of two or three junks typically arrived in November and remained until the following spring. Mission envoys traveled to the capital to present tribute, while the larger crew and functionaries remained in Fuzhou to trade. The supercargoes of the Liuqiu ships furnished local brokers with manifests of imported trade goods and specified desired exports, which were obtained from as far as Canton and Suzhou. Figure 4.1 is the cargo manifest of goods exported by the final Liuqiu mission, which indicates the particular commodities obtained in Canton and Suzhou, including British manufactures, all textiles. These tribute missions were regularized trade ventures, formalized by ritual encounter and supervised by the Board of Rites, which privileged appropriate modes of conduct over direct trade and accumulation. Trade on Western terms, by contrast, would prioritize ideas about economic rights over ritual.

The south China coast was the region of international ports sought by

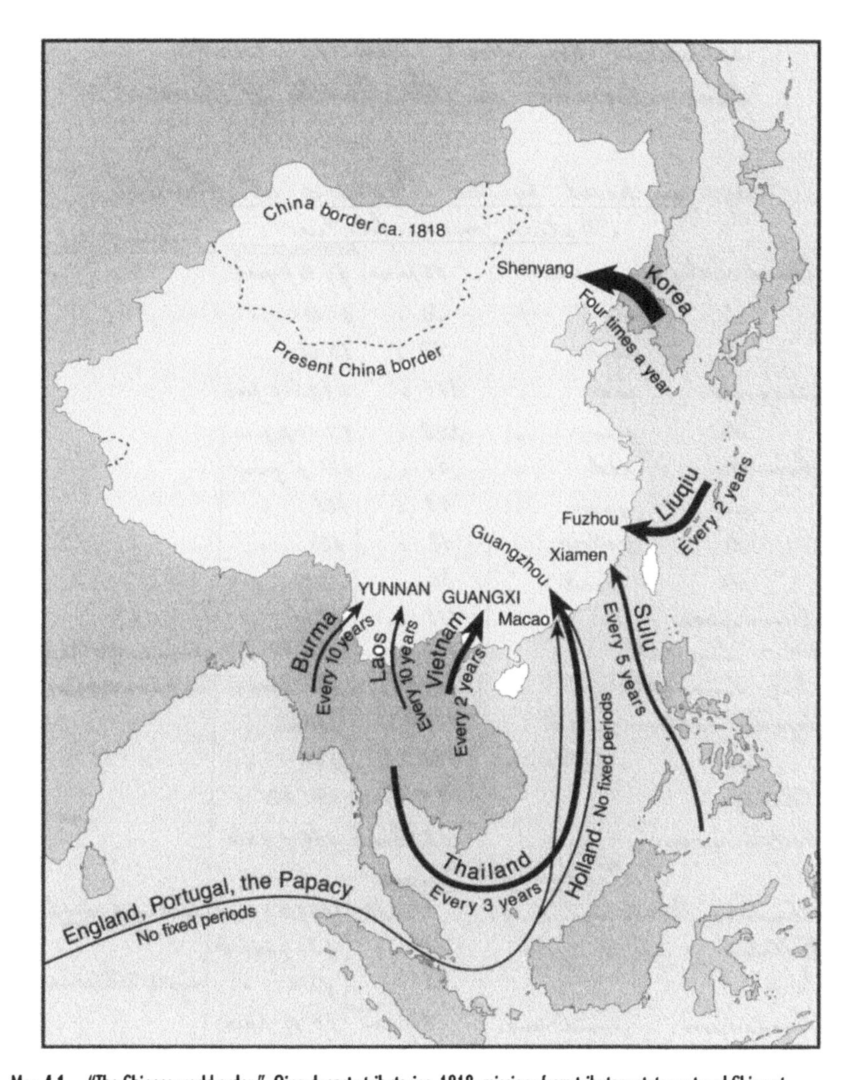

Map 4.1 "The Chinese world order." Qing dynasty tributaries, 1818; missions from tributary states entered China at designated ports and border provinces.
Source: Fairbank (1968); line work by Jane Sinclair.

all foreign merchants, from the earliest Islamic traders to the Western mercantilists. In the first encounters between China and Western powers, China sought to interpret the arrival of the West through the tribute system. [2] But Western imperialism fundamentally changed the terms of exchange and forced the Qing government to open the coastal cities to foreign residence and trade. The resulting "treaty system" carved out a

80

Return of Merchandise exported by two Tribute Junks from Loochoo in the Month of June 1851.

Cargo purchased by the Brokers at Canton.

1°. British manufacture:

Item		Quantity	average prices	observations
Broad cloth (plum colour)		50 pieces	$1.10 p yard	
do. (green)		30 "	$1.10 "	
do (black)		50 "	$0.95 "	
Longells, red,		400 "	@ $9.50 p. piece	
do, green,		200 "	$7.50 p. piece	
Camlets, red,		50 "	$25. p. piece	
do, green,		40 "	$25 "	
do, yellow,		10 "	$23. "	
do, black,		30 "	$23. "	
Bombazettes		100 "	$23 "	
White long cloths		3000 "	$3.60 "	each piece weigh[g] 6 catt. 12[m]
Grey long cloths		3000 "	$3.20 "	each piece weigh[g] 6 catt $1½
Dyed cottons, red,		1000 "	$5.20 "	
do, yellow,		500 "	$4.– "	
Chintz		600 "	$5.20 "	
Cotton yarn		30 Bales	$40. p bale	

2°. Produce of China

Item		Quantity	average prices	observations
Rhinoceros horns, med. drug,		6 piculs	$9 @ $12 p picul	from 8 inch[s] to 1 foot long without root
Buffalos do,	do,	10 "	$3. p picul	
Ivory		25 "	$150. "	weigh[g] 30 catties up to 100 each piece
Hartshorn,	med. drug.	20 pairs	$4. p. pair	
Caustic cuttings	do.	1000 cuttings	$2.20 p catty	
Putchuck	do.	5000 catties	$3.20 p 100 catt.	
[illegible]	do.	8000 do.	$–.10 p catty	
Ma-hwang	do.	3000 "	$2.50 p catty	
Kan-sung	do.	10.000 "	$10. p 100 catt.	
Liquarice	do.	20.000 "	$10. p 100 catt.	

Figure 4.1 Exports carried by the final Liuqiu tribute mission, 1851.
Source: FO 228/128, Fuzhou, 18 June 1851, to Bonham/from Sinclair.

Produce of China continued:

Rhubarb, med. drug.	8000 catties	$7. p 100 catt^s	
China root, do.	40.000 "	$1.60 p 100 "	
Miscellaneous drugs.	6000 "	total 750.000 cash	
Cloves	2000 "	$14. p 100 catt.	
Tortoise back, shell	50 "	$8. p catty	
do feet, "	200 "	$8. p catty	
do belly, "	200 "	$2.50 "	
Vermillion	3000 "	$1.20 "	
Quicksilver	3000 "	$1.10 "	
Sub borate of Soda	3000 "	$68. p 100 catt^s	used in medicine
Snake Skins	10 skins	400 cash each	
Silk of the white worm	2 piculs	$500. p. picul	finest quality, used for fine embroidery - is produced in the Tsing quen dist. in Canton Province - - -
H do	20 "	$350. "	
Hainan Grasscloth	60 bolts	$12. p bolt	is close to Canton

3° Foreign imports:

Foreign Ginseng 1st quality	1000 catties	$800. p 100 catt^s
do. 2nd do	1000 "	$500. "
Sapan Wood	20000 "	$2.20 @ $3. p picul

N. B. One of the ten Foochow Brokers, specially appointed for the Loochoo Trade, was deputed to Canton to purchase the above goods; they were chiefly carried overland from Canton to Chang chow foo, near Amoy - at the rate of 4000 cash each coolie, for a load averaging 90 to 100 catties. From Chang chow foo, these goods were sent to Foochow in Junks. The most bulky were brought up direct in Canton Junks.

Cargo purchased by the Brokers at Soochow.

1° Crape, red and white.	400 pieces	480 cash p Tael	each piece weighs 9 taels
Silk piece goods	400 "	430 " "	" " 8 @ 9 "
Hangchow bombazettes	20 "	480 " "	" " 25 "
White Lutestring (piece goods)	20 "	320 " "	" " 5^t. 6^m.
Velvet	300 "	$2.30 p. piece	

Silk piece goods and Satins for the King & Court for 20.000 dollars.

Figure 4.1 (cont.)

Cargo purchased at Soochow continued:

Hangchow fans	10.000 fans	240 cash each	
Coarse fans.	12.000 "	50 cash each	
Red felt carpets	600 carpets	$1. each.	manufactured at Soochow
Porcelain tea caddies	1600 caddies	140 @ 200 cash each	fabricated at E. king in Kiang, near Pow.
Red dye	70.000 leaves	$42. p 10.000 leaves	also a lip salve for the fair sex

2° British manufacture:

Longcloths (white)	2500 leaves	$ 3. p piece

N.B.
The above cargo was purchased, partly at Soochow, partly at Shanghai, by one of the Loochoo-trade Brokers sent for the purpose. They were brought down in native Junks to Foochow.

Cargo purchased at Foochowfoo.

1° British manufacture:

White longcloths.	4.500 pieces	$3.10 @ $3.30 p p.
Longells (red)	120 "	$ 8.40 p piece

N.B. In this quantity are included 2000 pieces bought by the Broker at Amoy on his way to Canton. 2000 pieces purchased from the native hongs here, payable in 2 months, and the rest of this cargo either bought from Mr. Compton's Hong, or from the Dealers in the City.

2° Produce of China:

Fuhkien paper	400 bundles	1500 cash p bund.	
Coarse do.	500 reams	250 cash p ream	
Writing paper	400 bundles	1355 " p bundle	
do do.	500 "	1500 " "	
Color'd paper	10.000 sheets	50 cash each	used chiefly for inscriptions and ornamental writing.
Oil paper	3000 sheets	200 @ 300 cash p sheet	
White sugar	10.000 catties	5600 cash p 100 catt.	
White loaf-sugar	5000 "	6800 " "	
Sugar candy	4000 "	9600 " "	
Writing brushes	25000 brushes	16. up to 80 cash each	made at Foochow

Figure 4.1 (cont.)

Indian Ink	10.000* catties	480 cash p. catty
Bamboo Combs	6000 combs	10 cash each
Bamboo Chopsticks	20.000 pairs	16 cash p. pair
Varnished trays &c	4.600 pieces	70 " each
Varnished wooden trunks	100 trunks	400 " "
Course porcelain cups &c.	6000 sets	70 cash p. set
Porcelain flower Vases	100 vases	800 " each

N.B. These are manufactured at the District town of Tih-hwa in Fokien, distant about 8 days journey from Foochow.

Pewter tea canisters	700 canisters	300 cash each
Kittysols	6000 kittysols	200 " "
Second hand clothes	100 garments	total 451.000 cash
Needles	60.000 needles	3000 cash p. every 10.000
Candied Oranges	600 catties	8000 cash p. 100 catties
Preserves and sweetmeats	1000 "	9600 " "
Small sized drums	100 drums	100 cash each
Tea - black	20.000 catties	320 a 960 cash p. catty
Tea - green	6000 "	$9. p. 100 catt.
Foochow grasscloth	2000 bolts	$2.50 p. bolt
Hemp	1400 catties	160 cash p. catty
Cassia	20000 "	$4.50 p. 100 catt.
Red dye²	20.000 "	$77. p. 100 catties ² a red flower
Incense sticks	102 cases	@ 6 cash a bundle each case cont.g 500 bund.

3° Bird's nests	10 catties	$40. p. catty

Sugar and Drugs bought with the Funds sent by the King and Court amounting to $10.000.-

British Consulate
Foochow. 18th June 1851
Chas. A. Sinclair

Figure 4.1 (cont.)

geography of open ports and mercantile access, and led to the growth and internationalization of two of the largest economic centers in Asia, Hong Kong and Shanghai. Despite their historic significance, Ningbo, Fuzhou, and Xiamen, the three smallest of the first five open ports, did not become leading centers of international trade but maintained local positions in the regional trading networks. What is interesting about these cities is the degree of continuity in social and economic practices that characterized them in the face of foreign imperialism, and how local social organizations and economic activities resisted the exigencies of the treaty system. The historiography of the treaty port era, typically written as a chronology of events between world powers, is also a geography of incremental imperialism, which was, despite its formulaic institutional character, quite unevenly received in the cities of the south China coast. This chapter assesses forms of continuity and resistance in the regional formation through local conditions in Ningbo, Fuzhou, and Xiamen during the treaty port era, and in scaled perspective, from the cities to national-scale intersections with processes unfolding in the larger world economy.

"Semicolonialism"

Understanding contemporary China is complicated by the experience of "semicolonialism" that began with the Opium War, 1839–42.[3] If China is a developing country at the start of this millennium, and some analyses characterize it that way, it does not share the profile of colonial history experienced in much of Asia, Africa, and Latin America. Neither can China's developmental history be limited to the terms of transitional economies in post-socialist states, since communism in China must also be understood against the historic experience of semicolonialism as well as the socialist revolution. "Semi" is an awkward modifier which signals how foreign powers practiced forms of colonial-style imperialism in China but never colonized the country at large, and that China's historic losses of sovereignty were geographically concentrated in a group of cities and their hinterlands, incrementally opened to foreign merchants and missionaries by a series of treaties.

Differences between Chinese and Western worldviews about trade and diplomacy culminated in the Opium War between China and Great Britain, and were underscored in a series of subsequent wars and military conflicts with different foreign powers. From Western perspectives, these conflicts initially centered on China's implementation of tribute system-style controls over mercantile relations, which sought to limit trade to just a few ports (map 4.1). Such geographically circumscribed terms of economic relations clashed with nascent ideologies about free trade, and Westerners wanted to call at any mercantile port where they could anchor a ship. The ritual protocols of the tribute system, in performing the kowtow (*ketou*), are especially well known as the focus of ire of especially the British envoys, but understanding the kowtow as embodied spatial practice, and as a representation of the Chinese world order, changes the terms of the debate. The rituals of imperial reception prescribed that the tribute envoy perform "genuflections and prostrations" in front of the seated emperor, which constituted kneeling and bending over to knock one's head on the floor. Rather than the gesture of degrading subservience it is often represented to be, in the imperial worldview "the kowtow was merely a part of the universal order of Confucian ceremony which symbolized all the relationships of life. The emperor performed the kowtow to Heaven and to his parents, the highest officials of the empire performed it to the emperor, and friends or dignitaries might even perform it mutually to each other. From a tributary envoy it was, therefore, no more than good manners" (Fairbank, 1953, p. 29). James Hevia's (1995) reinterpretation of the tribute system emphasized how the British understood these rituals as critical to establishing mutual power relations. Yet Western mercantile goals were fundamentally economic, and British envoys may have continued to perform the kowtow if they had free access to the ports they desired. What they were really piqued about was the Canton system. After 1759, China limited foreign trade with the West through Canton, which evolved to restrict Western traders to deal through designated imperial merchants, known as the Cohong. As ever, a good deal of informal trade took place outside the system, especially at Macao and Xiamen (Hao, 1986, pp. 17–20), and the Cohong was not the complete imperial monopoly Western traders claimed it to be. Nevertheless, Western mercantile powers maintained that the system unreasonably restricted trade, and after a series of debacles over opium trade and other issues through the 1830s, Britain moved to settle matters by force and sent a military fleet to the China coast, which initiated the Opium War. The fleet initially blockaded major ports, and, in another round, gunboats blasted Guangzhou, Xiamen, and Ningbo. China was militarily unprepared to defend the cities and capitulated.[4]

The treaty settlements

The Treaty of Nanjing, the Treaty of Tianjin, and the Treaty of Shimonoseki are three of the most significant of more than a dozen major treaties between China and foreign powers that opened China to foreign residence and Western mercantilism. The first treaty, the Treaty of Nanjing, signed in 1842, settled the Opium War.[5] The treaty provided for the opening of the ports by allowing "British subjects, with their families and establishments . . . to reside, for the purpose of carrying on their mercantile pursuits, without molestation or restraint, in the cities of Canton, Amoy, Foochowfoo, Ningpo, and Shanghai" (Hertslet, 1908, pp. 7–8). It gave over to Britain the island of Hong Kong in perpetuity, and abolished the Cohong system. The treaties also often required that China pay indemnities to foreign powers, and the Treaty of Nanjing required China to pay $6 million for opium, confiscated by imperial commissioner Lin Zexu in Canton, $3 million for debts owed to foreign merchants by the Cohong, and another $12 million to reimburse the British for their expenses incurred in the military confrontation. This last element of the agreement, in which China was compelled to pay for its own destruction, would be repeated in subsequent treaties. In a supplementary treaty the following year, the British obtained the important "most favored nation" clause, which granted specific treaty privileges between China and any one country to all other foreign powers. In this way Western powers shared with each other the fruits of their separate imperialist programs in China, and prevented China from maintaining distinct diplomatic positions with individual countries.

In 1858, the Treaty of Tianjin settled a second war between China and Britain, allied with the French, over continued tensions in the coastal trade. The provisions of the Treaty of Tianjin significantly enlarged the scope of foreign presence across the Chinese landscape by allowing foreigners holding passports, including missionaries, to travel anywhere in the interior. It ultimately opened ten more ports to foreign trade, including Shantou in Guangdong, two ports in Taiwan, and four on the Yangzi River, and specified residence of a British ambassador in Beijing. The Beijing residence provision, however, in the heart of empire, was predictably resisted by the Qing government. British and French troops forced the issue, and marched on the capital and beyond to the Yuan Ming Yuan, the imperial summer palace in the suburbs north of the city – and burned it to the ground. Destruction of the Yuan Ming Yuan signaled a terrible loss of control over the ability of Qing troops to protect imperial patrimony, and compelled the Qing to sign the Convention of Beijing, in 1860, which affirmed the articles of the Treaty of Tianjin and extracted further land and money resources. The Convention ceded the tip of the Kowloon peninsula, part of the mainland, to the British colonial territory in Hong Kong, opened the port of Tianjin, the traditional maritime gateway to the capital, and permitted British ships (and all other foreign ships) to carry Chinese emigrants. Thus these treaties not only carved out a geography of imperialism but set in motion diverse spatial processes that simultaneously tied the coastal cities into an imperialist world economy and fundamentally constrained China's participation in it.

The Treaty of Shimonoseki in 1895, which concluded the war between China and Japan, further eroded China's territorial sovereignty, most distinctively by making Taiwan a Japanese colony. This treaty authorized 200 million taels in indemnities, ceded Taiwan, the Pescadores, and the Liaodong peninsula to Japan (of which the Liaodong peninsula was later relinquished for 30 million taels), opened four more ports, and yielded to foreign interests the right to set up manufacturing plants and warehousing in the interior. The opportunity to engage in fixed capital investment also prompted infrastructural development in waterworks, electrical, telephone, and telegraph systems, and railroads and roadways. Western firms bid to construct and finance railway projects, with large high interest loans, and foreign railroad construction became a focus of serious national debate. In response, many local elite groups, in the mode of "self-strengthening" and nascent nationalism, were motivated to finance large-scale infrastructural development with domestic capital, and the "rights recovery movement" emerged to wrest infrastructural development rights away from foreign control (Wright, 1968). The urban merchant class with local gentry support created an institutional alliance known as *guandu shangban* (official supervision and merchant management) to manage a range of manufacturing and service industries (Feuerwerker, 1958; Fewsmith, 1985; Rankin, 1986). It was especially active in the open ports where the extremes of foreign demands were well understood. The movement compelled a US interest to yield control over the Guangzhou–Hankou railway, and won the return of construction rights from a British company for the Shanghai–Hangzhou–Ningbo railway (Chan, 1977). In Fujian, provincial leaders sought to finance the Xiamen-Zhangzhou railway by selling shares to local merchants and gentry, and also to Fujianese overseas in the Nanyang (Godley, 1981). But at the national scale, the import of opium, indemnities paid to foreign powers, and debt service on foreign loans left China in a weak political and economic position.

Opium for tea

From a world economy perspective, the origins of the treaty ports are based in the opium trade. The textile products of Britain's industrial revolution did not form a sufficient basis from which to offset purchases of Chinese tea, porcelain, and silks, and Britain had a trade deficit with China.[6] The cost of the colonization of India further strained Britain's national accounts, and an improved balance of trade became a national priority. Frederic Wakeman's assessment of Britain's evolving world economy evinces the ideological basis of the economy at stake:

> After centuries of trade, the West had finally found something the Chinese would buy in large quantities. Moralists might feel a guilty twinge at the thought of the nature of this product, but was not the drug the staple of the Country trade? And was not the Country trade in turn the epitome of those values every Anglo-Saxon of the age valued most highly: self-help, free trade, commercial initiative? Thus were twinges ignored, moralists scorned, and doubters derided. If anything, the free traders felt more was due to them. Manchester was on the rise, and the Country traders chafed at restrictions the Select Committee almost took for granted. Jardine wrote to a friend: "The good people in England think of nothing connected with China but tea and the revenue derived from it, and to obtain these quietly will submit to any degradation." (Wakeman, 1978, p. 173)

Opium was planted in British India, from where it was traded to China by the agency houses of the Country trade, private trading firms licensed by the British East India Company to sail between India and China. Opium remained the leading Chinese import until 1890 and the Company taxed tea imports upwards of 100 percent, which, together, contributed to balancing Britain's national accounts (Feuerwerker, 1969).[7] During the period 1810–26, China maintained a total trade surplus of about $74,700,000, but from 1827 to 1849, $133,700,000 in silver flowed out of the country (Hao, 1986, pp. 122, 129).

In the early years of consumption tea was costly and a privileged beverage of the upper classes. It even occupied a symbolic space in late seventeenth-century English homes – locked in caddies to be kept safe from servants and common use. With increased supply through the eighteenth century, tea began to transform into a more common drink and entered realms of public consumption. Tea parties and tea gardens, as scenes of polite social encounter, emerged as sites of leisure activity. Tea also appeared on the menu in coffeehouses, including a coffeehouse bought by Thomas Twining in 1706, which brought tea out of the domestic arena and more squarely into the realm of an evolving public sphere (Ukers, 1935, pp. 40–6). Coffeehouses, opened widely in Restoration England, became sites of public information exchange, meeting places for the discussion of news and politics, and in London shipping and trade (Pincus, 1995). The theoretical work of Jürgen Habermas (1989), on the emergence of the public sphere as a basis for civil society, derives from the experience of England in this period and the role of places in which people of diverse class backgrounds pursued critical assessments of state and society. The coincident British temperance movement promoted the new stimulant beverages of the tropics in opposition to regular drinking of beer and ale. The important role of women and the clergy in the temperance movement made tea a beverage that facilitated new axes of social encounter between the private and public spheres, and its consumption grew with the expansion of the urban population during the Industrial Revolution. Tea was a symbolic and material civilizing medium as well as a basis of economy in modern England. China served as the main source of tea until the late nineteenth century, when British planters began to achieve mass production of tea in the Indian and Ceylon colonies.

The relationship between the tea trade and the opium trade underscored the value of these commodities. The Western merchants of the era were all trading opium, despite the series of imperial edicts forbidding its sale and consumption, and as with any drug economy of global proportions, government decrees could not fundamentally influence the political economy of an extremely valuable trade. William Jardine, whose writings on the China trade have left a distinct perspective on the political economy of the era, was the leading British merchant in the China trade. He founded the largest private foreign partnership in the China trade, Jardine, Matheson Company (which, in turn, became one of the largest multinational corporations in Asia, and was headquartered in Hong Kong until 1984). Jardine often represented merchants' interests in the debate over the opium trade, and he was called forth in response to the imperial anti-opium activities in Guangzhou. The most consequential anti-opium strategy, ordered by Fuzhou native Lin Zexu, court-appointed imperial commissioner with special powers to deal with the opium problem in Canton, precipitated the Opium War (Collis, 1946; Waley, 1958). In the spring of 1839 Commissioner Lin ordered confiscated and destroyed the foreign stocks of opium, which amounted to over 20,000 chests or three million pounds. The method of destruction – chests broken open, the opium dumped into trenches, wrecked with salt and lime, and flushed out to sea – was undertaken with all the boldness of a public spectacle. The foreign merchants even jointly funded Jardine's return to London to promote the cause of recovering the cost of the destroyed opium "and to ensure that the moral objections to the opium traffic being raised by various Protestant missionary societies did not gain too wide an audience" (Spence, 1990, p. 154).

The Open Ports

The treaty ports were port and borderland cities, on China's coast, rivers, and international boundaries, which were opened to foreign commerce and trade beginning with the Treaty of Nanjing.[8] Foreign presence in China concentrated in the treaty ports, and they were the essential bases from which scholars evolved the "impact–response" paradigm. What were the effects of the treaty ports on China? The two most common responses to the impact–response approach concluded that Western involvement led to change and modernization in society and economy; or that Western penetration inhibited or prevented Chinese social and economic development, particularly in the ports' rural hinterlands (see Esherick, 1972). Such dualistic accounts have given way in the face of more recent evidence that a small amount of economic growth occurred in China through the late nineteenth and early twentieth centuries.[9] Thomas Rawski (1989) has shown that certain Chinese industries, especially banking, transportation, and communications, experienced economic growth, albeit at a small rate around one percent, during the prewar period. This conclusion must reflect the fact that economic development was geographically uneven, and also variable by industrial sector. Still, the existing scholarship, in its focus on the idea of treaty ports, their impacts, and degree of economic growth, has inhibited broader understandings of these cities. "Open ports" (*kaifang gangkou*), "mercantile ports" (*shangpu*), and "trading ports" (*tongshang gangkou*) are arguably more appropriate terms for the ports open to foreign residence and trade (see Feuerwerker, 1976).[10] Treaties technically opened the ports, but the treaty port name has construed the notion in the Western literature that the ports owed their existence to a nineteenth-century Sino-Western political economic event, masking contextual geographies, marginalizing the realities of local social and economic processes, and belying the reality that opening the ports did not yield Western traders the practical conditions of widespread access to mercantile activity in China they desired.

On the basis of the "unequal treaties" – treaties signed under unequal power relations – it was not surprising that Westerners should have found subsequent settlement and mercantile pursuits in some of the ports especially hostile. In fact, the British were so disappointed with economic prospects in Fuzhou and Ningbo in the 1840s that they soon petitioned the Chinese government to effectively trade them in: they would give up trading rights at Ningbo and Fuzhou in exchange for opening other ports that might prove more commercially successful, such as Hangzhou and Suzhou.[11] From the 1850s to the 1870s, trade at Fuzhou revived somewhat, in response to lifting an interdiction against sending tea down the Min River. But the tea trade fell into decline by the 1870s and no other products replaced the demand for tea (Gardella, 1994). Foreign imports were similarly narrow. Opium made up the bulk of what foreigners had to offer, in addition to lead for lining tea chests, and a small quantity of shirtings. In 1876, the British Consul at Ningbo lamented, in a common refrain, "trade on the whole has been and still continues to be very dull. The whole trade of the port, with the exception of opium, is entirely in the hands of the Chinese."[12] In 1891, the British Consul in Fuzhou reported that most of the wholesale and retail trade was conducted by local Chinese firms (China, Imperial Maritime Customs, 1893, pp. 416–17). In 1896, the US consulate at Ningbo closed for lack of business. The foreign population in the ports did not necessarily decline, though, but changed in composition. Missionaries began to find more success than merchants in some cities, particularly Fuzhou, where the population of Western missionaries grew as merchants left. Why were these ports disappointingly unremunerative for foreign mercantilists during the "treaty port" era, when Shanghai and Hong Kong developed into two of the largest port cities in the world?

The three smallest of the first ports opened to foreign trade were different than Hong Kong, Shanghai, or Canton, and from each other. While in Hong Kong, the colony, and Shanghai, the center of China's mercantile economy, fortunes were made in foreign trade, the smaller ports remained local and regional centers, though still tied to distant shores and commercial networks by the lifepaths of emigrant merchants and their overseas legacies. The growth of Shanghai, by far the largest port, quickly overshadowed Ningbo, but Ningbo merchants underwrote a certain amount of Shanghai's success by establishing the financial industry of the larger city, the prewar center of China's banking industry. Fuzhou, the provincial capital of Fujian, and one-time imperial capital, remained as a south coast repository of anti-imperialist and later nationalist fervor, a center of strong local identity and fierce dedication to local rule in the face of Western occupation and a short-lived tea trade. Fuzhou especially did not welcome the treaty system, yet became one of the largest centers of Christian worship in the country. The Amoy region, its merchant class already networked around the Nanyang to centers of commercial maritime activity, was the area whose overseas community distinctively maintained a large remittance economy and returned home to invest in local infrastructure. In these ports local merchants prevailed in several mercantile sectors, in some cases even out-competing Western traders in Western products. Let us turn instead to some of the social and economic landscapes that characterized these places.

Foreign settlement

The open ports proved difficult to occupy and the first consuls regularly found their work impeded. Local resistance to foreign occupation of the ports emerged early in the contest over the space of foreign residence and trade. Establishing the consular facility itself was an indicator of things to come. Indeed, the Treaty of Nanjing seemed to have provided foreigners only the right to arrive at the port and transfer to the shore. Robert Thom, first British consul appointed to Ningbo, arrived in December 1843 aboard the steamer *Medusa*, whose commander simply landed Thom, his four consular staff, and suite of baggage – and left. Thom spent several days searching for a consular residence in the city, only to find that no one would rent to them. At last he managed to rent an "unoccupied woodyard" with assistance from local officials who reluctantly observed the provisions of the treaty.[13] G. Tradescant Lay, the first consul at Fuzhou, was deposited by ship from Hong Kong at the mouth of the Min River in July 1844. Any expectation of British pomp was nowhere in sight, and, as a later historian described the situation, "The Consul was allowed to find his way in the most undignified manner to the seat of his official post, with as little ceremony and respect as if he had been a common peddlar of bazaar jewelry" (Beard, 1925, p. 13). Local authorities provided the new British consular staff with a small, humble residence "constructed partly of thin boards."[14] Four months later Lay disparaged the place as "disreputable in appearance and seated in one of the lowest neighborhoods in the suburbs of Foo-chow."[15] Rutherford Alcock landed in June 1844 at Gulangyu, the small island facing the city of Amoy that was to become the principal area of foreign settlement. British troops had continued to occupy Gulangyu since the end of the Opium War two years earlier, and so Alcock's initial residence was secured upon arrival. His later choice for a consular residence was "a building situated within the walls adjoining the rampart and called the Library."[16] Alcock's "library" was the city's official examination hall, where local scholars periodically sat for China's civil service examination system. The examination hall, funded by voluntary subscriptions from local gentry, was not a multiple-use facility in need of more permanent occupation. In this episode, the British representative chose to misunderstand the importance and singular utility of a structure that represented one of the fundamental bonds between state and society, a building at once private and public in its symbolic meaning of place. To turn over the local examination hall to a foreign representative was a simple incredulity.

The space of foreign presence in the ports was not predetermined, and consuls, mercantilists, and missionaries all had to negotiate for residential and commercial property upon arrival. Local officials limited or blocked Western locational opportunities, and symbolic spatial elements of the city became the focus of debate. One of the most protracted contentions was the issue of foreign residence inside the walled cities. Power of place in traditional China was associated with government office and its means of attainment, through traditional education, and its rewards, in social status and wealth. These symbolic qualities of place dependably found spatial expression inside walled cities. Official buildings were located near the center of the city and gentry lived nearby. Foreign officials and missionaries, flush with assumptions about their own privileged positions, presumed to be able to occupy these places of significance and in order to exert influence with local officials. In Guangzhou and Fuzhou, provincial capitals with relatively large walled cities, foreigners contended early and long term to rent properties inside the walled city.[17] After four months at Fuzhou, Consul Lay requested quarters in the Temple of the White Pagoda, "or in some other convenient building inside the city."[18] He explained:

> the Consulate must be within the walls of the city, since it is the residence of not only all the civil and military functionaries, but of every person who lays claim to respectability. . . . In the ordinary phraseology of the natives, "within the city" is a term used to imply what is polished and correct, while "outside the city" is associated with every thing that is low, mean and ill-bred.[19]

Lay's dualistic rendering of the issue hinged on the symbolic properties of the wall, and his determination to reside within it. Power of place in China also found definition in historic temples, and in association with *fengshui*-defined sites.[20] Based on *fengshui* principles, houses, temples, graves, and other sites were selectively located in relation to topographic relief, waterways, trees, and southern vistas, all of which attract *qi*, the energy force of the *fengshui* belief system. On these cosmological terms, the Temple of the White Pagoda, on one of the three hills of Fuzhou and situated near the south gate, was one of the city's most significant places. In 1845, in the face of strong opposition, Consul Lay was able to obtain quarters in a disused Buddhist temple on Wushishan (Black Stone Hill), which afforded a view of the river and arriving ships (Carlson, 1974, p. 21), but a local movement against the consular residence would ultimately compel the British to abandon Wushishan.

Disputes over property locations resulted in foreign enclaves on the urban margins, and typically on lands that were physically separate from the larger city. In Amoy the foreign settlement at Gulangyu, a small island just seaward of Xiamen, followed in the wake of British occupation of the island after 1842. In Fuzhou too the foreign settlement was originally restricted to an island, Nantai, in the middle of the Min River. The foreign settlement at Ningbo lay on a tombolo-shaped piece of land inside the confluence of the Yong and Yuyao rivers, and at Canton it centered on Shamian Island. Shanghai of course had the largest foreign settlement, spread out over the west side of the Huangpu River and divided into two main areas of foreign residence, the International

Settlement and the French Settlement. Local people continued to live in the areas of foreign settlement, which later became residential areas of choice during the years of upheaval associated with the warlord governments and the Nationalist regime.

As enclaves of Western commercialism and culture, the open port cities were places where realities and representations of foreign culture, economy, policy, and ideology could be challenged directly. Tensions around the full range of differences between China and the West, fomented by continued foreign demands, led to the emergence of local movements in the port cities, especially in strikes over foreign economic practices and boycotts of foreign goods. In reality, what Western history has often recounted as disputes over terms of trade on the China coast is, in certain sectors, more accurately understood as organized resistance to Western economic power and early multinational corporate activity. Still, many foreign consuls understood the subtleties of local social and economic issues and the diverse problems of foreign imperialism, and in their place-based contexts, often wrote to support the perspectives of local people rather than Western powers and the forces of international political economy. From a scaled perspective we can view both the serious ethical problems of Western imperialist practices in the ports, and local ties between diverse people forged in the contexts of place.

The spirit of Min

In Fuzhou, a former capital of its own empire, meanings of place in legacies about political independence have been deeply embedded in local history. The imperial government opposed the inclusion of Fuzhou among the open ports, and Fuzhou gentry petitioned against foreign residence inside the walled city. It was one thing for the British consul to have obtained quarters inside the city, but when English missionaries of the Church Missionary Society gained quarters in a temple property on Wushishan in 1850, Fuzhou residents were stunned. Protesters organized against their occupation and new building construction, but the missionaries unadvisedly pressed on with construction. Demonstrations ensued and culminated in the infamous Wushishan riot of 1878, in which missionaries were injured and missionary property was destroyed (Carlson, 1974). Foreign display of dominance, whether through political and economic imperialism, racial superiority, or religious belief, would have to have been considered a particularly impolitic arrogance in Fuzhou. The prevailing Consul Sinclair saw through the missionary complaints and concluded that "The purchase of the houses so close to Wooshihshan was undoubtedly made purposeful to cause irritation."[21] The Wushishan incident, however, was not emblematic of the missionary presence in Fuzhou in general. How Fuzhou also became a notable center of Christian conversion, to the degree that contemporary Fuzhou remains distinguished by a significant Protestant population, raises questions about complex axes of social formation, in this case in ethical and anti-imperialist alliances against particular kinds of economic activity.

By 1850 three missionary societies had arrived in Fuzhou that would dominate Protestant evagelical work in the city through the twentieth century: the American Board Mission, the Methodist Episcopal Mission, and the Church Missionary Society, affiliated with the Church of England (Cochrane, 1913, p. 50). The work of the Protestants in Fuzhou was significant, not only for the strength of its numbers of converts, but for the degree to which local Protestants became involved in early civil society and nationalist movements.[22] Map 4.2a, by comparison to national population distribution, map 4.2b, shows that the coastal provinces experienced the greatest degrees of Protestant missionary activity, and, among them, that Fujian was the leading province of Protestant conversion relative to the total population. Map 4.3a demonstrates the degree to which this religious activity in Fujian was largely a coastal phenomenon, and, by comparison to map 4.3b, the degree to which Protestant activity concentrated around Fuzhou. By the turn of the century, Fuzhou Christians had become an ardent and organized group of protestors against a range of problems, including continued opium trade, foreign product monopolies, and the US Chinese Exclusion Act of 1905. Christianity in Fuzhou was embraced by people of diverse class backgrounds, including local elites, and Fuzhou Protestants held a disproportionate number of positions in the new provincial government after 1911 (Dunch, 1996, pp. 208–9).

The anti-opium cause in Fuzhou especially united the gentry and local reformers with missionaries to create an effective boycott. The Fuzhou branch of the national Anti-Opium League formed in 1906 under the chairmanship of Lin Bingzhang, the great-grandson of Lin Zexu. In this period of emergent nationalist movements, Lin Zexu had become both a local cult figure and a national hero, and the Fuzhou headquarters of the League was effectively a shrine to Lin Zexu (Wu and Lin, 1983). The League invoked his work against foreign opium trading at Guangzhou to galvanize local participation, especially from the foreign missionary population. An annual anti-opium parade assembled at the League headquarters and marched through the city, led by men carrying a large image of Lin, down to the custom's office on the river where a crowd of thousands gathered to watch the burning of seized opium paraphernalia. The parade was a public and nationalist spectacle, and, in the manner of territorial processions of cult festivals, ritually reenacted Lin Zexu's

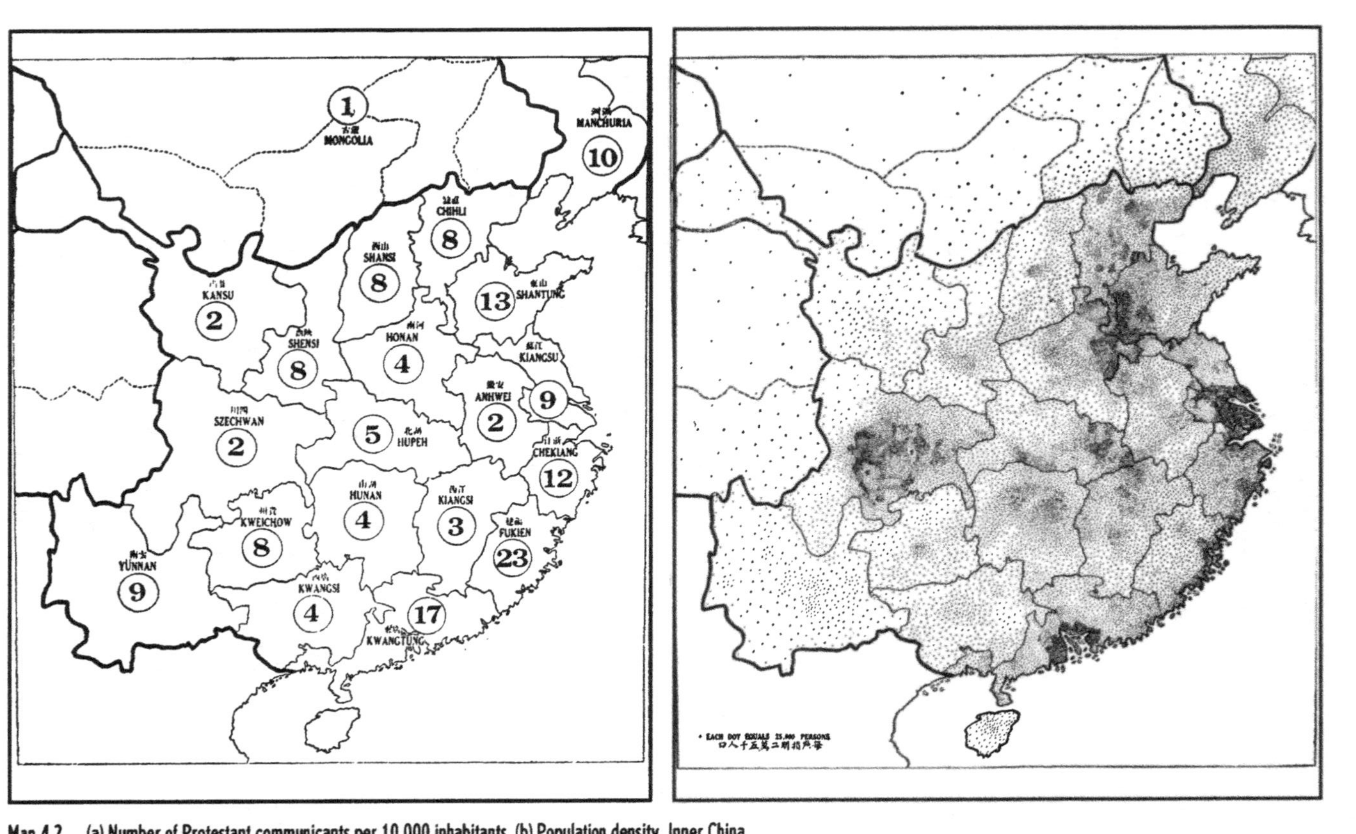

Map 4.2 (a) Number of Protestant communicants per 10,000 inhabitants. (b) Population density, Inner China.
Source: China Continuation Committee (1922).

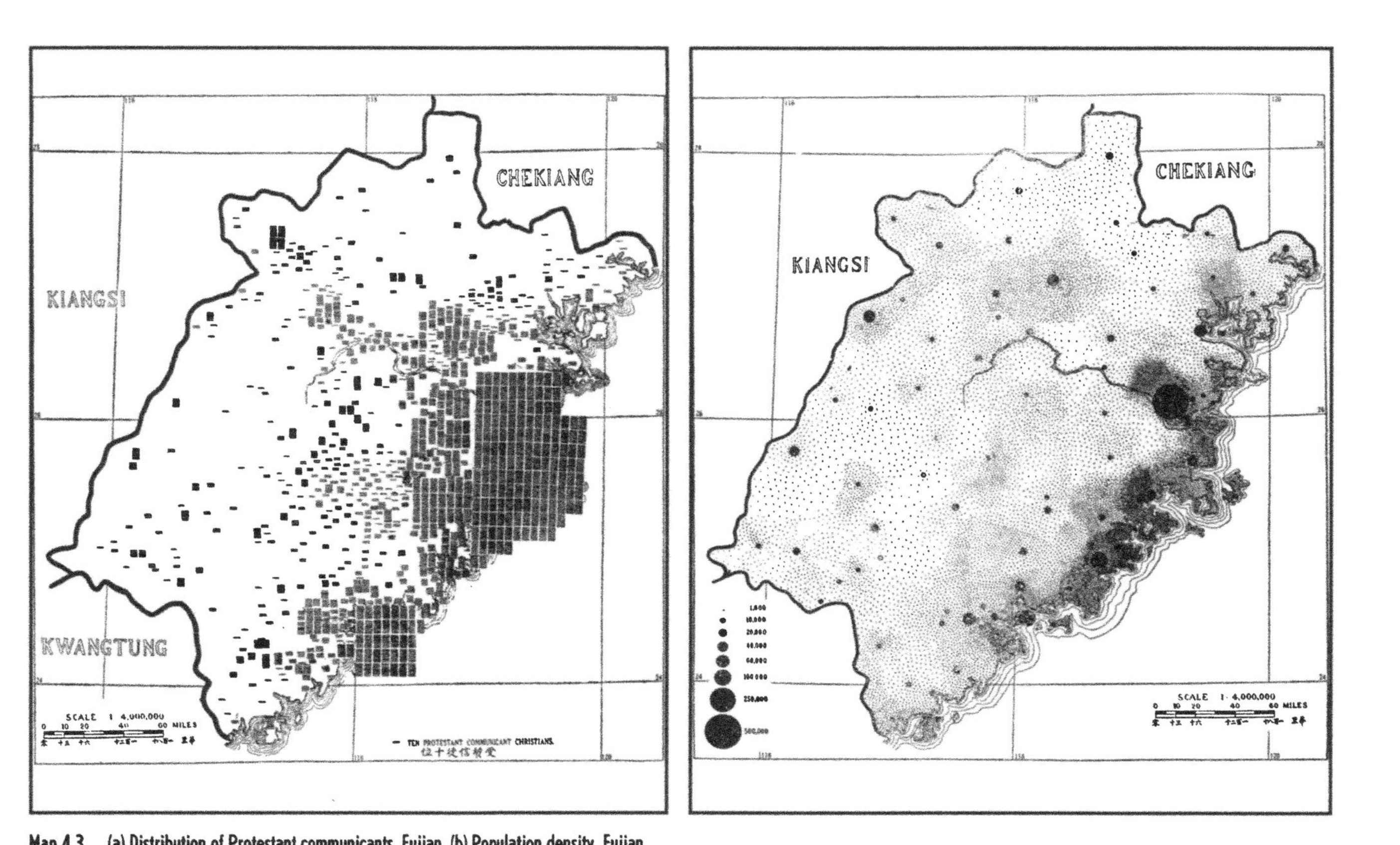

Map 4.3 (a) Distribution of Protestant communicants, Fujian. (b) Population density, Fujian.
Source: China Continuation Committee (1922).

campaign against opium and imbued Fuzhou with an anti-imperialist spirit.

The anti-opium campaign broadened in its support groups and target issues. In Fuzhou, the Methodist Church and Church Missionary Society avidly protested not only opium but cigarettes.[23] Local students, both Christian and non-Christian, joined the movement, and the cooperative character of the protest created a successful alliance between the local population and foreign missionaries in Fuzhou. The egregious quality of tobacco promotion in China no doubt motivated the alliance. In 1909 American Consul Samuel L. Gracey wrote: "You will understand the matter better if you are informed that one of the Greatest Tobacco Companies in the world just recently put to work here five foreigners who have posted an immense number of illustrated advertisements and are allowed to spend $1,000 a month in giving away cigarettes, and it was this special effort I believe, which aroused the Christian people to put forth the great efforts they have."[24] The firm in question was the British-American Tobacco Company, which, through its long-term strategies of market expansion, has become the largest transnational corporation in the world tobacco industry.

Alliances between local Protestants and missionaries in Fuzhou reflected a larger culture of learning and criticism that characterized the city. The missionaries established both boys' and girls' schools, and colleges for young adults, including one of the first women's colleges in China, the Fujian Hwa Nan Women's College. Foreign language learning was common, and the city's leaders engaged the national and international debates of the era. Newspapers proliferated in Fuzhou through the turn of the century, and the city was a center of many nationally known political reformers. One late dynastic perspective that recognized these conditions, from the perspective of the northern vantage, was the description of Fuzhou intellectuals as "stars twinkling in the dark" (Li, 1982, pp. 221–5).

Commercial Organization

Urban Chinese society revolved around a diverse array of social and economic organizations, but Western economic interests largely encountered the mercantile organizations, in highly articulated local trading systems on the south China coast. If Western merchants' reports about meager trade prospects in Ningbo and Fuzhou reflected a general condition (rather than a Western lament over limited foreign success) then we should expect to see evidence of social and economic decline more generally. The Taiping Rebellion (1850–64)[25] disturbed local economy in its wake, especially in the Jiangnan, but decline of the southern coastal ports is not the general case, as social movements in Fuzhou demonstrate, and the cities continued to transform and urbanize through the turn of the century. Merchant networks in the coastal cities supported the commercial economy from returns of trade, which made the circulation of merchant capital more significant in sustaining local economies than local landed pursuits. Western consuls complained about the absence of commercial activity in Fuzhou and Ningbo, and blamed their lack of opportunity on the power of local "guilds." The guilds were merchant organizations whose networked economic activities foreign merchants mistook for monopolies, antithetical to their ideas about free trade.[26] In reality, the superior competitiveness of local merchants was better understood as based in local knowledge systems and diverse contacts maintained through long distance trade networks and memberships in multiple merchant associations (Hamilton, 1977).

The Ming–Qing transition was a time of commercial expansion and increased interprovincial travel, and merchant organizations increased in numbers, especially in large cities. Mercantile activity was maintained in general through the Qing dynasty, and increased again in the second half of the nineteenth century. Greater numbers of merchant organizations appeared in cities of the south coast, and especially in the rapidly commercializing Jiangnan region (Fu, 1956, 1957; Ho, 1966). Between 1840 and 1911 the formation of merchant organizations in Shanghai more than tripled, especially those formed on the basis of trade in specific commodities and services (Yu Z., 1997). Increased mercantile organization reflected increased commerce, merchant mobility, and the need to maintain local markets in the face of foreign trade and increased retail activity. Merchants maintained a somewhat symbiotic relationship with the state in this era: merchant organizations depended on government recognition and support, and the state depended on merchant organizations to regulate commerce, determine the value of commodities, and maintain economic stability in general (Hamilton, 1977; Mann, 1987).

A systematic treatment of mercantile organizations in China has never been produced, which underscores their diversity and complexity. A preliminary typology of merchant organizations may be based on two categories, organizations by province or place, especially major cities, and organizations by trade.[27] Provincial and place-based organizations, the *tongxianghui* and *huiguan*, were both social and economic organizations. Variation in names such as *gongsuo* (public office) also appeared in the names of *huiguan*, such as the important Siming Gongsuo, the Ningbo *huiguan* in Shanghai during the open port era. While the terms *huiguan* and *gongsuo* were interchangeable to a degree, the different names also represented some differences in functional activities of the organizations. When they were different, *huiguan* were almost exclusively associated with places, while *gongsuo* were based on specific trades (see Xu and Wu, 2000, pp. 181–3). *Huiguan* supported economic activities, by providing business and employment information, regulation of commerce among members, and savings and loan functions. They offered temporary lodging to traveling fellow provincials, and passage home for the destitute. Their leadership commonly stepped in to adjudicate disputes among members or with other larger-scale institutions, such as foreign consulates or chambers of commerce. *Huiguan* also maintained cemeteries and temples, and observed cult festivals. Some also ran schools and arranged apprenticeships. They were the basis of social organization in sojourning society.[28] The term *hanghui* (occupational association) designated a commodity, trade, or service provision association. Commodity-specific associations formed around all items of long distance trade, from silk, porcelain, pearls and gems, tea, cotton, sugar, and rice, to fish, medicinals, indigo dye, and coal. Some of these associations more specifically supported trades, such as copper and iron smiths, and makers of cloth, furniture, paper, rope, mats, and candles. Commercial organizations for service providers, such as carpenters, metal workers, cloth pressers, bankers, doctors, pawn brokers, and other specialists, characterized large cities. These organizations, as *gongsuo*, could offer some of the same social services as *huiguan*, depending on their size and situation. The word *bang* (group) has also been used to designate a community of workers, based on place or origin or dialect.[29] *Bang* is also in common contemporary usage in Chinese overseas populations to differentiate among communities based on dialect and homeland area of origin, such as in Singapore (Cheng, 1985). In Singapore and Malaysia the largest Chinese population subgroup is the Hokkien *bang*, which is the dialect group of southern Fujian.

Peng Chang's (1957) analysis of the distribution of merchant organizations by province also observed merchant organizations in three categories of scale: rural merchants attending periodic markets, urban merchants with regular shops, and long distance merchants who wholesaled relatively high value commodities. Maps 4.4a–f are based on Chang's analysis of the location and distribution of provincial merchant groups engaged in long distance transregional trade. They depict the extent to which merchants developed commercial networks in China beyond their own provincial bases.[30] In rank order of activity and influence, merchants from Guangdong had more organizations of diverse types in more major cities in China than any other province, and Guangdong merchants dominated trade within the province. Merchants from Fujian generally were the second most active group of merchants

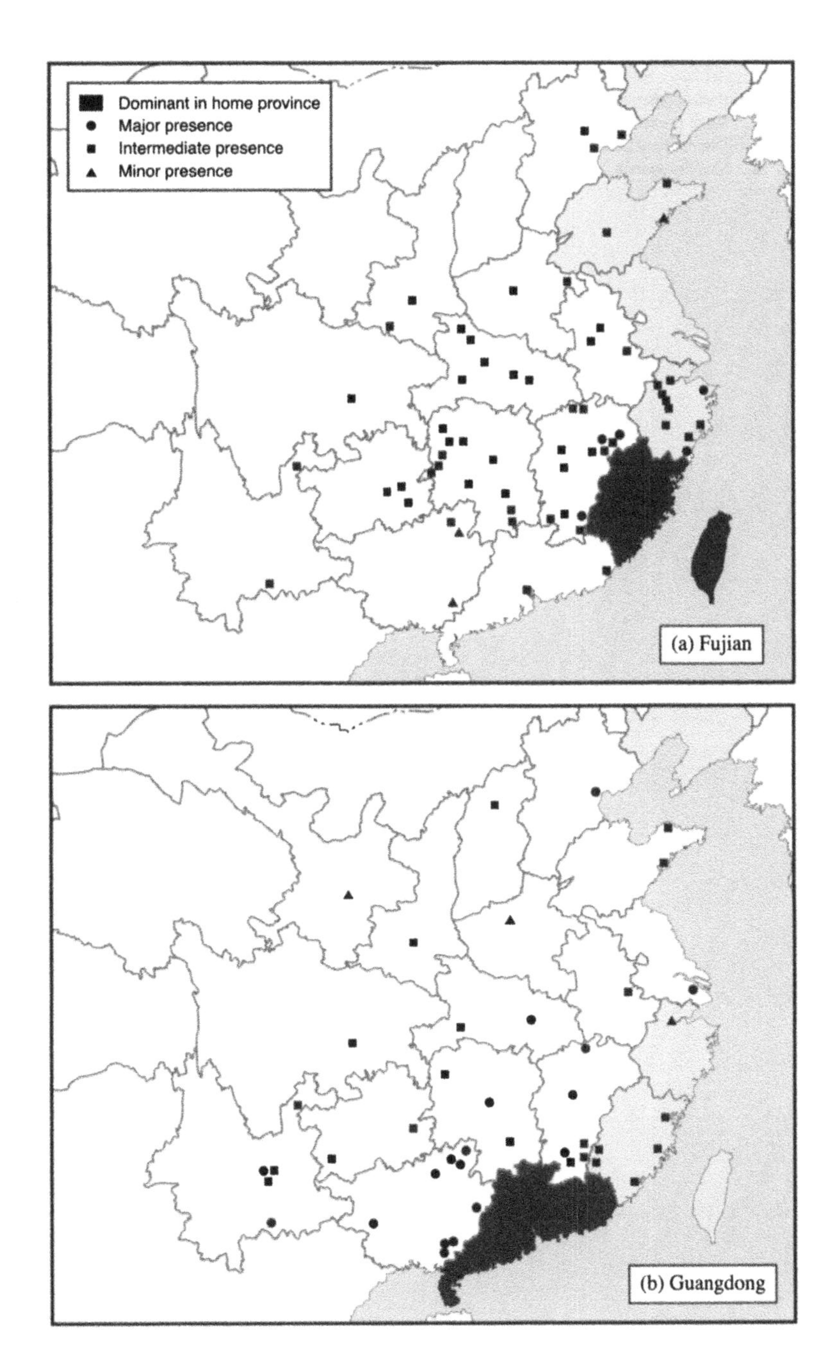

Map 4.4 Merchant organizations. (a) Fujian. (b) Guangdong. (c) Zhejiang. (d) Jiangxi. (e) Shandong. (f) Jiangsu.
Source: Peng Chang (1957); line work by Jane Sinclair.

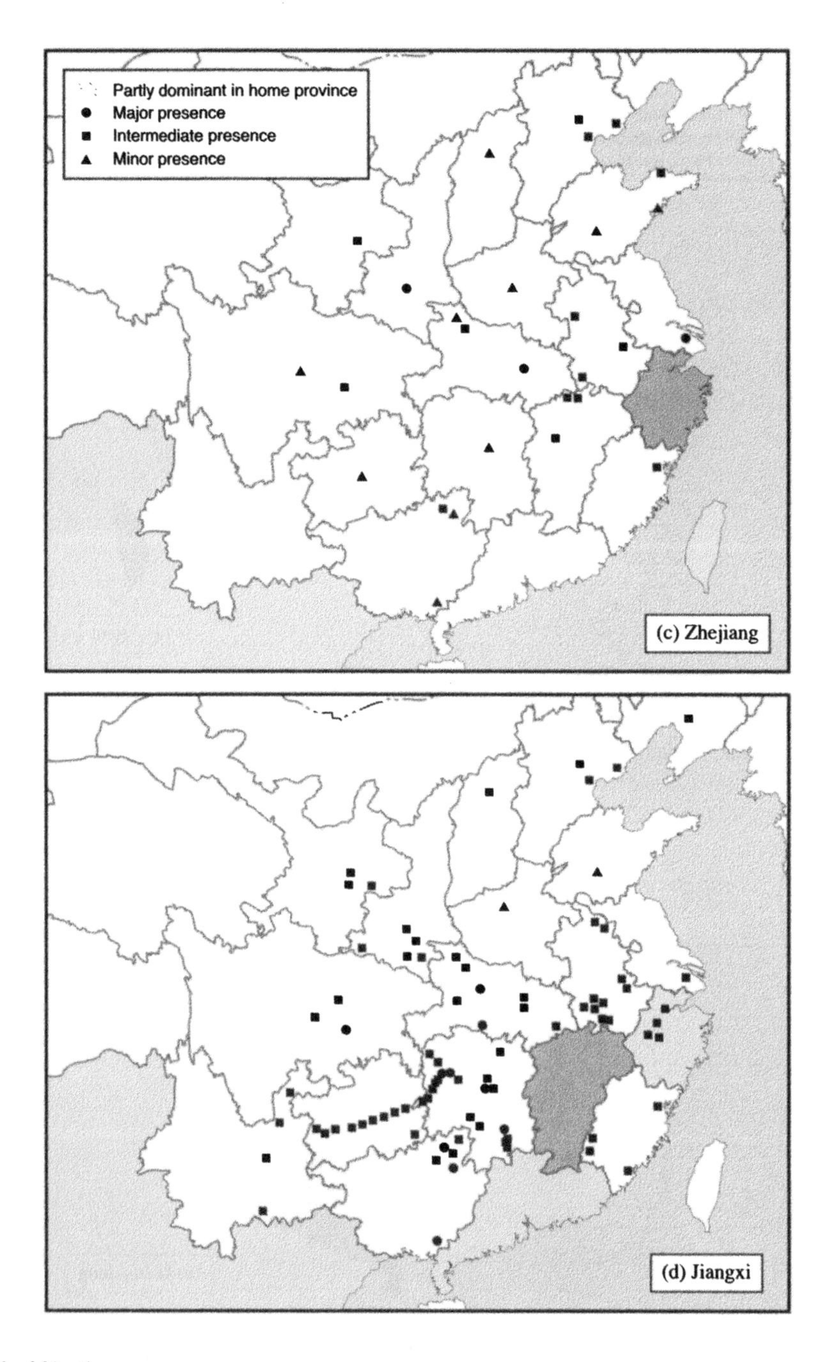
Partly dominant in home province
Major presence
Intermediate presence
Minor presence
(c) Zhejiang
(d) Jiangxi

Map 4.4 (cont.)

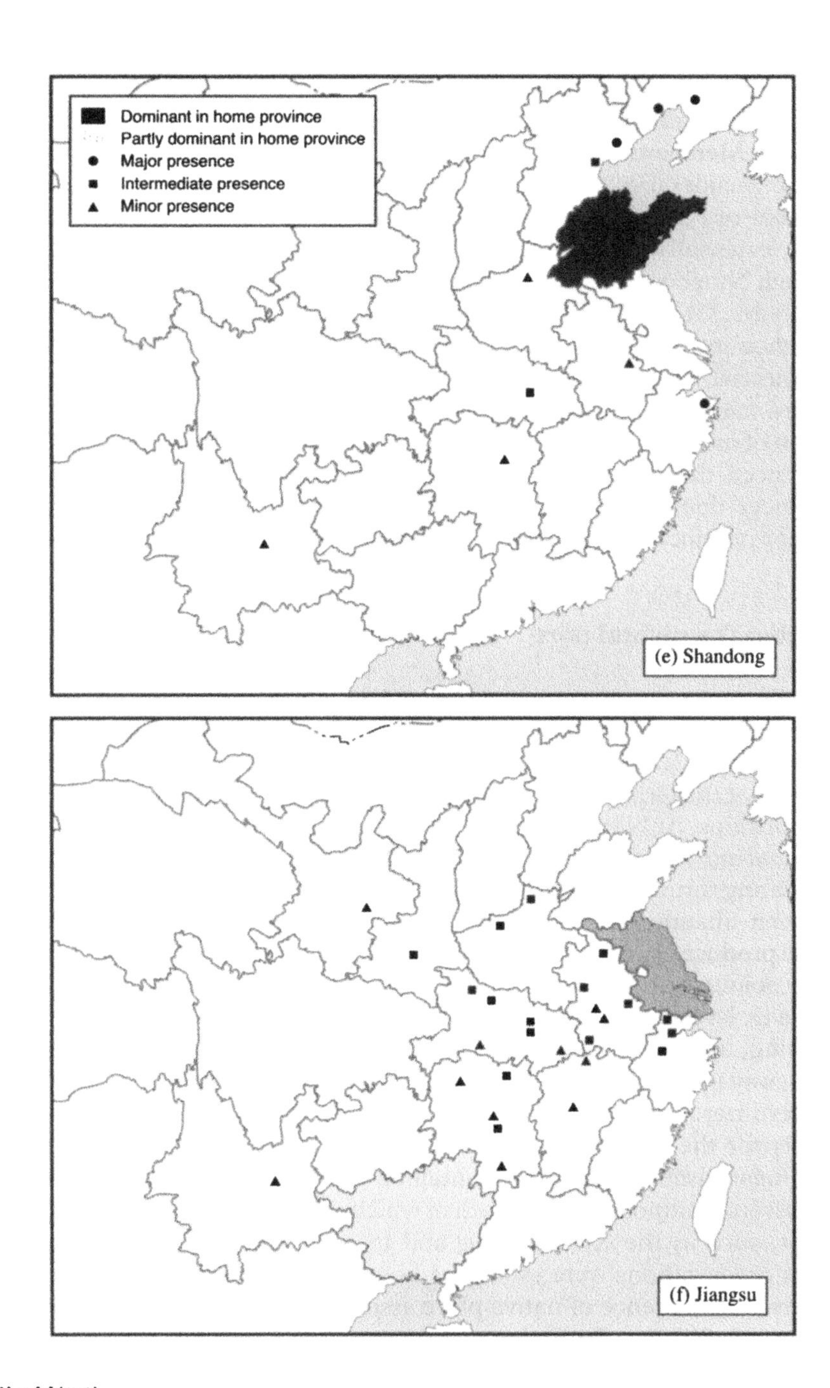

Dominant in home province
Partly dominant in home province
Major presence
Intermediate presence
Minor presence
(e) Shandong
(f) Jiangsu

Map 4.4 (cont.)

by province. Fujian merchants' groups could be found in fifteen provinces, and dominated commerce within Fujian and across the strait in Taiwan. Merchants from Zhejiang were the most prominent merchants in three major cities, Shanghai, Hankou, and Xi'an. Because of the pivotal role of Ningbo in the coasting trade, merchants from other provinces, especially Shandong and Fujian, played a major role in trade through Ningbo, and Zhejiang merchants only partly dominated provincial trade. Merchants from Jiangsu had a more concentrated sphere of influence in the Yangzi basin, and as in Zhejiang, the significance of commercial activity in the Grand Canal towns of southern Jiangsu attracted merchants from distant provinces. By comparison to the distribution of merchants from Jiangxi, it is clear that basic relative geographies influenced the range of mercantile presence. Merchants from coastal provinces dominated coastal mercantile activity, and merchants from interior provinces ranged more widely across central and western China.

Ningbo: the pivotal port

China's domestic trade also had regional divisions in northern and southern trading groups, and Ningbo served as the entrepôt for the coasting trade. Ningbo's central position on the coast, and location at the effective end of the Grand Canal, left it poised to capture a host of mercantile relationships. At Ningbo, locals referred to the *beibang* (northern group) and *nanbang* (southern group) to identify the two basic patterns of trade circulating through the port.[31] The *beibang* traders plied the coastal route between Shandong and Ningbo, carrying a regular trade in northern staple products, such as beancake, soybean oil, wheat, noodles, and hemp. They sojourned in Ningbo and returned with southern products. The *nanbang*, largely from Fujian, brought to Ningbo poles of pine for shipbuilding, timber for house construction, paper, sugar, and oranges, and took away products transhipped from the north, especially beancake, for soil fertilizer. Ningbo junks sailed to Fuzhou in ballast just to purchase timber for the shipbuilding industry.[32] In the mid-nineteenth century, the Fujian firms in Ningbo maintained a provincial-level organization and several component *bang*, each of which specialized in particular trade goods, such as the Xiamen *bang* and its trade in sugar.[33] Fujian merchant organizations were established especially early in Ningbo, before the era of emergence of native place associations. The earliest recorded centers of affiliation for Fujian mariners were not *huiguan* or *gongsuo* but rather Tianhou temples, which likely served as a basis for the later evolution of dedicated merchant organizations (Shiba, 1977, pp. 416–17).

But if Ningbo was such a strong entrepôt in the coastal carrying trade, even through the turn of the century, why did Western merchants find the place so commercially inhospitable? Local traders did dominate Ningbo, and it was difficult for foreigners to gain a foothold in the local market. Ningbo's economic strength was bound up with the role of merchant activity in the evolving economy of Shanghai. Commercial groups in the Ningbo area specialized in banking, and established financial services at Shanghai as early as the eighteenth century, when the city's role as the major trade center in the Jiangnan began to emerge (Mann, 1972). Ningbo banks were *qianzhuang* (native banks), which were one of two major types of regional banking institutions in China. The *qianzhuang* were more common in south China, relatively small in scale, independent, and specific to subregional contexts, such as the Yangzi delta or southern Fujian. (The northern type was the Shanxi bank, which was affiliated with the state and performed long distance interregional transactions.) Ningbo bankers and merchants formed the basis of the largest and most powerful *huiguan* in Shanghai, the Siming Gongsuo. The Gongsuo was located inside the French Concession and gained notice throughout China after successfully organizing two major protests, in 1874 and 1898, to maintain the adjacent Ningbo cemetery against French efforts to have it removed. During this period the Ningbo community concentrated in the French Concession and made up about half its population, an estimated 400,000 people. The protests effectively fended off further imperialist encroachment, and in Goodman's (1995a, p. 394) estimation of the issue and related events, such activity demonstrated how "early Chinese nationalism built upon native place sentiments in anti-imperialist movements." In such cases, economic practices, political activism, and meanings of place coalesced in *huiguan*, which demonstrates how cultural, economic, and political spheres were spatially constituted and entirely interwoven in imperial and early modern China.

The nature of business practice at Shanghai also held down trade at Ningbo.[34] Two types of commercial transactions in use – barter and cash – affected the ability of merchants to conduct trade at Ningbo. In Ningbo, and other smaller markets, cash price transactions prevailed. At Shanghai, merchants conducted high volume trade by barter pricing, but, at the same time, merchants also agreed on nominally higher cash transaction prices, which were reported in official trade notices. The reality of barter pricing affected all merchants who were unable to deal in high volumes. The real problem for the foreign merchant at Ningbo was that the volume of Western goods traded in large quantities under the barter price method permeated the lower Yangzi region. Because barter price was in reality lower than cash price, Ningbo merchants dealing in barter transactions in Shanghai could afford to transport British products as far as Ningbo and still turn a profit. On these terms, merchants from Ningbo could out-compete Western traders in Ningbo in Western products. Profits earned by Ningbo merchants in Shanghai recirculated back to Ningbo, and these economies were not entirely visible in the commerce and trade of Ningbo itself.[35] Outside the barter price system, the independent foreign merchant in Ningbo did not stand a strong chance to maintain a regular trade.

Xiamen: remittance economies

Except for the era of the tea trade in Fuzhou, foreign trade through Amoy was generally a bit more vibrant than in the other intermediate ports. Still, Amoy consuls also complained that the bulk of the trade was in the hands of local merchants. The British noted with some accuracy and regret the number of local mercantile firms in Amoy that served the coasting trade, the cross-strait trade, and the Nanyang trade.The 1880 trade report for Amoy counted over 40 firms engaged in trade with Shanghai, Ningbo, and other northern ports, while just as many were dedicated to cross-strait trade with Taiwan. Trade with the Nanyang depended on dedicated firms serving regional destinations: 11 local firms specialized in trade with Singapore, Melaka, and Penang, 15 firms called at the Dutch East Indies ports, 15 specialized in the Philippines trade, and nine firms handled trade to the French Indochina ports in Vietnam. All told, the consular office reported 183 local wholesaling firms compared to 27 foreign firms (China, Imperial Maritime Customs, 1880, pp. 212–22). Foreign firms dominated only the trade with Hong Kong, which amounted to about a third of Amoy's total trade. Even so, 16 local firms specialized in trade with Hong Kong. Overall, local merchants handled the bulk of trade through the port, which represented the continuing strength of the "Amoy network." Local merchants also shared the urban landscape with the cult of Tianhou: 26 Tianhou temples dotted the city, and the larger merchant houses maintained their own Tianhou temples.[36] The ubiquity of Tianhou temples must confirm not only the popularity of the goddess among merchant mariners, but the economic wherewithal to support the cult.

The history of maritime trade in southern Fujian made sojourning a way of life in Xiamen and an opportunity for people in the hinterland. Regular economic migration to the Nanyang ports took place side by side with contracted labor migration. By the early twentieth century upwards of half a million journeys to and from Xiamen took place each year (Dai, 1988, p. 38; *Xiamen huaqiao zhi*, 1991, pp. 15–30). Migrants regularly remitted money home, and sustained a circular migration across the South China Sea that created in its wake a circulation of merchant capital that fueled Xiamen's growth and development more than any one local industry or landed pursuit (Remer, 1968, pp. 186–7; Dai, 1996, p. 162).

From a world economy perspective, nineteenth-century colonial expansion created an international market in bonded laborers. Workers from China were often the least cost alternative for colonial economies, and a "coolie trade" developed on the China coast within a decade of opening the ports.[37] In the first decades of the open port system, Amoy gained a reputation with foreign labor contractors as the best place to fill a ship with Chinese laborers (Dai, 1988; *Xiamen huaqiao zhi*, 1991). While the majority of voluntary labor migrants from Amoy headed for the Straits Settlements and other ports in the Nanyang, British, French, Spanish, and American ships also transported workers to distant colonies, including Belize, Bourbon, Cuba, Guadalupe, Demerara, Havana, Maritius, Martinique, and Peru. The demand for workers often exceeded supply and "crimps" drew the men to the emigration ships through a range of fraudulent promises, and sometimes kidnapped them. Colonial practices also contributed to the organization of remittance economies. In Amoy the British soon established emigration agencies to organize laborers and process paperwork to arrange remittances. The workers could order a monthly allotment payable to a relative, who held a contract book which they presented at the office of the emigration agency on the last day of the month to claim their cash. Consuls characterized the bookholders as "elderly and aged" Chinese women, whose sons or brothers had promised a monthly remittance. In one instance, the British Governor at Belize wrote to Consul Swinhoe at Amoy that the "majority of coolies . . . desire to lay aside their obligations to their friends in China," in an apparent effort to staunch capital outflow.[38] Family members appealed to British Consul Swinhoe who recorded that he supported the women, and sought to put an end to the issue by demanding written declarations from the workers themselves.[39]

Xiamen, on average, claimed nearly one-fifth of all the remittances in China. Fully 90 percent of those remittances originated in the Nanyang, and by the 1930s, 50 percent of those came from the British Straits Settlements, Melaka, Penang, and Singapore (Zheng, 1940, pp. 30–3). An indicator of capital liquidity in Fujian was the proportion of native customs duties paid in currency. In the 1920s, Amoy and Fuzhou had the highest rates of cash payment of native customs duties among all the ports, at 50 and 46 percent respectively. Wenzhou had the next highest rate of cash payment at 38 percent, followed by Ningbo at 33 percent (Wright, 1927, pp. 12–19). A range of types of institutional arrangements managed the money flow, from large commercial banks as well as *qianzhuang* (native banks) and *minxin ju* (postal exchanges), which handled overseas letter services and money orders. During the period 1920–30, six modern commercial banks were established in Xiamen, all established by emigrant capital (Dai, 1996, p. 166). The increase in the number of local banks underscores the growth and ubiquity of the remittance economy. There were 23 *minxin ju* in Xiamen by the 1880s, a number which remained relatively contstant until the 1910s; then in the 1920s the postal exchanges tripled to over 60 (Wu, 1937, p. 196). By 1932 Xiamen had 105 postal exchange offices (*ZGYHQZ*, 1996, pp. 43–7). A survey conducted by the Bank of Taiwan in 1914, under the Japanese colonial government, enumerated 57 postal exchanges in Xiamen city in 1914 and noted that all but two were branches or agents of *minxin ju* in the Nanyang. The greatest number served Singapore and Penang, in addition to multiple branches in the Philippines, Rangoon, Saigon, Medan, and Java (Hicks, 1993, p. 87). The numbers of *qianzhuang* also increased, from 18 in the 1880s, to 33 in the 1890s, and to 45 in the 1910s. The big increase in *qianzhuang* came in the 1920s when the number in Xiamen nearly doubled to 84 offices (Wu, 1937, p. 224). Clearly, remittance activity in southern Fujian increased through the early twentieth century. Remittances were the leading source of support for many families in southern Fujian and bound local households to distant shores through the transboundary regional economy.

Infrastructural Development

Modernization of local infrastructure in the port cities depended upon returns from merchant capital, and in Xiamen it distinctively depended on investment by *huaqiao* or overseas Chinese, and *guiguo huaqiao*, returned Chinese sojourners, from the Nanyang.[40] In the 1920s and 1930s locals returned from overseas became the leaders of infrastructural modernization in Xiamen and remade the urban landscape. New roadways and public infrastructure, a redesigned commercial district, and housing in the urban core and on Gulangyu completely transformed the city.

Rebuilding Xiamen

Despite the political instability in China after 1911 and the years of warlord government, thousands of *huaqiao* returned home to settle and invest in Xiamen. People from Xiamen itself and those originally from *qiaoxiang* in the hinterland concentrated their investments in the safety of the city of Xiamen and the International Settlement on Gulangyu. The period 1927–

37 was the peak stage of overseas Chinese investment in Xiamen, and real estate was the leading sector (Lin and Zhuang, 1985, pp. 35–43, 55–6; Lin, 1988, pp. 24–5). The real estate boom engendered another round of land reclamation, and as one observer of the local transformation remarked, "reclamation of shore lands has been the most spectacular feature of this improvement program" (Ch'en, 1939, p. 205). Real estate companies proliferated, and by 1930 real estate values became some of the highest in the country, second only to the International and French Settlements at Shanghai (Lin, 1936, pp. 51–2). Organized rebuilding of Xiamen took place in the 1920s and 1930s, and plans for the commercial district reflected migrant lifepaths and years of experience in the British colonial Straits Settlements, especially Singapore. The new building form for the commercial district was the Straits Chinese-style shophouse, a syncretic architectural form that combines the traditional Chinese shophouse – a narrow, tall two- or three-story structure with housing above a shop – with Western beaux-arts or art deco style facade treatments (see Kohl, 1984, pp. 179–85). The ground floor of the shophouse is set back underneath the second story, which creates a covered walkway along the street. The Straits Chinese shophouse, functional and aesthetic, served downtown Xiamen the rest of the century (plate 4.1). The central commercial district in Xiamen is the only place in China where the Straits Chinese shophouse prevails in the urban landscape, and it is under conservation by the Xiamen municipal government.

The greatest concentration of elite housing was built on Gulangyu, the International Settlement, where property investment and internationalized lifestyle were protected. In the 1920s *guiguo huaqiao* built over 1,000 houses on Gulangyu, including imposing mansions in art deco style (Chen G., 1995). The mansions remain, but Chinese overseas families were marginalized during the Maoist era, and during the Cultural Revolution their properties were often split up and occupied by multiple households in pursuit of "spatial equality." Under reform, the government has initiated new housing assignments for the Cultural Revolution occupants and restoration of property to descendants of the original owners.[41] Gulangyu, closed to motorized transport, has not experienced the effects of rapid development otherwise characteristic of the larger Xiamen SEZ. The island is Xiamen's leading tourism destination and remains distinguished by its beach, gardens, Zheng Chenggong Museum, and rehabilitation of musical culture. In the era of the International Settlement, locals embraced Western classical music and piano playing was a popular elite past time. People again call the place "musical island" and "island of pianos," after resident devotion to classical music (Kraus, 1989, p. 129). The Gulangyu concert hall and new public library dominate the island's central district, and the ferry passenger terminal takes

Plate 4.1 Straits Chinese style architecture, Xiamen
(Photograph: Carolyn Cartier.)

the shape of a grand piano. In China under reform, the coastal cities have reemerged as centers of internationalization and cosmopolitan ideals, and the open port era appears less significant, one of several phases in the region's historical maritime economy.

Mercantile society and mercantile cities

Urbanization on the south China coast was based in forces that tied the region to transnational mercantile economies. Profits that drove urbanization, especially in Ningbo and Xiamen, were earned elsewhere and circulated back to the cities through merchant networks. Ningbo merchants and financiers in Shanghai, working on a barter price basis, could out-compete Western merchants in Western products. Overseas Chinese from southern Fujian practically commuted to cities in the Nanyang. This circular migration reflected a circuit of merchant capital so strong that the financial base of Xiamen lay well outside China. Processes of mercantile organization allowed Chinese to maintain commercial enterprise in the face of foreign competition and to manage capital returns at distant ports. Ningbo, Fuzhou, and Xiamen were different types of cities in the urban administrative hierarchy, and so the range of functions served by each city varied. But their regional economies accounted more for local variations in society and economy than administrative status. Ningbo was a prefectural-level city, second only to the provincial capital of Hangzhou. Yet in distinctive ways, Shanghai was Ningbo's hinterland, from which it borrowed a commercial dynamism that otherwise seemed absent in the city itself. Fuzhou, the provincial capital, was for Fujian what Shanghai was for central China: a city of significant proportions poised to capture the commerce of a river system and its hinterland. Fuzhou, though, arguably became more distinguished by social and cultural transformation than economic activity, and its large Christian population debated the events of the era on its own terms. Despite its economic significance, Xiamen was not formally designated a city in the adminstrative hierarchy until the 1930s, which Ng Chin-keong (1983, p. 67) has evaluated as "a result of the government's recognition of its foremost commercial and strategic importance as a maritime center." Southern Fujian was the leading area of emigration from the province, and people from the region traveled to and settled in the Nanyang more than other cities and regions to the north. Long distance economies tied to the Nanyang substantially transformed both coastal south China and maritime cities of Southeast Asia.

NOTES

1 See also FO 228/114, Fuzhou, 28 Nov. 1850, to Bonham/from Sinclair; FO 228/128, Fuzhou, 18 June 1851, to Bonham/from Sinclair.

2 See also John Fairbank's (1953, pp. 23–38) discussion of the tribute system and work by John Wills (1968, 1988), who has especially cautioned against totalizing interpretations of the tribute system and assumptions that the Chinese empire "fit" Western powers into the tribute system on terms that were the same as relations with other Asian states.

3 See Maurice Collis (1946), Fairbank (1953), and Michael Greenberg (1951) for accounts of the Opium War.

4 Historians have used the term "gunboat diplomacy" to describe the nature of foreign relations on the China coast during the nineteenth century. Indeed, foreign consuls regularly complained in their logs that local Chinese officials were more cooperative when a man-of-war lay in the harbor, and practiced sending Chinese officials ultimatums when an armed ship was in port; see, for example, USDS, Amoy, 10 Dec. 1855, and 30 May 1859, to Secretary of State/from Hyatt. Ellsworth Carlson (1974, p. 3) noted that the British made use of this perspective at Fuzhou.

5 For the text of the treaties see Godfrey Hertslet (1908, volume 1).

6 Woolen goods were an impractical material in the humid subtropical climate of the region; this was lost on Europeans of the era, who retained the notion that tropical disease vectors could be avoided by keeping the body covered in layers of clothing.

7 Because opium was also grown in China, the British argued that merchants should be able to continue to import opium unless China could prove complete domestic eradication. China began a national opium eradication campaign in 1906; Mary Wright (1968, p. 14) estimated China achieved an 80 percent reduction in opium cultivation.

8 The treaty port term was also used to distinguish from the three other categories of open ports, namely "ports voluntarily opened by China," "ports in the leased territories," and "ports of call"; see En-Sai Tai (1918, pp. vii–viii). No systematic scholarly account of treaty ports exists, and the first studies of individual ports focused on the larger ports in the era of the "impact-response" paradigm and through the lens of modernization theory. The classic account is John Fairbank (1953); see also Rhoads Murphey (1953) and Albert Feuerwerker (1976).

9 The Chinese Marxist scholarship emphasized the negative impact of foreign economic activity, especially on rural handicraft industries. See Jack Potter (1968) for a discussion of the scholarship and an argument that the effects of foreign enterprise were uneven, and the review article on the historic Chinese economy by Ramon Myers (1991).

10 I am indebted to Laurence Ma for suggesting *tongshang gangkou* or trading ports as a further alternative to the treaty port label. This issue of appropriate nomenclature is only initiated here and deserves a sustained discussion.

11 See for example FO 17/164, 3 Sept. 1850, to Bonham/from Palmerston; and references to further correspondence in FO 802/113, 24 Dec. 1849 and 18 July 1850.

12 FO 228/570, Ningbo, 16 May 1876, to Wade/from Warren.

13 FO 228/42, Ningbo, 30 June 1844, to Davis/from Thom.

14 FO 228/41, Fuzhou, 30 Aug. 1844, to Davis/from Lay.

15 FO 228/41, Fuzhou, 28 Oct. 1844, to Davis/from Lay.
16 FO 228/50, Amoy, 12 Feb. 1845 to Davis/from Alcock.
17 For a discussion of the debate about Guangzhou see Hosea Ballou Morse (1910, pp. 367–99).
18 FO 228/41, Fuzhou, 28 Oct. 1844, to Davis/from Lay.
19 Ibid.
20 The southeast coast was also the region of evolution of a leading school of *fengshui* practice during the eleventh century. J. J. M de Groot (1897, volume 3, Joseph Needham (1956–62), Stephen Feuchtwang (1974), have demonstrated the complex conceptual bases of *fengshui* and its origins in Chinese cosmological thought.
21 FO 228/650, Fuzhou, 7 May 1880, to Wade/from Sinclair. Consular records clearly demonstrate both Chinese and British consular frustrations over dealings with Western missionaries. In Amoy in 1878, British Consul Alabaster wrote that missionary complaints were "generally, greatly magnified and misrepresented," in FO 228/606, Amoy, 9 Oct. 1878, to Wade/from Alabaster.
22 See Richard Kraus (1999) and Mayfair Yang's (1999a) introduction to *Spaces of Their Own*, about the evolution of civil society in China and for an assessment of using Habermas's work in the Chinese context.
23 USDS, 14732/13, Fuzhou, 3 June 1909, to Rockhill/from Gracey.
24 Ibid.
25 The geography of the Taiping Rebellion is interwoven with the history of the Jiangnan region and Shanghai. The rebellion originated in the south, in Guangdong and Guangxi provinces, and moved north to the Yangzi delta region. It disrupted trade at Ningbo and across the lower Yangzi region in the early 1860s. Shanghai remained the only major city under dynastic control and foreign protection. Consequently, Shanghai received increasing shares of trade from around the Jiangnan and grew as a result of the effects of the Taiping Rebellion; see Wen-Hsin Yeh (1996, pp. 57–8). Albert Feuerwerker (1958) and others have also argued that the power of merchant groups distinctively increased after the Taiping Rebellion. About the Taiping Rebellion, see the account by Jonathan Spence (1996).
26 The term "guild" has been used as a generic translation for these and other Chinese organizations in the historic literature, but the connotations of guild are too limited in scope to characterize the complexity of these social and economic organizations in Chinese society.
27 See Bryna Goodman (1995b) and the materials in Brian Ranson (1997), for specific discussions of diverse types of merchant organizations and native place associations.
28 Some *huiguan* had a broad membership while others were more exclusive. The summary in Xu Dixin and Wu Chengming (2000, p. 183) also noted that *gongsuo* membership fees were higher for non-locals. Keith Schoppa (1982) also discussed how some *huiguan* welcomed only elite merchants, and maintained power by excluding potential newcomers of insufficient means or connection.
29 The original meaning of *bang* was a fleet of boats, the grain transport boats

used on the Grand Canal; it evolved to designate the group of laborers serving the boats, then organized water-transport workers, and, ultimately, a group of workers in general. See Susan Naquin and Evelyn Rawski (1987, p. 50.)

30 Peng Chang's (1957) study remains important for its contribution to mapping the distribution and "relative strength" of merchant groups by province. However, Chang's study cannot be considered a complete representation of merchant group activity. Certain obvious omissions may be detected in the research, such as the absence of data on Fujian merchants in Ningbo. I have added the major presence of Fujian merchants in Ninbgo to the Fujian map reproduced here.

31 FO 228/757, Ningbo, 19 July 1884, to Parkes/from Cooper; FO 228/3280, Ningbo, 14 Apr. 1919, to Jordan from Shaw. See also Shiba (1977).

32 FO 228/62, Fuzhou, 12 Feb. 1846, to Davis/from Alcock.

33 FO 228/3280, Ningbo, 14 Apr. 1919, to Jordan from Shaw.

34 See FO 228/63, Ningbo, 10 Jan. 1846, to Bart/from Thom.

35 See Susan Mann's (1987) discussion of the recession at Ningbo during the 1860s.

36 *Xiamen zhi* (1839, chapter 2, 45b/50).

37 Consuls and traders in China were well aware of global labor costs. See, for example, FO 228/71, 16 July 1847, Amoy, to Davis/from Layton.

38 FO 228/405, Amoy, 19 May 1866, to Alcock/from Swinhoe; FO 228/405, Amoy, 18 Apr. 1866, to Boyd/from Swinhoe.

39 Ibid.

40 See Wang Gungwu (1981), for discussion of the meaning of *huaqiao*, which is typically translated as "Overseas Chinese." Because *qiao* means journey of some limited proportions, and thus return to China, the term is used here to refer to migrants in the era of sojourning and during the colonial period in Southeast Asia. "Chinese overseas" is adopted to refer to contemporary ethnic Chinese populations in diverse countries around the world.

41 Interview, Xiamen, 1995. Return of housing to descendants of pre-revolution owners is variable and appears to depend at least in part on the owners' relationships with the government.

5

Revolution and Diaspora

> [B]eing Chinese in South China has involved continuous negotiation of cultural identity and history in order to establish and legitimate position in a volatile but all encompassing state order.
> (Helen Siu, 1993, p. 22)

Twentieth-century revolutionary movements in China had important bases of activism in the cities of the south China coast and networks of support among the overseas Chinese. China's political transformation, from empire to nation-state, was bound up in the region of diaspora, even as the transformation was a territorial process that ultimately loosened ties to the diasporic population and the transboundary cultural economy on the south coast. In the late dynastic era and during the Republican period, the government treated Chinese abroad as essential to the nascent modernization project and enlisted overseas Chinese capital to support national infrastructure projects. During the 1950s, the Chinese Communist Party evolved a contradictory line that alternatively courted, disdained, and disenfranchised the overseas Chinese. The Communists marginalized people with overseas connections, while decolonization in Southeast Asia after the Second World War compelled Chinese immigrants to adopt local citizenship. These simultaneous events transformed relations between the China and diasporic populations, and led to the evolution of new identities for Chinese communities in postcolonial Southeast Asia. Still, the legacies of China's overseas policies, combined with postcolonial state-making projects, have fueled nationalistic positions on the ethnic Chinese in Southeast Asian states. How diasporic communities have negotiated these kinds of transformations, sustained distinctive communities, and evolved complex translocal identities, based in both local places and homeland regions, are the subjects of this chapter.

The Southern Revolutionary Axis: From Guangzhou to Shanghai

The cores of revolutionary activity in the early twentieth century were in Guangzhou and Shanghai (Dirlik, 1989, 1991; Chan, 1996b; Bergère, 1998). Many revolutionary leaders were based in these cities, and, having been educated abroad, maintained connections between political communities in Guangdong and Shanghai and international centers of activism in Japan, the Nanyang, and the West. Sun Yat-sen led the country's first revolutionary movement, and was by several measures the most important revolutionary leader of the first half of the twentieth century even as he spent much of his political life outside China (Schiffrin, 1968, 1980). Sun was born into a peasant family in the Zhujiang delta, educated at a Western missionary school in Hawaii, and later in Hong Kong at the College of Medicine for Chinese. After leading a failed uprising in Guangzhou in 1895, he retreated overseas, and instead organized support for a Chinese republic in Europe and Japan. In Tokyo in 1905 he formed the Tongmeng hui (Revolutionary Alliance), which attracted international sponsorship and substantial membership among Chinese students abroad, and especially students from southern and coastal provinces – Guangdong, Hunan, Jiangsu, and Zhejiang (Bergère, 1998, p. 104). Sun also popularized the Alliance in the Nanyang and generated financial support particularly among the Hokkien community in Singapore (Wang, 1959, p. 31; 1976, p. 408). In early 1912, in the wake of the collapse of the dynastic era, Sun returned to China to become the first provisional president of the nascent Chinese republic, installed in Nanjing. Forced to leave China again in 1913 after the northern warlord Yuan Shikai gained control of Beijing and wrested the office of the presidency from the deposed Qing emperor, Sun returned again to Shanghai and Guangzhou in 1916 and reorganized his political platform in the Guomindang (GMD) or Nationalist Party. His biographers have attributed his successes in part to his identity formation, and its basis in the cosmpolitan geography of the south China coast. Marie-Claire Bergère contextualizes his character in the following terms:

> Sun Yat-sen, a Cantonese raised in Hawaii and Hong Kong, was a pure product of maritime China, the China of the coastal provinces and overseas communities, open to foreign influences. The travels, the encounters, and education that the young peasant received in missionary schools initiated him into the modern world and aroused in him a desire to give China a rank and role worth of it in that world. (Bergère, 1998, p. 3)
>
> His extreme geographic mobility nurtured his equally great versatility of mind and temperament. He could cross cultural boundaries as easily as

> geographical ones, adapt to all societies, all types of men. Some critics have seen this flexibility and plasticity as a mark of inconsistency or even duplicity. But Sun Yat-sen . . . knew that in order to convince you need to speak the language of the person you are dealing with. He was as capable of operating in missionary circles as in the lodges of secret societies,[1] in merchant guilds as in students' cultural societies, and was as active in Tokyo, London, and San Francisco as he was in Hong Kong, Hanoi, and Singapore. (Bergère, 1998, p. 6)

These geographical contexts of identity formation are emplaced regional experiences, formed in the reach of social relations that have characterized the regional cultural economy in south China and the distant places its mobile population connects. Memories of Sun Yat-sen also bridge the contemporary Taiwan Strait. Sun has been revered by Taiwan's political leadership, and by contrast to the inheritors of the Nationalist Party he founded, Sun Yat-sen has also been celebrated in late twentieth-century China as a revolutionary leader. In 1979, Mao Zedong's portrait in Tiananmen Square was replaced by one of Sun Yat-sen, alongside those of Engels, Lenin, and Marx. Sun Yat-sen's internationalist philosophy for modernizing China suited Deng Xiaoping's Four Modernizations platform, which prompted revisionist study of Sun Yat-sen thought and served to situate the opening of China to the world economy in revolutionary history.

Communism emerged as a revolutionary movement in China in the same era, and the first plenary meeting of the Chinese Communist Party (CCP) took place in the summer of 1921 in Shanghai, in the French Concession. Joseph Stalin and the Comintern were carefully following the Chinese revolution and promoted an alliance between the CCP and the GMD, which resulted in the announcement of a United Front in 1924 (Van Slyke, 1967, pp. 15–20). But Sun Yat-sen died the following year. The United Front led to the establishment of the Whampoa Military Academy near Guangzhou, where Chiang Kai-shek, who emerged to succeed Sun, directed the Academy. By contrast to Sun Yat-sen's background, Chiang Kai-shek came from a salt merchant family, in Fenghua near Ningbo, and had originally worked as a stockbroker in Shanghai. He instead made the military his career and left for Tokyo, where he joined the cause of the Revolutionary Alliance. In 1926, Chiang led the Northern Expedition from Guangzhou to Shanghai with the goal of reuniting China from warlord rule. Shanghai was by then under the control of labor unions, partly organized by the CCP, which, significantly, as Elizabeth Perry (1993) has argued, were built on existing trade and mercantile organizations and, thus, native place networks, rather than class alliances.[2] Shanghai in the 1920s was also the center of Chinese anarchist activity, which was focused on organizing urban workers

(Dirlik, 1991, p. 237). Under Stalin's advice, the CCP and the unions welcomed the advancing GMD troops (Brandt, 1958, p. 76).[3] But Chiang boldly betrayed the alliance and seized control of Shanghai: his troops attacked union headquarters and executed CCP members across the city. Additional victories in the north gave Chiang control over national leadership and in 1927 the Nationalists moved the capital from Beijing to Nanjing. During the Nanjing decade, 1927–37, the regime evolved fiscal strategies that would ultimately contribute to its demise, specifically by extorting money from Shanghai capitalists, practices which were based on Chiang's long-term mercantile connections to organized crime and opium trafficking in the city (Perry, 1993, pp. 88–103). But Chiang was bent on staving off the Communists, even as Japan began to attack the country. Chinese elites pressed Chiang to negotiate a national anti-Japanese strategy with the CCP, but this second united front was not organized early enough to prevent Japan's massive invasion of central China in 1937, and crimes of war against the population of Nanjing.[4] After the US bombings of Hiroshima and Nagasaki, the Nationalists received the Japanese surrender, but their popular support continued to erode. The US attempted to broker a peace agreement between the CCP and the GMD, but talks broke down and civil war ensued.

The geography of each regime's base of support illuminates politicized distinctions between the interior and the coastal cities, distinctions which became entrenched in the planned space economy of the Maoist era. The Communists maintained their prominent wartime base in Yan'an (Shaanxi province) and depended on the allegiance of rural populations. As early as 1941 the CCP was already engaged in breaking down the class structure in areas under their control, by identifying landlords and empowering poor peasants (Chen, 1986, 128–33). The Nationalists, by contrast, extended dominance over the coastal cities and had real and symbolic alliances with Western democracies and international economic activity. The Nationalists received considerable US assistance yet their government faltered, fundamentally burdened by financial corruption and extraordinary inflation (Chang, 1958). The CCP began to win the country and the penultimate battle of the civil war closed when the Communists marched over Guangzhou. At that point, the only major city left under Nationalist control was Xiamen, which was held open as a maritime gateway for the retreat to Taiwan.[5]

The port cities of the south China coast represented ideological problems for the new Communist regime, even as it desired to maintain the foreign exchange accumulation of the overseas Chinese remittance economies. The People's Republic of China established national politices of socialist reconstruction that defined the coastal cities of the "treaty port" experience as parasitic places of Western-style capitalism and

Guomindang sympathies. As Mao (1954, p. 32) viewed the cities during the war, "The capture of the cities now serving as the enemy's main bases is the final objective of the revolution." The Communist regime later relegated the status of China's most economically developed cities, in their alliances with transnational capital and symbolism of bourgeois consumption, to the lowest level of geographical significance in the new country (Murphey, 1980, pp. 25–33). The Maoist platform focused on turning "consumer cities" into "producer cities," revitalizing the peasantry, and building up the industrial infrastructure of the interior provinces. As a result, when economic reforms opened the coastal cities to foreign investment in the early 1980s, the urban infrastructure in Shanghai, Xiamen, Ningbo, and other former open ports existed much as it had in the 1930s and 1940s.

Revolution and the Overseas Chinese

From the first phase of the revolution in China, leading up to 1911, and through the decade of Nationalist rule, similar state perspectives on the overseas Chinese prevailed: *huaqiao* were sojourning Chinese citizens the revolutionary regimes courted for political and financial support. Policy shifted with the Communist victory. During the agrarian reform movement in 1950–2, the investigation of class background categorically classified overseas Chinese families and thus named them as different from the rest of the population. They were encouraged to participate in socialist reconstruction, change their "wasteful habits," and join the mass organizations. The CCP government also innovated a system of overseas Chinese farms, which became principal areas for settling the returned population without relatives, as well as administrative settlements where *guiguo huaqiao* could be quarantined until appropriately reeducated. The first overseas Chinese farm was established on Hainan Island in 1951. By the 1960s, Fujian had 16 state-sponsored farms for returned overseas Chinese and Guangdong had 17; several also existed in Guizhou and Yunnan (S. Fitzgerald, 1972, pp. 70–1, 207–9). Between 1949 and 1966, almost half a million overseas Chinese returned to settle in China, and in 1960 nearly 100,000 *huaqiao* returned from Indonesia after the end of the dual nationality system, and imposition of discriminatory legislation toward non-citizens (Mackie, 1976). During the Cultural Revolution, 1966–76, intense political campaigns for eradicating "bourgeois" elements turned the *guiguo huaqiao* and their relatives into a vilified class. In 1967 a riot at the Overseas Chinese Affairs Commission in the capital led to its closure, and subsequently Red Guards also ruled over matters concerning overseas Chinese. In a somewhat bizarre transhistorical moment, Red Guards transported to Hainan Island a thousand *guiguo huaqiao* (Godley, 1989, p. 347), which invoked the dynastic era mode of punishment for banished officials.

In revolutionary China and postcolonial Southeast Asia the word *huaqiao* itself became a politicized trope. On one hand, the GMD and later the CCP used it to encourage overseas Chinese loyalties; on the other hand, Southeast Asian governments viewed it as a designator of a population group loyal to another country. As Wang Gungwu (1981, 1996) has repeatedly pointed out, the act of sojourning itself is not politicized; how the nation-state chooses to represent *huaqiao* status makes it a politicized position. In 1909 the Chinese government legally defined "Chinese" to include all Chinese abroad irrespective of actual citizenship status, which gave rise to certain concerns in the Nanyang under colonial rule. For example, after 1911 China began to evolve a national language policy that was also promoted overseas. By the 1920s, hundreds of Chinese-language teachers were working at Chinese schools, especially in Malaya and Singapore. They taught the new national language, the Mandarin dialect, and urged the overseas Chinese to help China (Wang, 1959, pp. 34–5). In such ways the rise of a modern educational system in the Nanyang and the language medium of instruction became fulcrums for debate in future nationalist movements.

Politics, though, did not encourage remittances. Remitted funds dropped precipitously from overseas Chinese, particularly as news spread that local level party cadres had appropriated remittances from recipients in the name of socialist reconstruction. The consequent losses of foreign exchange prompted much more tolerant policies (Wu, 1967, pp. 40–77; S. Fitzgerald, 1972; Godley, 1989). From 1954 to 1957 the Party downgraded the status of *huaqiao*-connected households from landlords and rich peasants to peasant households, and urged landlords who fled after 1949 to return. In 1955, the state guaranteed independent receipt of remittances, as well as the opportunity for *huaqiao* households to purchase goods otherwise unavailable to the general population. This privileged system included the establishment of special stores, and permission to use remitted funds even for "feudal superstitious practices," including cult festivals, *fengshui* consultations, and maintenance of ancestral graves (S. Fitzgerald, 1972, pp. 58–63). Such policies were covered extensively in the overseas Chinese press. The special privilege system varied by province, but generated widespread national complaint, and in 1958 the system was just as quickly canceled. Instead the state mandated that remittances could only be used to meet basic subsistence needs or be invested in state-sponsored programs.

By comparison to the early twentieth century, when Chinese leadership in the open ports was at the vanguard of social movements and nascent public sphere, Maoist ideologies rejected much of the same population as antithetical to the goals of the revolution. By the end of the 1950s, the process of decolonization in Southeast Asia instituted new forms of citizenship and the CCP advocated *huaqiao* choose citizenship in the Southeast Asian states. Still, the CCP continued schemes for overseas Chinese investment. Existing overseas Chinese companies in Guangdong and Fujian were merged into large-scale overseas Chinese investment companies, which guaranteed 8 percent annual dividends and right of capital ownership. This policy encouraged remittance investment, and by the end of 1959, roughly 6,000 households had invested in the Guangdong company, and around 10,000 households had remitted funds to the one in Fujian (Wu, 1967, pp. 54–5). In the early 1960s, branches of these investment companies were even opened in other cities. In these ways the Party maintained a privileged albeit irregular special perspective on overseas Chinese economic connections, which was ultimately tapped again in China under reform.

Southeast Asian Chinese Communities

The geography of the Chinese diaspora and its concentration in Southeast Asia is one of the great demographic subjects of the modern era. By the end of the twentieth century, over 30 million people of Chinese descent lived around the world, and over three-quarters were in the Southeast Asian region (Poston and Yu, 1990; Ma, 2002). The homeland origins of the overwhelming majority of Chinese overseas are Fujian and Guangdong provinces, and these places of origin are further concentrated in particular counties, especially in southern Fujian, in the Xiamen hinterland, and in the Zhujiang delta of Guangdong (Purcell, 1965; Fang and Xie, 1993). Historic patterns of migrant destinations have also been geographically specific. People with Fujian ancestry migrated in greatest numbers to Singapore and the port cities of Indonesia, Malaysia, and the Philippines. People from Guangdong, from both the Zhujiang delta and the Shantou area, have formed larger groups in Thailand and Vietnam, and are also significant but smaller groups in Malaysia and Singapore. Geographical associations between places of origin and destination significantly characterized the historic diaspora to the degree that migrants from a common region tended to settle in a common destination. These geographical patterns have been based on dialect groups, and the importance of common language as the basis of community organization and individual and group identity formation. The patterns of population distribution by dialect group in south China and in Southeast Asia confirm the historic ties between the two regions (maps 5.1 and 5.2).

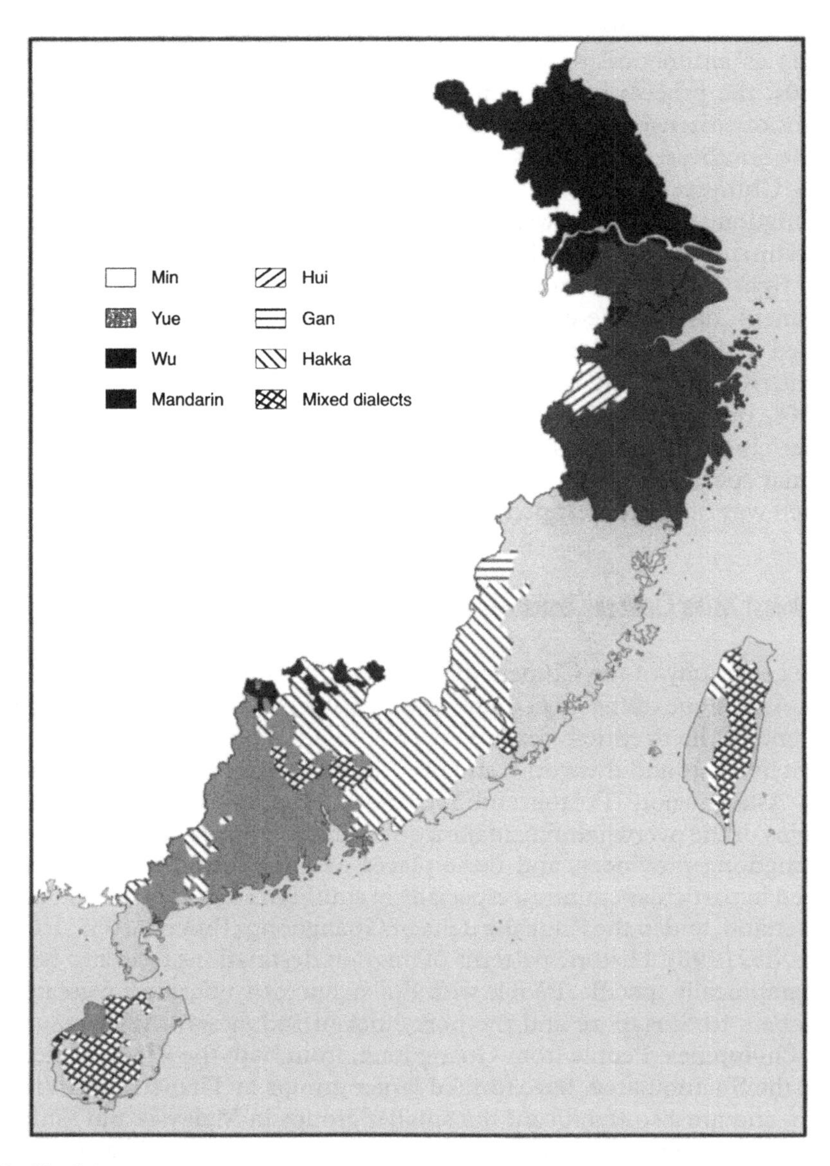

Map 5.1 Dialect regions.
Source: Language Atlas of China (1988); line work by Jane Sinclair.

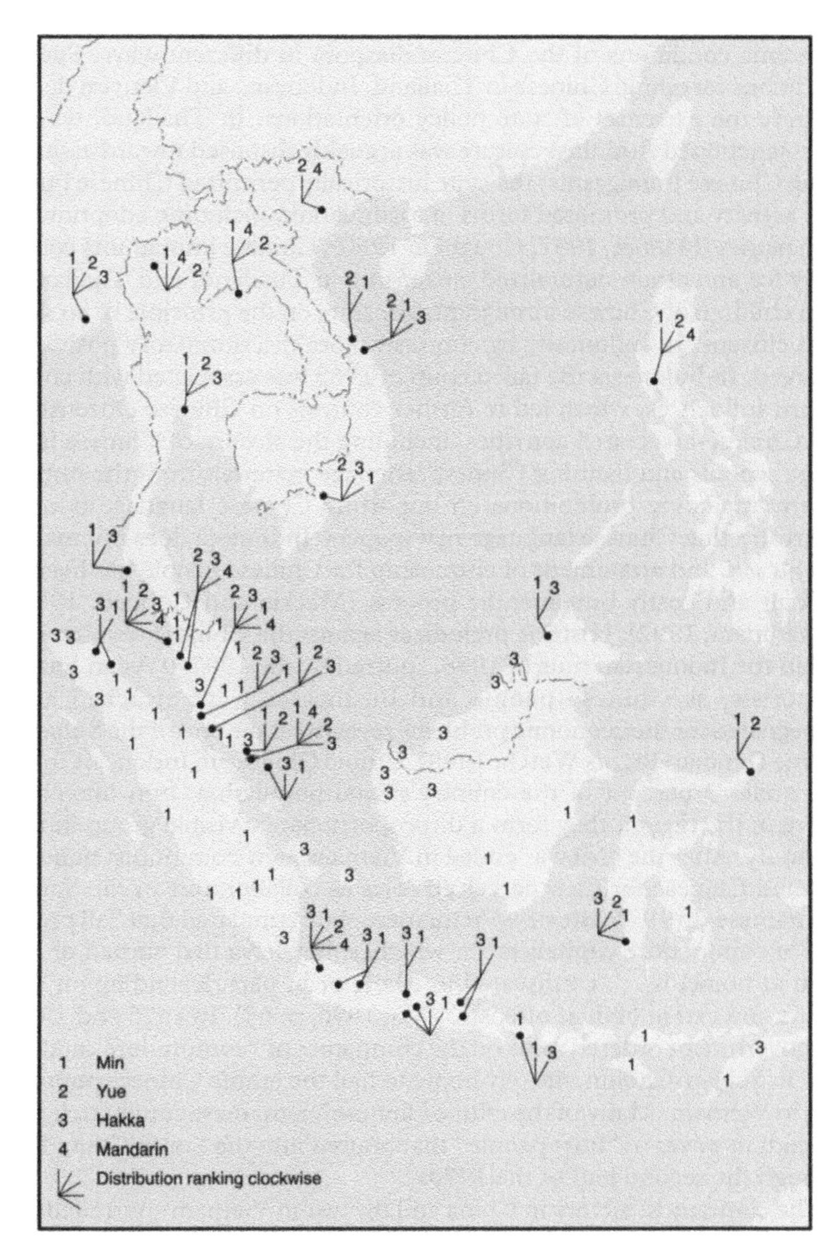

Map 5.2 Chinese overseas populations by dialect group. (Map reflects limited data for mainland Southeast Asia and the Philippines.)
Source: Language Atlas of China (1988); line work by Jane Sinclair.

In Southeast Asia, the postcolonial state has met the social and economic conditions of the Chinese diaspora in different ways. Social conditions for ethnic Chinese in Thailand, Indonesia, and Vietnam demonstrate the extremes of state policy orientations. In Thailand, where state-sanctioned Buddhist culture was arguably disposed toward assimilating Chinese immigrants, the state historically permitted Chinese business activity and promoted forms of assimilation, including adoption of Thai names (Skinner, 1957; Coughlin, 1960). Chinese immigrants could apply for and attain naturalized citizenship in Thailand, and Thailand-born children of Chinese immigrants became, on the principle of *jus soli*, Thai citizens. In Indonesia, by contrast, more discriminatory positions emerged. In Indonesia the failed coup of 1965 was associated with communist influences, which led to further controls on Chinese citizenship and Chinese-associated activities, including the closure of Chinese language schools and banning Chinese language materials from the public sphere, including prohibitions on importing Chinese language books, and restricting Chinese language newspapers. Indonesia does not maintain *jus soli*, and attainment of citizenship for Chinese people has been a difficult and costly bureaucratic process (Mackie and Coppell, 1976; Suryadinata, 1992). Historic prejudices against the Chinese were revivified in the Indonesian riots of 1998, spurred by the 1997–9 Asian financial crisis, as Chinese people and businesses were attacked and scapegoated for the economic problems revealed at the end of the Suharto regime (Human Rights Watch, 1998). Ethnic Chinese in Indonesia form just under 3 percent of the country's total population, but, like elsewhere in the region, they form a disproportionately visible group in the economy. After the civil war ended in Vietnam, new communist policies targeted Chinese for their perceived roles as collaborators of the South Vietnamese. In 1975, the new Vietnamese state stipulated that "all property of compradore capitalists . . . whether they have fled abroad or remain at home, is . . . confiscated in whole or in part depending on the nature and extent of their offenses" (Vo, 1990, p. 65). In 1975 and 1976 the government ordered raids on the companies of "compradore capitalists" in Saigon-Cholon, the city home to half the ethnic Chinese population in Vietnam. Many of the ethnic Chinese Vietnamese sought to leave instead, in waves of "boat people" that poured into the South China Sea through the second half of the 1970s.

The communist victory in China and the war in Vietnam spurred ideas about Cold War geopolitics in Asia. In 1967, at the height of the Cold War, the democratic market economies of Southeast Asia – Singapore, Malaysia, Indonesia, the Philippines, and Thailand – formed the Association of Southeast Asian Nations (ASEAN) to promote regional peace and security, and as a partial counter to communist governments in the region. Among the countries of Southeast Asia that became founding members of ASEAN, Malaysia was the one where the communist movement had threatened the stability of the state (Pye, 1956; Taylor, 1974). The Malayan Communist Party (MCP) had its origins in the 1920s as a left-wing group of the Malayan Guomindang, and emerged independently in 1930. It was the most active force of local resistance against the Japanese occupation, paralleling anti-Japanese sentiments in China (Means, 1970, pp. 68–9). The MCP claimed some pan-racial representation, and had a significant number of Malay members, even as the British colonial government sought to portray the MCP as predominantly Chinese (Cheah, 1979, p. 63). At the end of the Second World War the MCP was the most organized political force in the country, and in 1948 began an armed struggle against the colonial government. British troops put down the guerilla movement, labeled the "Malayan Emergency," but not completely until 1960. Even then, the MCP endured, gaining some following among labor unions in Singapore, and eking out an existence from remote forest bases. The MCP had no constituency among the larger Malaysian population, but endured nevertheless and did not officially abandon its platform of violent struggle until 1989. In the run-up to independence in 1957, new ethnically based political parties constituted a coalition to form the national government, and the Malaysian Chinese Association formed to become the legitimate and largest forum for political representation of Chinese people in Malaysia. The peninsular states of Malaya became independent in 1957, and Sabah and Sarawak joined the Federation of Malaysia in 1963. Singapore also joined the federation in 1963, but after two years of political tensions – centered on concern that Singapore's majority Chinese population would tip the balance of power in the new country – Singapore was compelled to leave the Federation and became an independent country.

The political geography of colonial Malaya had its origins in the British Straits Settlements – Melaka, Penang, and Singapore – which were the most popular destinations for Chinese migrants in the nineteenth and early twentieth centuries. As a result of the migration demographics, Malaysia has the largest Chinese minority population on the world scale. Partly because Malaysia's Chinese population is so large and is politically active, state policies have had to be accomodating. In postcolonial national politics, Malays dominate the national government, the Barisan Nasional, which is a coalition of several political parties that defines the Malaysian parliament. Its central members are the ethnically based United Malays National Organization (UMNO) and the Malaysian Chinese Association (MCA); the Malaysian Indian Congress forms a smaller party. The Democratic Action Party (DAP), largely Chinese, and Parti Islam SeMalaysia are the main opposition parties. Despite significant political representation, the memory of the Malayan Emergency, the organization of government based on ethnically affiliated political parties, and an uneasy history of race relations have made identity formation for Chinese communities in Malaysia an uncertain enterprise, and an ongoing process of negotiation. In addition, Malaysia did not normalize relations with China until the 1980s. Malaysia's head of state, Prime Minister Mahathir Mohamad, visited Beijing for the first time only in 1985, and during a domestic economic recession in part to stimulate growth by developing economic relations with China. The recession made the middle of the 1980s a tense time for the Malaysian leadership, and the state initiated unprecedented efforts to rescue what appeared to be a faltering industrialization drive. The plan to develop *Bukit China* (Chinese Hill) in Melaka was one of the most controversial development schemes of the era. The balance of this chapter examines Chinese community organization and identity formation through the national culture debates in Malaysia and the social movement to preserve Bukit China, the largest traditional Chinese cemetery in the world. What especially emerges from the analysis is how diverse Chinese communities in Malaysia forged alliances in the Bukit China movement based on place identities in Melaka, the importance of Melaka in diasporic settlement history, and ideas about homeland ties to the region of south China.

Malaysia and the National Culture Debates

Malaysia's official religion is Islam but the country remains a secular state and promotes some religious and "other" cultural tolerance in order to negotiate a population that is almost 10 percent Indian and over one-quarter Chinese. As a result of the political economy of the colonial period, the geography of Chinese settlement resulted in a distinctively urban pattern dominating the port towns of the west coast of the peninsula (Sidhu, 1976), and uneven capital accumulation in favor of colonial capitalists and the Chinese mercantile class (Jesudason, 1989). The uneven distribution of wealth by ethnic group became a significant issue in postcolonial national politics, and the state responded with a plan of social and economic restructuring that has increased political economic opportunities for Malays, in addition to emphasizing Malay cultural forms (Jesudason, 1989). Thus, by contrast to many former colonial societies, in Malaysia debates over postcolonial nationalism have not focused on the impacts of the colonial period, and instead have revolved around race and ethnicity, and the accommodation of the Chinese and Indian diasporic populations under Malay-dominated rule. The systematic approach to granting Malays special rights was originally legitimated in the constitution, and has been carried out under the state's New Economic Policy, 1970–90 (Cho, 1990). State promotion of the position of Malays found legitimacy in the memory of the riots after the 1969 federal election, in which a complex set of social, political, and economic tensions reduced to Chinese–Malay antagonism (Oo, 1991; Crouch, 1992). The state first formulated a national culture policy emphasizing Malay heritage in 1971, as part of a wider response to the events of 1969. Its implementation began to show substantial effects toward the end of the decade, and Chinese institutions and community leaders had to organize to attempt to maintain basic institutions and spheres of Chinese cultural activity. Controversial issues of the era were the establishment of a Chinese-language university, practice of the lion dance, and official recognition of Yap Ah Loy as the founder of Kuala Lumpur.

At the turn of the 1970s, Chinese leaders in education were embroiled in controversy over the plan for a privately funded Chinese-language university. The state ultimately forbade its establishment by law, even as by 1977 it had reduced non-Malay university admissions to less than 25 percent, and had already established an Islamic university (Means, 1991, p. 60). The goal to establish a Chinese-language university and the state's position against it has reflected the general policy to install the Malay language, *Bahasa Malaysia*, as the medium of instruction in state-funded schools. Such efforts to eviscerate one of the important bases of community formation, Chinese-language schools, spurred the organization of the United Chinese School Teachers Association (UCSTA) and important campaigns to maintain Chinese-language education. The UCSTA has successfully maintained Chinese medium primary schools within the national system, in part by working carefully with the MCA and UMNO to negotiate a sphere of legitimacy (Tan, 2000). But during the postcolonial period many secondary schools were compelled to convert to English medium instruction and, after 1969, to Malay medium, or lose federal funding. Still, Malaysia is the only country in the region where a system of state-funded Chinese primary schools continues to exist.

National controversy over the lion dance erupted in 1979, when a Malaysian state official advised that "foreign" cultural practices, such as the Chinese lion dance, could not be accepted as part of Malaysian national culture (Kua, 1990, p. 11). Chinese associations instead promoted the lion dance nationwide. Popularization of the lion dance, whose troupes had traditionally drawn on working-class membership (Carstens, 1999, p. 42), affiliated people of different class backgrounds and gave working-class Chinese a more prominent symbolic place in national culture. Combined with its politicized notoriety, the lion dance became an enhanced symbol of collective Chinese identity in Malaysia during the 1990s. State assault on Chinese cultural practices took another turn in

1981, when the Ministry of Culture, Youth, and Sports announced that Yap Ah Loy, the reputed founder of Kuala Lumpur, would be replaced in official histories by his nineteenth-century Malay counterpart, Raja Abdullah, a local sultan (Carstens, 1988, 1999). In response, newspapers ran long articles about Yap Ah Loy, explaining his important role in the early development of Kuala Lumpur. Even though the state prevailed in this case, the barrage of information on Yap Ah Loy left people newly aware of the identity of the historic founder of capital – and of ways in which landscapes and their histories are contested in postcolonial Malaysian society. In contemporary cultural politics, the Bukit China issue emerged as literal ground, a kind of geographic tableau, of the larger national culture debates in which a diverse Chinese community located the opportunity to reinscribe its position in Malaysian history.

Melaka:[6] The Bukit China Movement

Locals and regional specialists know Bukit China as the largest remaining traditional Chinese cemetery in the world (plate 5.1). Approximately 12,500 graves are scattered across the 42 km^2 hill, which is closed to new burials and serves as a local park. Bukit China is a monumental historic landscape overlooking the town of Melaka, capital of the state of the same name and a rapidly developing regional service center (map 5.3). In 1989 Melaka became Malaysia's singular designated historic settlement, a status the federal government granted to promote tourism and seek UNESCO world heritage site status for the town center. Melaka was the first major settlement on the Malayan peninsula, the first sultanate in the East Asian realm, a thriving port at the pivot of the monsoon, and home base for a diverse population of Asian traders and merchants, including a sojourning Chinese mercantile community. The Lusitanian colonial project gave Melaka another claim on regional history, as the first bastion of colonialism east of India. The Portuguese conquered Melaka by force in 1511 and some of their constructions still stand in Melaka in the company of buildings from the subsequent Dutch and British colonial eras. As a result, a legacy of edifices, from each regime that held Melaka as a key point of control in the Asian maritime trade, creates a landscape palimpsest of 500 years of mercantile history in Melaka's central historic district. Although the Malaysian application for world heritage site status was unsuccessful – because the Melakan state government obliterated the key site of local history in "reclaiming" the waterfront harbor – the national recognition conferred upon Melaka has resulted in significant deployment of state funds to develop heritage site tourism in the city. Bukit China, however, has not been featured in tourism promotion plans. Instead, in the middle

Plate 5.1 Bukit China, aerial view.
(Photograph: Ee Tiang Heng.)

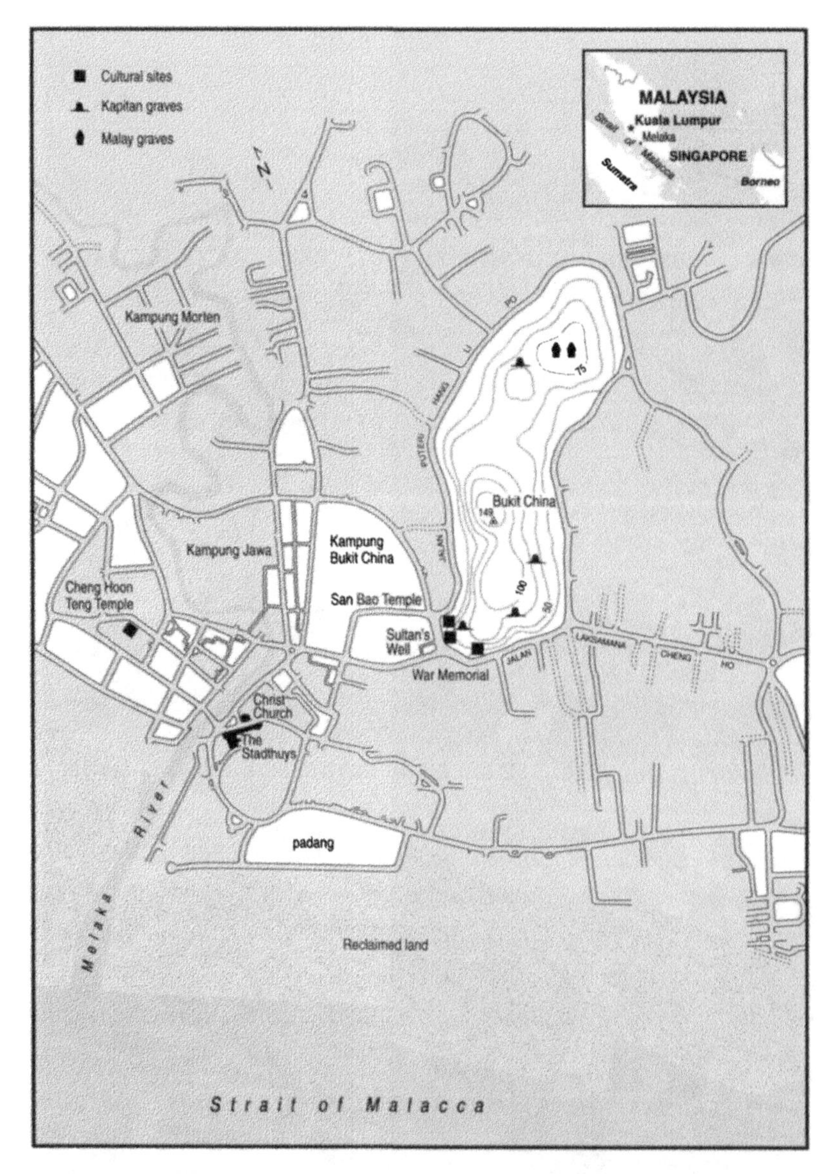

Map 5.3 Melaka town and Bukit China
Line work by Jane Sinclair.

of the 1980s, either one of two competing development proposals for the Bukit China site would have leveled the hill for real estate development and a cultural theme park. In response, a preservation movement rose in Melaka, expanded nationally, and ultimately influenced the state to cancel development plans. The preservation movement based its legitimacy on the idea of the importance of the hill as an unparalleled landscape of Malaysian national history, and therefore warranting state protection.

Rise of a social movement

In April 1984 the Melaka state government made public a development plan that would transform Bukit China into a mixed land use site of housing, commercial properties, and public amenities, such as a library and theatre. The government plan allowed for the preservation of a few late Ming dynasty graves and Muslim graves, but otherwise called for development of nearly 80 percent of the hill surface. A private Chinese land developer and leader of a faction of the MCA, Tan Koon Swan, countered the government proposal with his own: turn three-quarters of the hill into a cultural–historical theme park, complete with a "time-tunnel" and multistorey car park. The private proposal was advocated heavily around town, especially through a promotional propaganda tract, *Bukit China: The Road to Survival*. Published by a Tan Koon Swan organization, the booklet publicized plans for "the biggest historic cultural park in the world" (*MHHW*, 1984). What sounded like hyperbole people took seriously: Koon Swan had already helped develop Genting Highlands Resort, Malaysia's only gambling center, situated high in the mountain rainforests of the central peninsula.

These proposals inspired public interest in Bukit China and generated an eleven-month preservation drive that drew support from individuals and groups around the country. While early regional media reports identified the Bukit China controversy as a Chinese–Malay issue (Clad, 1984), preservationist and pro-development interests were not neatly divided by ethnic group, as demonstrated by the pro-development faction of the MCA (figure 5.1). The main agents of the preservation movement were the Cheng Hoon Teng Temple, the DAP, and the MCA. Complicating the debate, some individuals and parties adopted situational identities, alternatively supporting preservation or development, in negotiation of different roles they encountered in their daily lives in Melaka and beyond. One of the successes of the preservation movement, then, was its ability to transcend these potential divides and work as a platform for multiple identities and voices within Melaka and the larger Malaysian Chinese community.

Figure 5.1 "The war over *San bao shan* [Bukit China]." Cartoon images of Neo Yee Pan (left), chairman of the MCA, and Lim Kit Siang (right), chairman of the DAP, stand guard over Bukit China, while Rahim Tamby Chik, Chief Minister of the State of Melaka (left), and Tan Koon Swan (right) plot development of the hill. The Cheng Hoon Teng Temple lies at the center of the debate, in the image at the base of the hill.
Source: Jianguo ribao (November 14, 1984).

The Bukit China preservation movement affiliated diverse people and organizations, including the major and minor political parties, non-governmental organizations, Chinese community and cultural groups, and prominent individuals. The preservation movement originated in Melaka as a local effort of the Cheng Hoon Teng Temple, which holds the land title to Bukit China, and ultimately enlarged to the national scale through the participation of the DAP and the MCA. The Cheng Hoon Teng Temple is a Buddhist Guanyin temple and the oldest Chinese temple in Malaysia. The Temple's incorporation ordinance defined Bukit China as a "gift of land" made by founders of the Melaka Chinese community "for the promotion, propagation, and observance of the doctrines, ceremonies, rites and customs of the Buddhist and other religions" in Melaka (Cheng Hoon Teng Temple, 1949, p. 1), and named the trustees to protect the property from land use change. The state's development scheme for Bukit China challenged the authority of the Cheng Hoon Teng Temple, which is itself an important Buddhist site and a place of significant community formation. The Cheng Hoon Teng Temple focused its efforts on court action to demonstrate legal title to the property and the unreasonable nature of any development plans for Bukit China. The main body of the MCA founded a Save Bukit China Fund to collect donations for landscaping improvements and a state-threatened quit rent payment, and the DAP sponsored a series of activities across west Malaysia that evolved to create a national-scale movement.

The preservation movement had to organize in response to both the development proposals for Bukit China and the state's politicization of racial issues in attempts to thwart the anti-development movement. At various points, the state used the memory of the Malayan Emergency and historic interracial problems in order to attempt to intimidate preservation organizers. The Melaka state Chief Minister, Abdul Rahim Thamby Chik, asserted that development of the hill was in the interests of the people and necessary for local economic growth, but moreover, he deployed the insidious threat that opposing development was an act of treason. Preservationists were labeled "anti-government," "anti-national," and "anti-development" (*The Star*, 1984a), thereby equating development with nationalism and constructing preservationists as antagonists of the state. In a Melaka State Assembly meeting, the Chief Minister actually charged the outlawed Communist Party of Malaya with using the Bukit China issue to overthrow the government (*The Star*, 1984d).

The preservation movement worked to discredit both the agents of development and their development plans among several potential constituencies: the community of Melaka, the country's intelligensia, and the national political arena. The quick rise of the preservation movement owed considerably to the efforts of the DAP and its leader Lim Kit Siang, a resident of Melaka, who became the most active force in the movement by staging rallies and speeches, organizing marches and jogathons, and holding regular press conferences. The DAP's most prominent activities were signature and poster campaigns, which were conducted in four languages and launched in every state in the peninsula, assemblies at Bukit China, a fund raising drive, and direct challenges to the state and the Tan Koon Swan faction for community accountability. The DAP distributed across Malaysia over 160,000 posters proclaiming "Save Bukit China as National Heritage" (*The Star*, 1985), and sponsored a "Jog to Save Bukit China" which drew about 800 participants and resulted in numerous arrests for unlawful assembly (Pandiyan, 1984). These types of activities heightened awareness of the debate over Bukit China well beyond Melaka town.

The problematic attitude of the Melaka government and the extent of the DAP activities motivated other groups to articulate their positions. Local bodies, especially the Melaka Chinese Chamber of Commerce, the Melaka Chinese Assembly Hall, and a group newly formed for the movement, Graduates and Professionals of Melaka, gave considerable support to the Cheng Hoon Teng Temple. In October 1984 Chinese organizations representing every state in Malaysia, including the East Malayisan states of Sabah and Sarawak and alumni associations of Taiwan universities and Nanyang University, Singapore, submitted a joint memorandum to the Chief Minister of Melaka for the preservation of Bukit China (*The Star*, 1984c). The document invoked UNESCO guidelines for historical and cultural heritage preservation, including recommendations from the UNESCO-sponsored World Conference on Cultural Policies, which heightened local and national perceptions of legitimacy (*The Star*, 1984c, p. 2).

Simultaneously, the Temple drew up and published in key newspapers a survey form soliciting public response to the development proposal. Anti-development responses were received from 553 Chinese associations and *huiguan*, and 14,451 individuals, which represented a total estimated 294,000 people; only 73 individuals expressed their approval for the government's plans (*The Star*, 1984b). The support of prominent individuals, especially Tunku Abdul Rahman, Malaysia's former Prime Minister, lent a certain national level credibility to the preservation movement, which worked to undermine the power of the development discourse. At year's end, representatives of political parties called on the office of the Prime Minister to intervene and settle the Bukit China crisis (*The Star*, 1984e). The Bukit China preservation campaign achieved diverse ethnic and class representation and became a national scale movement, the first grassroots movement to grow to such proportions in the history of the country.

Region as homeland

The construction of a community identity position in the Bukit China landscape fueled the preservation movement, and based on the reality that burials and monuments on Bukit China represent diverse identity groups in Melaka town. Multiple meanings of Chinese identity have characterized the Chinese community of Melaka (Clammer, 1979, 1983; Tan, 1983, 1988; Cartier, 1993), based on language subgroups, lineage groups, class, and family history in Melaka. The Hokkien (southern Min) subgroup is by far the largest in Melaka, and Hakka, Cantonese, Hainanese, and Teochiu (the Shantou area) form smaller but significant groups. In addition, Melaka is the hearth of Baba Chinese culture, a syncretic Chinese-Malay cultural form that evolved in the early colonial era as Chinese men married local women and adopted aspects of Malay culture (Tan, 1983, 1988). In British Straits Settlements society, the Baba were British subjects, which allowed them to integrate into higher level social and economic positions in the colonial economy. Their local birth combined with complex identity positions – simultaneously Chinese, Malay, and English – distinguished the Baba from later Chinese migrants. Baba Chinese culture has remained a distinctively Chinese subethnic identity with Chinese and Malay cultural roots. Bukit China has represented these and other identity groups in diversity: unlike in other Chinese burial grounds in the region, which were exclusive to dialect or lineage groups (Yeoh, 1991), burials at Bukit China were not restricted by subethnic or clan group, and there was no particular descent relationship between persons buried on the hill and preservation activists. Bukit China has symbolized multiple identities, through burials of individuals from diverse groups, as well as through symbolic local history, by tying Melaka to a common region of origin in south China.

Bukit China represents the fundamental historic tie to the land of origin in its grave sites, where character inscriptions on each and every monument mark the person's place of origin in Fujian or Guangdong, and often down to the county level. A few of the most prominent graves on Bukit China are monuments to local Chinese associations or *huiguan*, which designate sites of collective worship and communal power. In response to the call for the beautification and landscaping of Bukit China, some associations refurbished or replaced their monuments and thus reclaimed symbolic place on the hill. The most prominently located grave, recently restored, lies at the main summit overlooking the Strait of Melaka (plate 5.2). This grave marks the Eng Choon Association, which is the largest contemporary Fujian province *huiguan* in Melaka. Eng Choon is Yongchun, a prefectural capital inland from Quanzhou in Fujian. In this representation of space, the summit of Bukit China becomes a place in Fujian, and a symbolic monument to collective identity borne of the homeland region. Some of the largest graves on Bukit China belong to prominent Chinese community leaders, known as *kapitan*, or leaders of ethnic groups during the colonial period. Five graves of Chinese *kapitans* create a symbolic concentration of historic Chinese community leadership; one of these is the largest grave on the hill. *Kapitan cina* Li Wei King is acknowledged in local history for buying Bukit China from the

Plate 5.2 Bukit China summit, Eng Choon Association monument.
(Photograph: Carolyn Cartier.)

Dutch in the middle of the seventeenth century. The British abolished the *kapitan* offices, but the role of Chinese community leadership in Melaka reconstituted in the chair of the board of the Cheng Hoon Teng Temple. Thus the Cheng Hoon Teng has inherited the mantle of community leadership. More than symbolic power, Heng Pek Koon (1988, p. 5) has evaluated the *kapitan* system as the functional foundation of contemporary Chinese political institutions in Malaysia.

In addition to the gravesites, a landscape element on Bukit China has suggested local identifications with representations of Zheng He. In the early 1990s a volunteer group installed in poured concrete on the western slope of hill three Chinese characters, about six meters high, that read *San Bao Shan* (三保山, three protections hill), a Chinese name for Bukit China.[7] The *bao* (保) character on the hillside is also a homonym for the *bao* (寶) meaning treasure or jewel – the same character used in the name of the small temple at the base of the hill. The *san bao* of the temple name, the three-jeweled, is Zheng He's sobriquet. Thus one cultural reading of the characters is that they code a place-based cultural identity, formed in the reach of social relations, between local Melakan identities and the homeland region on the south China coast, trans-historically tied in the reality and memory of Zheng He's travels. Like the Chinese of Melaka, Zheng He's ships set sail from the south China coast, and three of his voyages called at Melaka. In this landscape reading, San Bao Shan is a representation of a translocal human geography and a potential focus of community identification (Cartier, 1997).

Yet in the late 1990s, at the edge a car park across the street from Bukit China, a large granite statue of Zheng He lay on its side encased in the scaffolding of a wooden shipping crate, apparently abandoned. The statue's uncelebrated arrival in Melaka resulted from a local economic mission to Fujian province. No doubt the statue's larger relation – the Zheng He monument on Gulangyu overlooking Xiamen harbor – inspired the Melaka mission to order one home. The Zheng He statue in the car park was to be emplaced on Bukit China, but the arrival of the statue in Melaka generated a controversy over appropriate representations of Zheng He, and whether placing the statue on Bukit China could lead to undermining conservation of the hill. Some members of the Melaka community feared that the installation of the Zheng He statue would lead to pressures for further land development, such as a Zheng He museum and related tourist enterprises. Would respresenting Bukit China's significance as a Zheng He site compromise the meaning of the Buddhist landscape, and invite commercialization? Melakans and the Cheng Hoon Teng Temple are compelled to weigh such concerns in the continuing effort to perserve Bukit China, in a global era of tourism development projects and waning public disinterest in authentic landscapes.

Laurel Means (1994, p. xiv) has assessed the idea of common identity about homeland origin in south China in work of diasporic writers, especially evinced in the poem "Bukit China," by native Melakan Shirley Geok-lin Lim (1994):

> Bless me, spirits, I am returning.
> Stone marking my father's bones,
> I light the joss. A dead land.
> On noon steepness smoke ascends
> Briefly. Country is important,
> Is important. This knowledge I know
> If it will rise with smoke, with the dead.

In her memoir, Lim (1996, p. 20) recalled the "moment [that] imprinted on me the sense of Melaka as my home, a sense I have never been able to recover anywhere else in the world," in the experience of accompanying her paternal grandfather's funeral procession to Bukit China. "Every other place," Lim concluded, "is foreign after this moment." Lim's participation in the ritual journey to Bukit China invoked the memories of translocal experience that bound her into a Melakan community and a diasporic world, and the embodied, ritual emplacement of that world at Bukit China. In the poem itself, the word "Country" is interestingly ambiguous. Our first reaction is to find it synonymous with nation, but on reflection it more likely codes understanding some combination of getting close to land, place, and region. In all these ways, from actual gravesites and *huiguan* monuments, to ideas about Zheng He and representations in the diasporic literature, Bukit China has represented a diverse community, local political leadership, and ties to the homeland between Melaka and south China.

Negotiating absolute space

How is it appropriate to theorize the place of a burial ground that is also a symbolic landscape, the embodiment of a significant cultural politics in the national order? Henri Lefebvre's writings on spatial practices and types of space problematize the relationship between abstract space and human-modified places, particularly with regard to ways societies have evolved control over space. The Lefebvrian spatial matrix allows a dialectical negotiation of the formation of place from space, and makes rare provision for conceptualizing the actual space of tombs and cemeteries in "absolute space," which Lefebvre (1991, p. 235) defines as "above all else the space of death."[8] Lefebvre's examples of absolute space are temples and the Greek *polis*, spaces that preceded or survived the Roman era of land privatization. In the Lefebvrian spatial triad, absolute space is "space of representation," space that represents cultural practices. Bukit China is a space of representation, real and symbolic, where activities of daily life are reconstituted, and where *fengshui* has been practiced to select burial sites. Spaces of death in the cultural worlds of Buddhism, ancestor worship, and *fengshui* practice are also sites of emplaced cultural experience necessary to the well-being of the living, where descendants return to perform ritual activities.[9] Attending gravesites of ancestors, from the burial sites of the common family to the imperial tombs, works to protect fundamental community well-being. The Chinese emperor's performance of ancestral rites ensured the stability of the empire itself (Rawski, 1988a, b). In other words, "Ancestors, not the individual or a legal entity, are symbolic representations of collective authority and rights. These rights do not come naturally but rather are created and maintained through manipulations of politico-economic interests on the ground" (Liu, 1995, p. 36). The ritual attending of graves is a set of embodied practices that serves to bind families with memories of the ancestors, whose appropriately propitiated legitimacy is then received to recreate beneficial familial and community relations. Representations of space, by contrast, are conceived and abstract spaces, the space of planners, developers, and the like who seek to control and administer space, to make it productive. In Bukit China, the state and private land developers sought to exert hegemony over the representational space of ethnic and community groups by abstracting it – planning and transforming it into a commodified and "productive" function.

In the Lefebvrian matrix, religious elites selected sites in the natural environment, such as mountaintops, caves, springs, and rivers, to create absolute spaces, centers of symbolic religious and political power. Through such spatial practices, "natural space was soon populated by political forces" (Lefebvre, 1991, p. 48). Bukit China is the mountaintop, selected and kept by the early leadership of the Melaka Chinese community; the Chinese belief system of *fengshui* imbues the natural topography with spiritual forces; and the contemporary practice of this belief system ensures its ritual significance. Thus absolute space "continues to be perceived as part of nature. Much more than that, its mystery and its sacred (or cursed) character are attributed to the forces of nature, even though it is the exercise of political power therein which has in fact wrenched the area from its natural context" (Lefebvre, 1991, p. 234). In the world of imaginary/real geographies where forms of knowledge prevail in ritualized belief, the space of Bukit China is a distinctive natural site that symbolizes historic and contemporary cultural practice and political leadership. Because *fengshui* is also a representational system "which subsumes ideology and knowledge within its practice" (Merrifield, 1993, p. 523), the landscape may also be thought of as a representation *of* space. Although such spaces are commonly the space of capital, in this case *fengshui* ideology is also a conceived space that accounts for social order, in intergenerational and localized prosperity and political–religious symbolism. Bukit China is then a *place* of cultural practice and a *space* of political symbolism for the Chinese communities, at local and national levels. Lefebvre held that under certain privileged conditions absolute space constitutes the "unity of a representation of space with a space of representation," which occurs when "gestural systems create spaces which are representational and lived, functional and symbolic" (Stewart, 1995, p. 612). Lefevbre's concern with gestural systems encompasses the embodiment of ritual aspects of spatial practices. In the practice of *fengshui* at Bukit China, *fengshui xiansheng,* or geomancers, have connected the burial, in the remains of the body, by a ritualized spatiality in the location and placement of the gravesite, to an ideology of idealized environment. This embodied act of recreating the social order and reinscribing traditional cosmology at selected sites on the landscape is a transhistorical force at Bukit China. Bukit China may thus be seen as a remnant paragon form of absolute space in which "space, time, and the body have not yet become practically or conceptually inimical" (Stewart, 1995, p. 612).

The social construction of place

Peter Jackson and Jan Penrose (1993, p. 2) have utilized a constructivist approach to analyze place-based nationalism, in which "race" and "nation" are located, socially constructed positions, the products of specific historical and geographical forces. My point of departure from their approach is that in Melaka, subethnic constructions of cultural identity within one "racial" category, the Chinese, take precedence over "race" as the conceptual force mediating interpretations of place and place-based versions of national culture. The subethnic category is significant here, since bridging diverse cultural identities in the Melaka Chinese community was the first step in the evolution of a successful preservation movement. Once established, community activism gave the Melakan population an opportunity to review and reconstitute the landscape history of Bukit China.

Bukit China has been the site of significant events, whether in chronicled history or hagiographic account, representing each period in the history of the city. At the national level, preservation activists reconstituted Bukit China's history in microcosmic representational form to demonstrate how the land embodied a history of Malaysia itself. Lefebvre's (1991, p. 237) theorizations about absolute space as "microcosms of the universe" capture the national: "Hence absolute space cannot be understood in terms of a collection of sites and signs; to view it thus is to misapprehend it in the most fundamental way. Rather, it is indeed a space, at once indistinguishably mental and social, which *comprehends* the entire existence of the group concerned (i.e. for our present purposes, the city state), and it must be so understood" (Lefebvre, 1991, p. 240, emphasis in original text). In this, Bukit China as absolute space represents the platform of an emergent national culture, and also the "power of place" (Agnew and Duncan, 1989). Bukit China as place is also a representation of "social and political order," which "serves to mediate between the everyday lives of individuals on the one hand, and the national and supra-national institutions which constrain and enable those lives" (Agnew and Duncan, 1989, p. 7).

The dialectic mediating the theoretical category of absolute space and the place of Bukit China is the set of social processes that construct human meaning about the land. In becoming place, the actual site of Bukit China and people's imaginary geographies of the hill are both contexts for human action and landscapes that people interpret in reflecting on and formulating human identity. Ideas about spaces of representation, like meanings of place, tend to evolve from "less formal, or more local forms of knowing" and are "geographically and historically contingent . . . frequently the result of socially specific spatial practices" (Stewart, 1995, p. 611). Further, "spaces of representation are the places that become sites of resistance, and of counter-discourses which have not been grasped by apparatuses of power, or which 'refuse to acknowledge power'" (Stewart, 1995, p. 611). In Bukit China the landscape became the literal and metaphorical site of struggle, through which political parties and grassroots community groups negotiated its fate, and redefined community identity and national symbolism. Massey's (1993, p. 66) interpretation of place is similarly based on the realization that place "is constructed out of a particular constellation of relations, articulated together at a particular locus," where the social processes at work, in this case in a preservation movement, may be constructed at different scales, and connect to distant places or regions that transcend the location of the place itself.

Constructing the nationscape

The preservation movement generated myriad verbal exchanges and media reports about Bukit China and its proposed development. For over a year, preservation activities spilled over into daily life and the Bukit China issue became the hot topic on the street in Melaka. The social construction of meaning about national culture in the landscape of the cemetery required the diffusion of knowledge about historic landscape events, which occurred through the activities of the preservation movement and media reporting. The discourse of the movement, as medium for social interaction, served to construct local community identity. Hundreds of news articles in both Chinese- and Malay-language newspapers summarized and amplified the events. Several major articles recounted the indigenous, immigrant, and colonial histories of the landscape, and their headlines upheld the central message of preservationists: "Bukit China: its past and present" (Tan Pek Leng, 1984), "Bukit China belongs to the people" (Kua, 1984), "Preserving Malacca's historical heritage" (Kancil, 1984), and "Bukit China – a hill steeped in history" (Tan Chee Koon, 1984). Two key documents, The Cheng Hoon Teng Temple survey and the Joint Memorandum of Melakan graduates and professionals to the Chief Minister of Melaka, crystallized the substance of the community response to the movement.

The Joint Memorandum, a two full-page newspaper advertisement, detailed in eight sections the context of the development controversy, and the goals of the Chinese community of Malaysia to preserve the hill in its entirety: as a cultural heritage monument of the Chinese community, as a tribute to positive Chinese–Malay relations in Malaysia, and as a public park land. The document repeated the history of Bukit China, and in addition to the visits of Zheng He and the role of Bukit China narrated in the first indigenous Malay text, discussed below, it focused on the symbolism of the hill as "an intimate bond between the Chinese community and Malaysia," and how "amidst the communalist politics that has frequently clouded inter-community relations in this country, Bukit China serves as a reminder of the bond of friendship between the Malay and Chinese communities that already existed in the early days of the Melaka Sultanate" (*The Star*, 1984c). One of the conclusive points listed in the Joint Memorandum was the protection of Bukit China under colonial rule. The British colonial government in Melaka apparently made several attempts to acquire Bukit China. The final attempt in 1920, which involved using earth from the hill to reclaim land in the Strait, led to the case being brought before the Privy Council in England – which turned down the colonial government's plan (*The Star*, 1984c). Thus even under colonial rule authorities prevented the acquisition and development of Bukit China.

The formation of a community view on the Bukit China landscape facilitated the social construction of local representations of national culture, about indigenous, immigrant, and colonial histories, in the landscape of the hill. But geographies of national culture are often both real and imaginary, real in the material landscape, and, at Bukit China, in imaginary geographies about legendary historic events known to have occurred at the site. The earliest indigenous Malay text, the *Sejarah Melayu*, narrates the genealogical history of the Melaka sultanate and describes how when the Chinese emperor discovered the importance of Melaka he sought to forge an alliance by marrying his daughter with Sultan Mansur Shah. She arrived amidst an entourage of hundreds, who were "bidden to take up their abode at Bukit China: and the place goes by that name to this day" (Brown, 1952, p. 91). Chinese chronicles of events relating to the tribute system, however, do not corroborate the account in the *Sejarah Melayu* (Yeh, 1936, p. 76; Purcell, 1947, p. 119; Sandhu, 1983, p. 95), and scholars have considered it an apocryphal tale to explain the existence of a Chinese wife in the sultan's seraglio, one of many vignettes in the *Sejarah Melayu* to express the greatness of Melaka in its relations with other empires (Brown, 1952, pp. 1–11; Sandhu and Wheatley, 1983, pp. 497–8). What matters in contemporary Melaka is that people widely recount this history, and with good reason, since a series of panels in The Stadthuys, the state Museum of Ethnography and History, includes a depiction of the marriage of Sultan

Mansur Shah to a Chinese princess. The *Melaka Map and Guide*, published by the state Tourist Development Corporation of Malaysia (TDCM, 1991) retells this account under a section, "Where it all began . . ." An analysis of the Bukit China case in *Asiaweek* (1985) reported this version of history as basic information; so did a short article in *Malaysian Business* (Selvarani, 1984). The value and meaning of these concerns in Melaka, and their function in a scholarly realm, is perhaps best bridged by Pierre Bourdieu (1977, p. 156): "rites and myths which were 'acted out' in the mode of belief and fulfilled a practical function as collective instruments of symbolic action on the natural world and above all on the group, receive from learned reflection a function which is not their own but that which they have for scholars." In Melaka such narratives have functioned to reinstate multicultural worldviews and a kind of local cosmopolitanism that has otherwise been undermined by the state in the era of the New Economic Policy.

Forms of local knowledge in folk belief and myth – and especially how people choose to operationalize that knowledge – were consequential elements of the discourse in the social construction of the Bukit China landscape. Significant situated and partial knowledges hold the potential to create alternative geographies of landscape in the same physical space, a condition which requires any analytical component of interpretation to capture and transcend the distanciation between the space of "real" geographies of the visual and material landscape and imagined geographies of representation. Recognition that "the imagination has become an organised field of social practices . . . and a form of negotiation between sites of agency ('individuals') and globally defined fields of possibility" (Appadurai, 1990, p. 5) yields a conceptual viewpoint about how people can create simultaneous and different meanings of the same landscape. As Cosgrove (1993, p. 281) has expressed, "landscape is able to contain and convey multiple and often conflicting discursive fields or narratives purporting to represent specific human experience. . . . Myths may both shape and be shaped by landscapes." The landscape of Bukit China has conveyed simultaneous historical, contemporary, and transhistorical narratives: local and national cultural heritage site, *fengshui* hill, internationally important cemetery, ethnic monument, weedy eyesore, prime undeveloped real estate, green lung, open space, and local park. It was the work of the preservation movement to address the dissonance among these views in order to realign a general public representation of landscape in favor of the preservationist narrative: a national monument and historic open space.

The success of the preservation movement ultimately registered in February 1985, when the Melaka state government announced its intention to cancel development plans. The representations of space expressed through the preservation movement had transformed common perception of the cemetery from a tired Chinese burial ground to the singular site in Malaysia to represent the history of the nation. As a result of the preservation movement, Bukit China once again became understood for what it has been: a sacred Chinese landscape at the site of the region's first sultanate, the nationscape of the multicultural Malaysian political order.

Place and national culture

The role of cultural identity in pan-ethnic nationalism is a critical one, since state nationalist ideology gels when builders of cultural forces align themselves with the paradigm of the state (Anderson, 1983; Hobsbawm, 1990). Since the 1980s, Malaysia has been constructing self-consciously Malay-centric national culture, which has left the Chinese and other ethnic groups to reclaim and assert their cultural forms and identities on the national scene. In preservationist representations of Bukit China, activists articulated a place-based version of national culture that symbolizes a total history of the Malay peninsula, which recaptured definitions of nationalism that prevailed before the New Economic Policy, including state acknowledgment of the significance of the Chinese community (Means, 1991, p. 9) and the importance of cooperation among ethnic groups (Abdul Rahman, 1969). These are versions of nationalist ideology that continue to be upheld in Melaka even as the contemporary state has steered a Malay-centric course. While the social constructions of meaning about history of nation in the Melaka case served to maintain a general category of nation (see Penrose, 1993, pp. 28–9), supposedly accessible to all racial groups, recapturing the earlier statist narrative was more desirable than allowing new alternative versions that would efface Chinese culture, in symbolic and geographical forms. Like analyses offered by anthropologists Bruce Kapferer (1988, p. 4) and Richard Fox (1990, p. 7), such a nationalist ideology rises from a local cultural context, rather than a state-level universal form. In this, the nationalist ideology symbolized in the social construction of the Bukit China landscape is a broader conception of national culture than the one currently under promotion by the state. Further, this case demonstrates Fox's (1990, p. 3) point about "a false rigidity to our conception of culture [in which we] artificially fortify our belief that cultural productions can be classified as either racial, or ethnic, or nationalist." This situation, in which different possibilities in the formulation of nationalism arise, has been anticipated by Partha Chatterjee (1993, p. 11), who has called for recording alternatives in community formation, and their intersections with national culture, that will not be swamped by the state.

The Bukit China preservation movement rose in response to a local crisis over place identity, a situated crisis in the collision of local and global forces over the potential loss of a monumental landscape and the histories of place and region it represents. Kernial Sandhu and Paul Wheatley (1983, p. v) foregrounded this larger set of problems: "What happens when a city of the first order of importance, in this case the capital of a thalassocracy dominating the sea lanes between East Asia and Europe, progressively declines over three centuries or so until it finally becomes a regional service center in a modern federal state?" Malaysia has restructured its economy, originally based on primary extraction industries inherited from the British colonial period, to promote both secondary and tertiary industry. The prescription for Melaka in the state-led industrialization drive has been tourism development (Cartier, 1996, 1998), an economic agenda which has set up conditions to contravene local experience of place. While preservationists likened Bukit China to Egyptian tombs and historic sites in Jerusalem (Li F., 1984), developers planned to transform the authentic cultural landscape into a theme park. Structural economic problems resulting from the mid-1980s recession only promoted such large-scale rapid development schemes (Means, 1991, pp. 172–4), and heritage landscapes also came under threat of development in Kuala Lumpur and Penang. It is common to examine such contemporary development problems in rapidly developing regions as economic issues; yet they are always also dynamic place-based processes where the cultural landscape ruptures to reveal stark questions about political economic hegemonies and forces of social and cultural change. In such circumstances, at the interstices of economic, political, and cultural forces, constructions of meanings about place, cultural identity, and expressions of nationalism, and the deployment of discourse and discursive practices by contending agents, reveal considerably how people understand place and decide what landscape means.

In the case of the Bukit China incident in Melaka, a diverse Chinese community formed a successful preservation movement by forging alliances at multiple scales, at local and national scales in Malaysia, and in the global arena by invoking international standards of heritage conservation. By contrast to more simplistic notions of community formation, based on ideas about ethnic homgeneity or race-based alliances, the basis of social movement formation in the Bukit China movement depended not on "being Chinese," but on place-constitued ties formed in Melaka, and the ways in which those ties framed larger-scale connections to multicultural visions at the national level in Malaysia, and historic imaginaries about a common homeland of origin in south China. The perspective on the region as homeland of origin is an important one for the ways in which it both realistically and ideologically elides the national imaginary of China and Chineseness as primary containers of interest, and thus association by race of all Chinese people as necessarily fixed within that interest. In the regional imaginary of homeland origin, Chinese people in Malaysia may both claim their roots and maintain positions of citizenship and nation orientation in Malaysian society. Negotiating identity through the complexities of place and region, in scaled formation, and in relation to different nationalisms, allows ways of seeing through to the realities of individual and group identity formation and avoids some of the intellectual problems of deterritorialized conceptualizations, such as the extremes of the "networked Chinese" approach, visited in chapter 2. The real problems of losing located, scaled perspectives and reverting to a "pure" network approach (cf. Latour, 1993) are arguably witnessed in totalizing associations of people of Chinese descent with a singular national imaginary of China/Chineseness, and the horrific rehersing of racism such positions can bring, as in Indonesia in 1998.

Diasporic identities, formed along lifepaths of high mobility in the transboundary cultural economy, are comparatively flexible identity formations. Bergère found these contextual conditions in the lifepath of Sun Yat-sen. In the Maoist era, the CCP could not maintain a stable policy on the diasporic communities and instead revealed the internal contradictions of the revolutionary struggle in its efforts to alternatively affiliate and marginalize the overseas Chinese, and still appropriate their capital. These contradictions reflect the complexities of the transboundary cultural economy in south China, in its transnational conditions, and the attempts of nation-states to contain them. Whether in China under communism or in Southeast Asia within the ASEAN, the modernization project has sought to command nationalistic loyalties and nation-state-based economies, and so has attached little importance to thinking about the region, and especially transboundary regions. Partly as a result, much more research on "ethnic groups" has examined ideas about conflict, separatism, and autonomy, aspects of population formation antithetical to the state, rather than questions about negotiation, resistance, and redefinition of identity formations and their regional contexts. It is not surprising then, as the nation-state has begun to become destabilized under forces of globlization, that alternatives to state/anti-state discourses have opened up, and that we have begun to theorize the constitutive relationships in these emergent geographies.

ACKNOWLEDGMENT

The extract from Shirley Geok-lin Lim, "Bukit China," in *Monsoon History: Selected Poems* (Skoob: London, 1994) is reproduced by permission.

NOTES

1 Labor organization among Chinese in the Nanyang depended on network association through *hui* (association or brotherhood), historically known in the West by the term "secret societies." *Hui*, according to Dian Murray (1993, p. 177), were mutual assistance networks that originated in the south China coast, "directly linked to the socioeconomic circumstances and migration patterns of China's lower classes," who did not have access to land. Murray (p. 179) has described *hui* as the "poor man's *huiguan*." These associations diffused with the migrants to overseas communities and were the primary form of labor and social organization for Chinese men in early colonial society.

2 Perry's argument is significant in that it questions understandings of the worker solidarity in Shanghai based on class formation and class alliances. She argues that union organization was an outgrowth of existing forms of social organization, especially native place organizations, rather than a particular evolution of class consciousness.

3 Historian Conrad Brandt (1958) reviewed the Soviet influence at this historical juncture as "Stalin's greatest failure."

4 The atrocity of Japanese warfare in Nanjing is well known around the world – except in Japan, where textbooks have been regularly censored to camouflage the virulent intensity of Japanese war crimes in China. See Iris Chang (1997). In 1997 Nanjing residents brought the first legal case on war crimes against the Japanese government.

5 It is not well known outside the Chinese culture world that the Nationalists also moved to Taiwan the greatest collection of China's material culture and fine arts, which is housed in the National Palace Museum on the eastern fringe of Taipei. For this reason alone the PRC's platform of national reunification with Taiwan is an important cultural-political goal.

6 The standardized Bahasa Malaysia spelling is "Melaka." "Malacca" is the historic rendition of the settlement name and remains in active use in the press and tourism industry literature.

7 The Chinese name for Bukit China is not a translation of the appellation of the state, and as with Chinese places the world over, the toponymy of the state arises from the simple association of place with the ethnicity Chinese.

8 This is not the generalized notion of absolute space that Neil Smith (1993, p. 99) writes about "as . . . what geographers, physicists and philosophers all recognize."

9 Only in the West have I encountered challenges about the significance of the Bukit China site, at least in part, no doubt, because most previous geographical research on cemeteries has treated them by descriptive and essentially symbolic accounts of the material landscape, in addition to the relative absence of annual ritualized social activity in Western cemeteries, combined with the Western tendency to separate cemeteries from other landscapes of importance and daily life.

6

Gendered Industrialization

> Future historians will remember the 1980s in China as a period of utopian vision on the one hand and an era of emergent crisis on the other.
> (Jing Wang, 1996, p. 1)

Economic reform in China continues to be a far-reaching program of national modernization and development. The reform program has encompassed distinctive domestic policies and internationalized development strategies in the arena of macroeconomic policy associated with neoliberal capitalism, including privatization of state industry and diminished welfare benefits. Reform has also become synonymous with the legitimacy of the state in the post-Mao era and forms the basis of contemporary nationalism: it is critical that the achievements of reform outweigh the problems produced in the course of economic restructuring. Typical analyses of reform have adopted a combination of political and economic approaches in attempts to explain the organization of production and economic growth. These analyses tend to assess the characteristics and transformations of state–enterprise relationships and diverse economic sectors, and some attempt to address the regional conditions of economic restructuring, but in general they do not address the social and spatial conditions of the production regime. This chapter examines some of the prevailing treatments of economic reform, and then considers instead the gendered conditions of industrialization and their regional and emplaced consequences in south China. It examines the geography of development to assess how a suite of reform policies has worked to produce a gendered landscape of development in the transboundary cultural economy – how the reforms, as national development strategies in scaled relations with regional economies, have borne gendered values and produced gendered results.

The relationship between industrial modernization and gender is a central axis of connection between state and society in China, around which the state has deployed representations of gender in support of modernization and development (Schein, 1996; Wang Q., 1999). As

Duara (1998, p. 296) has written, representations of women have been "a very significant site upon which regimes and elites in China responsible for charting the destiny of the nation have sought to locate the unchanging essence and moral purity of that nation." Through the waves of twentieth century modernization, it has been the representational role of Chinese women to be imbued with nationalistic values of each regime. In the first half of the century, representations of women typically evoked traditional subject positions in the "eternal Chinese civilizational virtues of self-sacrifice and loyalty" (Duara, p. 287). During the Maoist era, state feminist ideologies portrayed women's dedication to economic modernization through women's equality and the erasure of gender differences. Women entered the labor force in unprecedented numbers, occupied non-traditional employment positions, and wore unisex clothing. Communist policies raised the status of women by certain measures, and challenged repressive ideologies about women's roles in the traditional family, in which women were essentially productive and reproductive resources for male descent groups. But China's program of communist transformation did not fundamentally dilute gender bias embedded in the political economy (Gilmartin, 1993). Mayfair Yang (1999a, b) points out that the "erasure of gender" in the Maoist era really was about enhancing production with female workers, while circumscribing women's subjectivity and potential for identity formation. Maoism masculinized women and "obliterated sex difference, making it impossible for women to be women" (Ruo and Feng, 1987, cited in Barlow, 1994, p. 347). The CCP also established the All-China Women's Federation to articulate a uniform outlook on conditions affecting women and how women should serve the state. The Women's Federation, with a network of offices scaled down to the township level, has continued to serve as the state's voice on women's issues (Zhang, 1994; Howell, 2000).

These two axes of representation for women during the Maoist era, the unitary nationalist model and the way that it worked to cloak female sexuality, also symbolized the communist planned economy and its repression of human agency (Rofel, 1999, p. 117). In China under reform, as a result, shedding ideologies of the past has also meant recovering gender difference and sexuality: if the CCP promoted "a stiffly antibody, antiflesh, and antisexuality attitude" (Zha, 1995, p. 139), then the reform era could only witness an emergence of popular culture imbued with ideas about recovering the body – sexed. But rather than opening up onto a horizon of diverse points of gendered identity formation, patriarchy has reemerged under reform as the prevailing ideological position of difference between the sexes (Barlow, 1994, pp. 347–8). The advent of the market economy has led quickly to the rise of a male-gendered entrepreneurial culture and, as a result of economic restructuring, disproportionately high female unemployment. Still, the new urban commodity culture has broken open the nationalistic hold on representations of appropriate female roles in society. The market economy has bound ideas about sex and gender to commodity desire, and, in people's interest to seek things, provoked tensions about appropriate gendered social conduct (Schein, 1996, pp. 201–2). These complexities have emerged in debates on topics once subjects of comparatively private concern: issues about gender, the status of women, sex, and sexuality unprecedentedly surged into popular discussion in the 1980s and 1990s (Gilmartin et al., 1994; Evans, 1995; Yang, 1999b).[1] As Harriet Evans (1995, p. 157) has described the situation, "Sex – in some form or another – has emerged from obscurity to occupy a position of unprecedented prominence in public life in the People's Republic." In the 1980s and 1990s, and in association with the United Nations Fourth World Conference on Women, held in Beijing in 1994, academic research institutes and conferences on women increased, as did non-state women's organizations (Yang, 1999b). This combination of events under reform has simultaneously promoted male bias in the labor market, opened up new sites of identity formation, and led to a new sexualized cultural economy, especially in the transboundary region in south China.

The entry of women into the labor force in the twentieth century is arguably the most consequential social transformation in China's long history. Yet in contemporary China, the processes of industrialization and economic restructuring have actually reversed the trend of women's entry into the workforce and by some measures diminished the status of women. Trends produced by the complex interactions among especially five reform measures have contributed to these reversals: the open door policy and the emphasis on export-oriented industrialization, which depends on low-wage women's labor for manufacturing production; the rural household responsibility system, which has anchored many rural women's productive and reproductive activities in the household economy; the birth planning policy emphasizing one child, which tipped the birth rate in favor of males; the loosening up of the *hukou* or household registration system, which has given rise to gendered migration patterns; and the restructuring of state-owned enterprises (SOEs) resulting in significant layoffs, which has limited particularly women's employment options.[2] This chapter considers the effects of these reforms, and focuses on the labor regime of export-oriented industrialization to discuss how the transformations wrought through reform have especially implicated conditions for women. Understanding the gendered consequences of these reforms also depends on sieving histories of gendered social structures in China, to consider how contemporary economic reforms, colliding with local social norms and traditions, have reproduced new forms of patriarchal power and control.

In anticipation of a gendered analysis of reform, it is important to remember that the meaning of the term gender is often contested, and confusions around the term have bracketed its utility. "Gender," as Joan Scott (1999, p. ix) has reminded us, was a controversial term at the United Nations Women's Conference in Beijing, and based on disagreement between those who assume gender as based on biological sex, and those who have understood gender as socially constituted and variable. This comparison is complicated in China because of historical slippages in meanings around the Chinese terms for "woman" and "gender." In contemporary China, the common translation for the word gender, *xingbie*, is also literally translated as "sex difference," and used in that way to relive tensions left by the legacy of "gender erasure" during the Maoist era. Thus *xingbie*, depending on how it is used, can mean both difference based on sex and gender difference in the historically constituted sense (Barlow, 1994, p. 348; Cornue, 1999, pp. 80–1). In this analysis, I both assess issues around "sex difference" in post-Mao China and adopt a contructivist position on gender in order to make some preliminary suggestions about how to emplace gender dynamics in the regional cultural economy in south China.[3] At the core of Scott's definition of gender is a set of interrelated propositions: "gender is a constitutive element of social relationships based on perceived differences between the sexes, and gender is a primary way of signifying relationships of power" (Scott, 1999, p. 42). The focus on "perception" and "power" suggests the dynamic of gender formation, and its potential for difference and transformation. Scott's polyvalent approach examines gender relations through more discrete subjects, including cultural symbols and their gendered representations, state and institutional discourses that seek to fix gender ideologies, the political role of social institutions and organizations in constituting gender meanings, and subjective identities of individuals and how they may negotiate gender norms and barriers. Each of these subjects has spatial and place-based contexts in Chinese society, and in scaled relations between local and national levels of social and economic activity.

Most economic analyses of gender and development concentrate on women because of the male gendered condition of the patriarchal state and the development process. Further, this analysis concentrates on the gendered economic conditions faced by women because, no matter the urban or rural circumstances, research on the gendered effects of reform (cf. Davin, 1991; Judd, 1994; Summerfield, 1994a, b, 1997; Lin, 1995; Meng and Miller, 1995; Kelkar and Wang, 1997) substantiates the position, taken earliest and most visibly by Elizabeth Croll (1978, 1983) and

Judith Stacey (1983), that the position of women in the social hierarchy in general, and their state represenations, have reverted to more traditional role-based expectations under reform. A downward trend in the overall status of women in any society is a negative indicator of the broad spectrum of development standards in general.[4] Is the status of women declining in China?[5] What does a declining status of women indicate about the overall reform project? In the coastal cities, women encounter the extremes of greater opportunities to earn unprecedented levels of income and professional employment, and, on the other hand, relatively low-wage jobs in manufacturing or low-status service sector jobs in restaurants, domestic service, and prostitution. Thus the goal here is not examine to the reform process as fundamentally limiting women's opportunities by merely treating women as "victims" of state patriarchy and globalizing capitalism (see Cartier and Rothenberg-Alami, 1999); it is important to understand the diversity of women's experiences under reform. Rather, the objective is to assess how economic development and the patriarchal state have narrowed women's options in an era of state-defined excessive population, rural labor surplus, and surging urban unemployment, and to explore the arenas in which women have contested these issues and carved out spheres of interest where new social possibilities may emerge and alternative identities may be formed.

Who Cares?

How does normative economic theory treat the gendered conditions of economic growth? The concept of the transitional economy has been the normative approach used to explain the transformation of the Chinese economy under reform (see Naughton, 1995). The idea of the transitional economy is based on both theoretical explanations and policy recommendations for the market-oriented transformation of a number of so-called post-communist countries, especially in Eastern Europe, and, in Asia, China and Vietnam. The word "transition" implies the inevitability of transformation from communist planned economies into capitalist market economies. Ideologically, it invokes the history of the post-Cold War period as a definitive end to an era of communism and represents the political economic platform of the neoliberal regime, tailored for socialist governments. Economic policy recommendations of the transition approach seek to liberate "natural markets" and market forces, promote economic growth and development, and, often as a corollary, democratic governing ideologies. The idea of China as a transitional economy has posed challenges to the definitions of transition theory because China, by several measures, appears not to have followed the blueprints of its policy recommendations.

If China is experiencing a transitional economy, "the contrast between China's transitional economy and those in Eastern Europe and the former Soviet Union could not be more striking" (Walder, 1996, p. 1). One of the main differences in the China case is comparatively high economic growth. China has achieved reasonably sustained high growth at the national scale under reform without following some of the basic recommendations of transition policy. China has avoided the "big bang" or "shock therapy" policies used to privatize state assests and rapidly remove price controls, subsidies, and restrictions on foreign investment and competition.[6] In China, by contrast, through the first decade of reform, the state maintained control over the allocation of production in many key industries and social services. People and institutions do not change overnight, and China's strategies would appear the pragmatic response to what is otherwise in international transition policy less than realistic attention to conditions of local institutions and their dynamics.

Institutional conditions of Chinese society have arguably made the critical difference in the success of the reform experience (Oi, 1996, 1999; G. Lin, 1997; White, 1998a, b). As George Lin has argued, the role of the central state in the post-reform development of the Zhujiang delta has not been a major factor, and instead local organizations at the county, town, and village levels have articulated the economic activities at the basis of regional growth. Jean Oi has also identified the local state as the main agent of economic growth and development. In Oi's (1999, p. 12) analysis, the local state is "the activist state that has determined the outcome of reform in China." While Oi allows that the central state initiated the reform process, Lynn White (1998a) finds the origins of reform in regional transformation before 1978. White's thesis allows that the reforms were both central and local, but that the origins of reform can be traced to local and regional transformations, which central government leaders later sanctioned and implemented as national-scale policies. Thus reform processes, according to White, began in name only in 1978, and really had their origins in the period 1971–4, especially in Shanghai. This reinterpretation of reform, as a regional event in motion before Deng Xiaoping emerged to lead the Party, questions several points of conceptualization about reform, including the typical periodizations of reform history, the idea of the central government implementing reform from the top down, and the origins of reform in the capital. Understanding reform as a regional process opens up ideas about different sets of relations between the regions and the center, and suggests a dialectic between region and nation in ways that require the nation to acknowledge regional social formations and allow regions to articulate alternative national identities. The concluding section of the chapter engages this region/nation dialectic to situate contested images of mainland women in the transboundary region.

Regional economy

The emergence of distinctive regional economies in China has suggested different analytical approaches, especially in coastal south China. Would research on industrial districts as regional economies, based on the experiences of post-Fordist restructuring and flexible production systems in the industrialized West, help explain the realities of regional development in south China? The flexible specialization literature seeks to explain the emergence of industrial districts characterized by vertically disintegrated networks of small and medium-sized firms which produce small batches of goods in response to changes in styles and consumer demands (Piore and Sabel, 1984; Scott, 1988). Based on the model of the Third Italy, the regime of garment and fashion accessories production in northern Italy, Brad Christerson and Constance Lever-Tracy (1997) concluded that south China has already become an important center of high fashion garment production for the international market. Their data derive from survey research in Nanhai and Panyu in Guangdong province, Suzhou in Jiangsu, and Quanzhou in Fujian. They assume that the institutional characteristics of family firms in Italy, in trust-based working relations that allow quick changes in product lines, also exist in China, in trust relations based in ethnic and family ties of Chinese business networks.[7] Their assumptions depend on a transnational geography of capital formation in which Chinese compatriots and Chinese overseas invest in factories in south China, and ideas from the literature of Chinese business networks.[8] But their analysis has not examined the actual character of investment relations, the nature of social relations within the firms, or regional labor markets. In the absence of data on the social relations of production and labor–management relations, such an argument is delinked from the realities of the regional production regime and local factory conditions. In northern Italy, the maintenance of unions and socialist labor organization has underpinned the longer-term stability of society and economy. In China, a migratory labor force largely staffs sewing and assembly jobs, and both firms and the state discourage worker organization. While export-oriented garment production in south China in some ways resembles the concept of the industrial district, the characteristics of the labor force and relations between owners and workers are substantially different. Workers in south China are part of a "family" only in that the factory with dormitories serves as the workers' local home. Otherwise, owners and managers use family metaphors to socialize young women workers to accept role-based expectations of deferential behavior: the "dutiful daughter" works long hours at low wages in the interests of the (nominal) family firm (Lee, 1995, 1998; Hsing, 1998). In this socially constructed context of differential power relations, the firm is able to maximize surplus appropriation.

Much of the literature about flexible specialization and low-wage manufacturing generally appears to subordinate or overlook questions about women's work – as if the entire process depended on women's silent productivty. Yet if economistic approaches to regional and national development concur that the main subject of examination is economic growth, why isn't the source of that growth a central subject of analysis? Growth originates in surplus value – the profits gained based on the difference between the cost of a product and the wages that produced it – which means that labor, as performed by workers, is a central determinant of growth. "The central act of capitalism, the extraction of surplus value from workers by capitalists, is obscured in the illusion for both parties. People thus learn to treat human activity – labor – as a marketable commodity and to see capital as an active, even a rational, agent in society" (Gates, 1989, p. 799). Based in this contradiction, capitalism "thingifies" people and creates the illusion that economic activity is decisively rational. The goal then is to demystify and foreground the subject of workers, and as embodied economic agents, rather than treat workers as labor and only as a factor of production.

Engendering Industrialization

To understand the realities of the laboring subject we should assess the experiences and situations of workers, and how power relations constituted in real and symbolic alliances between firms and the state circumscribe workers' opportunities and intervene in their abilities to negotiate alternative conditions. From a larger-scale perspective, we should also assess how the state has adopted patriarchal positions on the roles of women and women's labor, and in support of new organizations of production that depend on low-wage labor and a secondary labor force. What complicates this analysis is history – the reality that Maoist cultural politics did not obliterate traditional social formations – and the ways in which gendered characteristics of the reform program are imbued with practices of gendered social structures and policy perspectives from earlier eras. Working to understand economic transformation in China compels thinking beyond the typical historical divides in economic analysis because in China, as Mayfair Yang (1999a, p. 9) has observed, "the state as mode of production and cultural force is much more ancient and developed than capitalism."

Traditional practices inherited from historic social formations have structured men's and women's life paths in China. Some have reemerged under reform, and certainly unevenly. Of course the strictures of the Confucian cultural complex have increasingly fractured through the twentieth century, yet what does it mean that celebrations of "virtuous wives" have reappeared in the countryside? The imperial state, especially in the Qing dynasty, conferred honors on virtuous women, chaste widows and exemplary wives and daughters, whose lives symbolized proper social order and community honor (see Elvin, 1984). Contemporary virtuous wives, like those of the past, have unselfishly devoted themselves to care for ailing husbands, their children when widowed – the chaste widows of Confucian ideology – and, in the logic of reform, to the assistance of husbands to help them make a fortune (Rai, 1994b, p. 128). As the example of working to make a fortune suggests, the revival of apparently traditional practices is a reinterpretation of historical social formations suited to the logic of reform. Relations between gendering and the state are imbued with economic logics, even as they may be displayed in terms of moral relations between state and society. The state as a mode of production in China is also a set of patriarchal institutions whose traditional forms may be instrumentally revitalized and transformed through new situated intersections with global capitalism.

Male bias in the development process

Relations between the patriarchal state and the gendered organization of economic activity in China under reform exhibit what Diane Elson (1991, p. 3) has conceptualized as "male bias in the development process," which "operates in favor of men as a gender, and against women as a gender." The concept of male bias in the development process recognizes that many decisions about economic organization, from the level of the household to the nation-state, transfer forms of bias about men's and women's presumed contributions to income generation.[9] The obvious problems of male bias are unequal wages and uneven access to productive resources, which include unequal pay for equal work, unpaid household labor, and assumptions about men as primary income earners. Male bias in the development process is produced in different and complex ways, through practices and traditions of everyday life, power relations in institutional spheres, and applications of economic policy that support gendered regimes of accumulation.

The notion of "market rationality" in normative economic logic is one example of institutionalized male bias in the development process. Women in China are being particularly disadvantaged under reform, in both educational and employment opportunities, based on male-biased perspectives that women are less productive or more costly to employ because they must also attend to reproductive responsibilities, especially maternity leave, childcare, and care of the elderly (Rai, 1994a, b). But what is really at stake is that enterprises must now pay for such work-related costs and that men in general have not shared household responsibilities. Dismantling of the state sector and restructuring of SOEs have shifted the cost of worker benefits to the enterprises, which, instead of paying out benefits, have instead demonstrated a pattern of attempting to reduce benefit costs by not hiring women. Among unemployed young people of both sexes with tertiary education, the proportion of women unemployed rose to 61.5 percent even in 1986, a high growth year (Rai, 1994b, p. 125). Among all workers losing jobs as a result of restructuring in the first decade of reform, an estimated 70 percent were women, and the number of women who lost jobs became higher than the number in the workforce (Croll, 1995, p. 120). As Lynn White (1998b, p. 264) has commented, the decline in numbers of working women was "just the opposite trend from what might be expected in a period of quick modernization." The normative economic perspective on the status of women holds that their positions relative to men should rise with higher education and paid employment, but despite increased levels of education and labor market participation, data on women's employment have confirmed that women are facing increased male bias in the job market (Bauer et al., 1992; Summerfield, 1994b).

Employment trends from the 1990s underscore the extremity of the issues at stake. Based on a 1997 survey conducted by the women's department of the All-China Federation of Trade Unions, over 70 percent of managers surveyed in 14 provinces and cities said they would not hire women, even if they were better qualified than men (Liu and Zhang, 1997). In Shanghai and Guangdong, half of the qualified female workers seeking employment could not find jobs. Shanghai has had a higher proportion of women than any Chinese province, and women in Shanghai have also made up a larger portion of the labor force than in most other provinces (White, 1998b, p. 263). Historically, Shanghai has been the leading city for women's education, and a center of publishing on women's issues (Beahan, 1975). But even in Shanghai the numbers of unemployed women disproportionately increased, and the numbers of women employed in professional positions declined. University students in Shanghai have understood that employers bargain with university authorities to hire larger numbers of men and assign women graduates to lower status and lower paid jobs (Forward, 1991, cited in White, 1998b, p. 264; Ma, 1999).

The state has only indirectly addressed the problem of women's unemployment – by encouraging women to leave the labor force and resituating women's roles in the domestic sphere. One of the leading policy guidelines evolved in response to women's unemployment has been lengthened maternity leave for women, up to five years (Wu and Xiong, 1998). The Ministry of Labor and other government departments have promoted this "solution" to women's unemployment, while the Women's Federation has attempted to protest the plan (Howell, 2000, p. 358). Other solutions have been floated, including a "half-day working system" for women, or a "flexible employment system," in which women work part-time on a variable basis in order to maintain household responsibilities. These issues have been widely debated in print media through complex economistic and historical arguments that valorize women's return to the home: because it is in the best interests of the women themselves, their families, and the state. It also eases the state's burden of male unemployment. Most of these arguments echo views about men's and women's "natural" differences and their corresponding divisions in the workplace and the home.

> From the point of view of "subject development," "women going back to the home" does not confine and oppress women. On the contrary, it allows women to move from an abstract liberation to real liberation. Ignoring natural and social selection and forcing women out to work is not liberating women, but destroying them. For a woman to become a person, she must first become a woman. The major lesson to be learned from nearly forty years of trying to resolve the issue of women's liberation is that women have not been treated as women, and a very heavy price has been paid as a result. (Liu, 1997, p. 32)

The ideas expressed in this statement reflect the fall-out from the Maoist era of "gender erasure." But such views must also reflect a certain amount of anxiety over significantly increased awareness of sex in popular culture. The emergence of sex as a common subject creates new tensions around interactions between men and women, which can be relieved in part by sending women home.

Gendered Reform Policies

Reform policies affecting employment, migration, residence, birth control, and other social concerns all have significant impacts on people's daily lives in China. Contemporary policies are sometimes explicitly gendered, as in the case of the prolonged maternity leave policy promoted by the state, and in other cases are implicitly gendered through existing social structures that influence policy implementation. The one-child policy is the outstanding example of a reform measure that was not explicitly gendered by the state, but in its very goals – one-child households – implicated a full range of gendered social processes and especially male bias in child preference. As a national plan and an interventionist reform at the household level, the one-child policy has particularly bound the home and the state in dialectical relations along women's life paths (Wong, 1997).

In 1979–80 China implemented the one-child policy with the goals of reducing demands on the resource base and promoting rapid industrialization by attaining zero population growth by the middle of the twenty-first century. It was fundamentally a national scale economic development policy (plate 6.1). Although national in scope, the one-child policy has been geographically uneven in implementation, acceptance, and results.[10] In reality the state has only kept urban families to the one-child restriction. After about 1986, and widespread opposition, rural farming families were allowed to have two children, especially when the first child was a girl, and other exceptions have been allowed. One unusual exception has involved the Chinese overseas and Guangdong province. Deng Xiaoping reportedly sanctioned the "Guangdong exception" for two children among peasant families in Guangdong from the start, because he "wanted to attract investment from overseas Chinese and they would be upset if their family roots were cut" (Becker, 1999). The two-child policy in Guangdong was a special case intended to demonstrate the comparatively liberal reform policies in the province, whose success, driven by foreign direct investment and transboundary investment, was critical to the first decade of the reform project. By comparison to the continued desire in rural areas to have more than one child, in major urban areas it has become more common for couples to want only one child. Daughter preference has also emerged in urban areas, as some parents perceive male children have become more caught up in the new cultures of reform and let go their family responsibilities.[11]

Despite policy exceptions in many areas, the one-child policy has lowered the number of births by an estimated 300 million people over the course of a generation. It has also resulted in lower than natural numbers of female births. Between 1982 and 1989 the number of boys born per 100 girls rose from 107 to 114, well above the normal level of 105 to 106 (Johansson and Nygren, 1991). The overall ratio was 100 to 111, based on the 1990 population census, and officially increased to 100 to 117 based on the 2000 census. New complex social patterns will continue to arise from these conditions. The perceived and real shortage of

Plate 6.1 Family planning billboard, Xiamen: "Times are different. It's the same to give birth to a boy or a girl." The one-child policy has been portrayed by the state as an essential element of modernization.
(Photograph: Carolyn Cartier.)

females, especially as the population enters marriageable age, is prominent among them, and the prospects for locating marriage partners have already begun to shift. New possibilities to migrate have, on the one hand, allowed women options to relocate for marriage to more distant and prosperous areas, especially in the coastal provinces (Fan and Huang, 1998). On the other hand, in more impoverished rural areas, the reorganization of farm labor in the household responsibility system, combined with the perceived and real shortage of marriage partners, has resulted in the revival of child betrothal arrangements (Croll, 1994, p. 169; Rai, 1994b, p. 125). The problems have also contributed to a commodity trade in women and children, a tortured form of market entrepreneurialism in which patriarchal power is abused to commoditize women as productive and reproductive resources (see Hirschon, 1984; Ocko, 1991; Evans, 1997; Biddulph and Cook, 1999). Estimates of the numbers of women and children kidnapped vary, but the total may be as high as 100,000 a year (Biddulph and Cook, 1999, p. 1445). Most studies examining the problem also point to the historic practice of selling women, but historic precedent alone cannot account for contemporary events. In the 1980s kidnapping of women was largely limited to poor areas, but by the 1990s kidnapping gangs had expanded in scope to urban areas, the transboundary region in south China, and international destinations. Problems wrought by contemporary reform measures and the patriarchal position of the state have contributed to this illegal trade. Impacts of the one-child policy in combination with the return to household farming, and new opportunities to migrate, are particularly implicated.[12]

Rural reform and the feminization of the agricultural labor force

The household responsibility system decollectivized agriculture and revived the household as the basic unit of agricultural production. Decollectivization of agriculture also gave rural families greater decision-making power and flexibility in the production process, by encouraging sideline production and other activities to boost household income. Rural incomes rose during the first half of the 1980s, but began to stagnate in the second half of the 1980s and through the 1990s. Households responded by reallocating labor to higher income earning enterprises. Where non-agricultural jobs were available, men increasingly sought wage labor in new township, village, and private enterprises, or by migrating to urban areas, especially for construction work (Jacka, 1997, p. 28). After the first decade of reform half of all male farm laborers in China had been employed in non-farm sectors, while fewer than one-fifth of female farm laborers had worked in non-farm sectors (Zhang, 1994, p. 83). This trend has increased to the point that, on average, women now perform over 60 percent of the farm labor (*China Daily*, 1999b), and in some areas, including parts of the Yangzi delta, women constitute 80–100 percent of the agricultural labor force (Croll, 1994, p. 167, 1995, p. 127; Zhang, 1994, p. 75; Jacka, 1997, pp. 128–9).[13] Comparative data on women's farm labor are limited, but among the regions, according to some observers, "South China led the way in this 'feminization' (*funühua*) of agriculture" (Jacka, 1997, p. 129). One of the main reasons for the maintenance of women's ties to the land is the contractual basis of the household responsibility system, which assigns families an additional quota of land when a bride enters the household. The bride must be visible in the village for the land transfer to be confirmed and to maintain rights to till the land (Huang Xiyis, 1999, p. 101). The Women's Federation has coordinated the state's main response to the feminization of agriculture, through campaigns oriented toward how to manage and improve agricultural production – what the state wants women to do –

rather than the realities women face (Rai and Zhang, 1994; Zhang, 1994). Gaps in services for women have been increasingly addressed by informal organizations, which have appeared especially at the township and county levels to assist women with a variety of concerns, from health care to creating new sources of income (Zhang, 1994).

Rural household labor contributions of girls and young women tend to be maximized through the patrilocal marriage system, which has resulted in the gendered problem of removing girls from school. Since the reforms, the dropout rate for both primary and secondary school age girls has been rising (Croll, 1994, p. 165; Rai, 1994b, p. 125). Boys continue to be educated, except in the most impoverished areas. A report undertaken by the Women's Federation sought to identify how reform has changed the lives of young rural women through sample sites in southern Jiangsu and Sichuan: the greatest change they identified is how levels of education have decreased (ACWF, 1993). In 1996 over 10 million school-age children, 6 to 14 years old, dropped out of school due to household financial problems, and two-thirds were girls (Shen and Wan, 1997). In the 1980s in Guangdong, over 88 percent of males were enrolled in primary school, compared to 64 percent of females (Peng, 1989). Even as normative economic theory holds that women's education will increase with higher incomes, in Guangdong and southern Jiangsu, where economic growth has been especially high, short-run surplus maximization can be achieved by adding labor to the household economy. The state has responded to these negative trends in female education by indirect means. In 1992 the state sanctioned the Spring Bud Plan (1997), a domestic and international campaign aimed at raising funds for girls' education.[14] Despite its successes, a solicitation scheme suggests how the state would consider girls' education a philanthropic project rather than a basic right and state responsibility, which reflects the neoliberal trend to diminish state provision of basic social services.

Restructuring in the rural sector, the feminization of agriculture, and new opportunities to migrate have created massive upheaval in many people's daily lives. Suicide rates among rural women are the most stark evidence of the problems. Among countries for which data were available in 1990, 56 percent of all women worldwide who committed suicide were Chinese, and most of them were in rural areas (Pritchard, 1996). The rural suicide rate was nearly three times the urban rate, and the rate among rural women was about 40 percent higher than among men. Moreover, these higher incidences of suicide and attempted suicide have emerged among young adult women in their twenties and thirties (Murray and Lopez, 1996, p. 824). What is remarkable about these data on a world scale is that in general men have committed suicide at a higher rate than women, the elderly have been more likely to commit suicide than younger adults, and suicide in urban areas has been more common than in rural areas. Research on reasons for the rise in female suicide in rural China points to a complex set of factors arising from transformations in society and economy, in "major psychosocial stressors" (Philips et al., 1999, p. 44) encountered in daily life under reform.

The practice of suicide as a reaction to extreme life circumstances has historical precedent in China (Hsieh and Spence, 1981). Should we attempt to understand female suicide against the backdrop of the past? Historic suicide took regional forms, and in Fujian it evolved into a social practice with all the overtones of a public spectacle (T'ien, 1988). The most notorious method of female suicide in Fujian during the Ming and Qing dynasties was a ritual hanging, sometimes organized for the family to witness, and on a larger scale staged for view of the entire community. The childless widow, demonstrating loyalty to her husband, was most likely to commit suicide. Critically, without a husband for protection, a woman might commit suicide to avoid in-laws or family members who would compel the women into remarriage in order to gain more reproductive opportunities or dowry wealth. In the literary terms of the state, suicide became locally constructed as the embodied statement of virtue and proper devotion, a male-biased interpretation of women's appropriate representation of *li* in traditional society. Historic female suicide also took place collectively, among women who practiced delayed transfer marriage in the Zhujiang delta (Stockard, 1989, pp. 118–22), and among diasporic communities. The Chinese massacre in Manila of 1603 apparently prompted a mass suicide among women in Quanzhou. "Almost one year after the 1603 massacre of the Chinese community in Manila, the news of the death of their beloved ones abroad finally reached the women and maidens left behind in Chu'üan-chou who then reportedly all decided to end their lives in a well-ordered manner" (T'ien, 1988, p. 49). The last recorded case of a female suicide by hanging in Fujian was as recent as 1850. Moreover, female suicide was also historically taking place in the context of imbalanced sex ratios. The male to female population ratio in parts of Fujian was low during the seventeenth and eighteenth centuries, and in some counties half or more of the men were bachelors (T'ien, 1988, p. 31). It is impossible to conclude that such historic practices account for contemporary patterns, but what appears to remain in China as a transhistorical value is "the acceptability of 'rational suicide' as a solution for a variety of social problems" (Philips et al., 1999, p. 43) women face under reform.

The feminization of agriculture has created new burdens for women, and is related to new opportunities to migrate as a result of relaxation of the *hukou* system in the early 1980s. Overall, men have been migrating more than women, which has left increasing numbers of women to manage on their own in rural areas. Based on the 1990 census, among Chinese migrants who moved between 1985 and 1990, 56 percent were men and 44 percent were women (Fan, 2000, p. 425). Reasons for migration also differ among men and women. Among the state's surveyed nine reasons for migration, the continued practice of patrilocal marriage in rural areas made migration for marriage the leading reason for women, at 28.3 percent of total female migrations. Women migrants were also somewhat more likely than men to come from rural areas, and migrate to rural areas. Men most commonly migrated for "industry or business" reasons, at 31.4 percent of total male migrations. Men also tended to migrate from rural to urban areas in greater numbers than women. Overall, 68.5 percent of men reportedly moved for economic reasons of various types, compared to 38.2 percent of women (Fan, 2000, p. 432). For both men and women, intra-provincial moves were more common than migration to another province, yet among inter-provincial migrants, more women were likely to engage in long distance migration from the interior to the coast.

Among the coastal provinces, Guangdong has been the leading recipient of long distance women migrants. Guangdong's intra-provincial migrants have also included proportionately more women. Data on total migrants to and within Guangdong province showed that migrants were nearly evenly divide by sex, with 50.8 percent men and 49.2 percent women (Fan, 1996, pp. 40–1). Guangdong province has also been different for the degree to which women migrants have identified economic reasons, rather than marriage, as their reason for moving to the province. Women in-migrants to Guangdong identified industry or business, at 56.4 percent of total migrations, as the reason for migrating to Guangdong almost as frequently as men, at 60.8 percent of total male migrations. This pattern reflects the sheer numbers of women migrants working in Guangdong province, and the importance of women's labor in Guangdong's export-oriented production regime.

The export-oriented industrialization regime

China's open door policy laid the basis of the industrialization drive in export-oriented development, and its reliance on low-wage manufacturing to produce consumer goods for the world market. The SEZs and open development regions on the south China coast became the centers of export-oriented industrialization and manufacturing work for women, in electronics and toy assembly, sewing in garment production, and mixed assembly and sewing in the footwear industry. Guangdong province, especially in the Zhujiang delta, has received more inter-provincial migrants, more FDI, and more investment from Hong Kong than any other coastal province, and so has become China's central place of articulation between labor migrants and the world economy. As the largest SEZ and a city established *de novo* as a center of export-oriented industrialization, Shenzhen has been built on the surplus value of migrant women workers. The average daily wage in the SEZ puts the production regime into perspective from the position of the laboring subject: migrants have made up 70 percent of Shenzhen's three million people and the average daily wage, for a 12-hour day in a toy factory, in the mid-1990s was under US$1.10 (Yi, 1998, pp. 28–9). Workers receive minimum wage and minimum overtime pay, they commonly must pay for meals and lodging at the factory, they pay fines for breaking factory rules, and they do not have dependable representation (A. Chan, 1997, 1998). Thus the labor regime compels maximum surplus appropriation: workers' daily lives are completely oriented toward factory production and tightly bound into the regional economy. The state supports the labor regime by disallowing local unionization and maintaining that the All-China Federation of Trade Unions is the only legitimate forum of worker representation (Howell, 2000).

In the coastal zone of rapid development, relations between workers and employers represent both the immediate need of manufacturing plants for large quantities of low wage laborers, and the insecurities young workers face in relocating long distances to live in factory dormitories. Hiring single young women serves several needs of management. While the bulk of the normative economic literature has characterized young women workers as relatively "docile" and "disciplined" with "nimble fingers," it is more realistic to understand their employment as affording management the greatest degree of power relations over a labor force. Compared to older women and male workers, young single women are the most susceptible to the authority and demands of management. The common manipulation of the "factory as family" metaphor by owners and managers suggests how workers should labor for some greater good on moral grounds rather than for decent compensation. Uneven power relations inside the factory also result in inappropriate demands from management for personal services from women workers, from hair washing to sex (Hsing, 1998, pp. 100–1). (The connection between these activities is that barber shops in the region of rapid development are coded as places of prostitution.) Despite difficult working conditions, manufacturing jobs remain viable choices for many women for whom other means of employment would result in even lower compensation or marginal social status (Davin, 1996a, b). Women migrants who return to home towns also gain social status and may be able to return with enough capital to start small businesses (Davin, 1999; Feng, 2000). On another hand, some rural migrant women who have returned home have also experienced increased spousal abuse, apparently associated with new perspectives on independence and alternative identity positions gained through their work experience (Liu and Chan, 2000). Assembly work, however, no matter the circumstances, does not offer long-term security. The prevailing pattern is that women will leave or be let go from assembly jobs after a few years, especially upon marriage or birth of a first child. Migrant workers especially are likely to be hired on limited term contracts, and supervisory jobs have typically remained the domain of men. These characteristics of the regional working environment are well known in the manufacturing zones and workers regularly seek knowledge about comparative factory conditions in order to avoid the worst of labor–management relations. While the degree to which workers have attempted to better their conditions remains largely undocumented, worker activism is emerging in the export-oriented sector. In the 1990s Guangdong experienced more than 10,000 labor disputes each year, and in 1997 alone 740,000 workdays were lost to strikes (*Xinhua*, 1998e).

The factory compounds and workplaces in the export-oriented sector are the primary places of experience for migrant workers, and the spaces within which women's laboring identities are formed and challenged. Primary pivots of identity formation are geographical markers, in rural and urban identities, provincial and home town origins, and linguistic affiliations. The significance of regionalism in Chinese society plays out in the factory compound, as managers and workers alike distinguish each other on the basis of place of origin and spoken dialect (Ngai, 1999). Women workers regularly affiliate based on dialect and provincial and home town ties, and sometimes act as intermediaries for bringing relatives and new migrant fellow provincials into the workplace. But in practice, this basis for identity differentiation also results in particular discourses of labor–management relations in which supervisors label individual workers by their provinces of origin, and admonish them through negative regional stereotypes. Inter-provincial migrants are referred to in general as *waisheng ren* (people from outside the province), and then by their province or region of origin, such as Guangxi *mei* (Guangxi girl), *beimei* (northern girl), and so forth. The largest-scale geographical identity categories divide workers into "northerners," from north of the Yangzi river, and "southerners," from south of the Yangzi. "Each group constructed the 'other' in derogatory terms: 'northerners' were bumpkins, silly, rude, and miserly; 'southerners' were cunning, promiscuous, dishonest, and spendthrifts" (Lee, 1995, p. 384). Migrants from rural areas are also continually reminded about their "coarse" and "backward" origins, which serves to construct the notion that rural women are less skilled workers, not fully assimilated into the industrial culture of the factory regime (Ngai, 1999, p. 4). Rural migrants are frequently assigned to less skilled, undesirable, and dirty jobs, whereas migrants from urban areas within the coastal provinces tend to be assigned to semi-skilled, relatively clean work (Huang X., 1999, p. 98). Workers from within Guangdong are divided linguistically on the basis of whether they come from Cantonese-speaking areas, Chaozhou-speaking areas in the Shantou region, or Hakka areas. In Shenzhen, people in management positions are typically Cantonese, who tend to favor Cantonese-speaking workers, with whom they have the easiest rapport, for desirable jobs and promotions. In all these ways, regional representations form the basis of identity formation in the emplaced circumstances of the factory workforce, and an essential basis of interpretation about daily life.

Women factory workers are known generally as *dagongmei* (working girls), young women migrants who experience a highly segmented labor market in informal and low-wage employment sectors. The *dagongmei* has become an element of contemporary popular culture in China. New cultures of work, gender, sex, and consumption have formed in the coastal zone around the *dagongmei* idea. Various print media, including magazines and newspapers, have negotiated the *dadongmei* identity, including a magazine simply entitled *Dagongmei*. Stories in popular magazines about the life of *dagongmei* feature the transformative themes migrants encounter, from struggling in the industrial world to changing attitudes toward love, sex, and marriage. In 2000, the release of the film *Durian, Durian*, about the experience of *beimei* working in Hong Kong (where everyone from north of Hong Kong is "northern"), by the director Fruit Chan Kuo, revealed the life of women from China who enter Hong Kong on three-month visas. The specific connotation of *beimei* in Hong Kong is "northern maiden," a euphemism for the prostitute. *Durian, Durian* follows the life of one *beimei* who has earned a small fortune in Hong Kong and returned home to the respect of family and friends, but they do not know about her working activities and she is not ever really comfortable with herself again. These symbolic qualities constructed around the *dagongmei* reflect the patriarchal conditions of the evolving market economy in China under reform, in which new sexualized identities for women are heightened, and which marginalize the importance of their labor contributions to the regional economy. Combined with the negative discourse about the working abilities of rural migrants, these images serve to construct low status for young women workers, which corresponds to low wages and maximum surplus appropriation.

The unusual demographics of the production regime have given rise to new patterns of household formation in south China. Chinese data on marriages show that Fujian, Guangdong, and Shanghai have the highest rates of mainlanders marrying non-mainland spouses, in which the typical union is a mainland woman marrying a Taiwanese or Hong Kong man. For some working-class men from Hong Kong and Taiwan, finding a wife and setting up a household in China has been an opportunity that was otherwise economically unattainable. The rate of mainlander–non-mainlander marriages increased four times over the first fifteen years of reform (Ye and Lin, 1996). Fujian has had a slightly higher rate of such marriages than Guangdong and Shanghai, and fully 85 percent of these marriages took place in Fuzhou, Xiamen, Quanzhou, and Zhangzhou – the cities longest opened under economic reform and with the greatest degree of cross-strait and overseas connections. In a recent innovation on cross-strait marriage schemes, mainland men have been obtaining residency in Taiwan by marrying elderly Taiwanese women. In 1999 over 300 Taiwanese women over the age of 65 married working age mainland men who desired residency rights for employment in Taiwan (Wu, 1999). The men pay the women monthly support based on a portion of their total earnings. Regulations allow a three and a half year stay for mainland spouses of Taiwanese over 65, whereas for the under-65 age group the residency period is only six months. Marriarge brokers arrange the unions, typically during trips women make with tourist and religious pilgrimage groups to visit Mazu sites in Fujian. By the end of the 1990s, the total number of cross-strait marriages of all types was over 50,000. These marriages, from legitimate unions to temporary relationships of convenience, are creating the basis for new types of linkages across the strait.

Informal or "secondary" households have also distinctively formed in the region. Hundreds of thousands of married men from Hong Kong and Taiwan have established households with so-called second wives in the manufacturing zones. The practice became the "the talk of the town" in Hong Kong in the run-up to 1997, especially after reports estimated that 300,000 Hong Kong men had established such households in Guangdong and their China-born children might be able to claim the right of abode in Hong Kong. The regional press reported the notion that peer pressure to take a mistress was so great that men increasingly related to other men with mistresses, so that business dealings for men without "second wives" could be affected as a result. In Hong Kong the male-gendered media message was female blame – blame that wives had lost their feminine powers and that they should make themselves more attractive and become more skillful in pleasing their husbands (Tam, 1996). The problem has led to increased divorce in Hong Kong and Taiwan, where support services have arisen for women whose husbands work in China.

In Taiwan, the practice was sensationalized in *One Country Two Wives* (Qiu and Lin, 1994), a book about the problems of mistress relationships and the rights of legal wives. Figure 6.1 (from the book) represents some of the popular notions about cross-strait relationships between Taiwan businessmen and mainland Chinese women. The images represent patriarchal objectifications of women, and about women's bodies, by depicting the mainland woman as younger, thinner, and more cheerful than her Taiwan counterpart, even child-like. The image of the man appears as if relatively powerful, and his behavior is clearly driving problems for women. The presence of large numbers of young mainland women in Taiwan, largely informal migrants, known as *dalumei* (mainland girls) has also fueled concerns about this problem. The *dalumei* may be a low-wage worker, but stereotypically is thought of in symbolic terms of the prostitute, in negative characteristics of a mainland woman who would seek economic gain through practically any means possible, and especially sex. In the terms of gender politics, ideas and realities about the politically unmediated desires of *dalumei* for Taiwan men provided opportunities for "native patriarchy . . . to further consolidate its arbitrary domination over native women" (Shi, 1999, p. 291). Taiwan's media discourse around the presence of mainland women heightened anxieties not only about gender and marriage relations in Taiwan, but about ideas of nation in an era of cross-strait tensions. One news article claimed that *dalumei* were known to have worked as spies in Chinese coastal cities frequented by Taiwan businessmen, and that some might have been sent to Taiwan to work as agents for the Chinese government (Shi, 1998, p. 300). Such representations treated mainland women as threats not only to the Taiwanese woman, the institution of the family, and the feminist movement, but to the security of the nation.

From perspectives in both Taiwan and Hong Kong, the arrival of the mobile mainland woman, young and single, in the transboundary region ignited patriarchal responses, and interpretations of the women in real and symbolic terms of invaded territory. Patriarchal forces objectified the mainland woman as a laboring resource and a sexual opportunity, in the way that labor is "thingified" under capitalism. In Hong Kong, people feared the invasion of hundreds of thousands of mainland children born to mainland mothers and Hong Kong fathers, and the consequent disruptions in society such a population influx would bring. In Taiwan too, the mainland woman has symbolized the potential to destabilize the family and undermine the coherence of society at large. These perspectives of "othering" mainland women from Hong Kong and Taiwan demonstrate what Louisa Schein (1996) has observed in internal boundary-making that is characteristic of nationalist tensions, and desires to maintain identity differences among different regions and sometimes within the same state. By contrast to earlier regimes in which women represented the civilizational qualities of nation, the mobile mainland

(a)

(b)

Figure 6.1 Cartoons from *One Country, Two Wives.* (a) The sign on the noodle stand to the right reads "Taiwan noodles 30 yuan per bowl"; the sign to the left reads "Mainland noodles 10 yuan per bowl." The male consumer reads the Taiwan sign and remarks, "these are too expensive." He turns to the mainland sign and concludes, "This side is good – cheaper and more." (b) The Taiwan woman stands and fumes while the businessman says to his mainland companion ,"Taiwan is so close and I'm very busy," to which she replies, "Yes, you've been gone too long."
Source: Qiu and Lin (1994).

women engages in the regional cultural economy on her own terms, seeking work, economic gains, and participation in the new commodity culture in China under reform. The idea of the new mobile woman in south China is both a site for independent identity formation and a symbolic reference point around which regional societies negotiate gendered power relations and the contradictions in the status of women in China under reform.

The new service sector: prostitution

Across Asia, industrial zones of assembly production are socially coded as places on the margins of normative society. They are places that exist as something like frontiers within the domestic space economy, and serve as liminal places of disengagement from home bound realities. In Southeast Asia the transboundary regions of the growth triangles contain known centers of prostitution, which operate seemingly beyond the reach of the law, on offshore islands and along international borders (Chan Hwee Leng, 1996). In China under reform, Shenzhen early emerged as a center of prostitution. The official Chinese press has blamed Hong Kong and Macao as sources of this "social disease," and the open coastal cities have been centers especially of prostitutes who have migrated from other provinces. Based on arrest records, over 90 percent of prostitutes in Shenzhen, Guangzhou, Xiamen, and Hainan were from outside provinces (Dong, 1996).[15]

Contemporary representations of prostitution in south China are also emblematic of contradictory perspectives on the position of prostitutes in Chinese society. The degree to which *waisheng* (from outside the province) prostitutes are held responsible for prostitution in general parallels anxieties about the mobile mainland woman in Hong Kong and Taiwan. The local state in south China always appears to construct prostitution as a geographically "othered" problem (see Schein, 1996). Based on contemporary narratives in the popular print media, Virgil Ho (1998) concludes that Guangdong authorities systematically blame *waisheng* prostitutes for illicit activity, while the apparently more occasional prostitute from Guangzhou receives a sympathetic response. The configuration of place of origin in prostitution also appears in historical examples. The first riot associated with the Siming Gongsuo cemetery debate in Shanghai in 1874 evolved from an incident of streetside verbal harassment when a group of Ningbo men yelled out at a Guangdong prostitute for serving French clients (Goodman, 1995a, pp. 391–2). In 1880, in an official communiqué to British Consul Giles on local social problems, the Daotai presiding over Xiamen wrote that "the prostitutes are

Cantonese who come up from Hong Kong."[16] In an example that invokes the dichotomy between land and sea and the virtues of landed society in the Chinese geographical imagination, British Consul Lay wrote in a despatch of 1844 about the places of prostitution in Fuzhou on riverside "pleasure boats." He described how families sold young women into such circumstances and how they worked to make their escape:

> having laid up a small peculium, the result of presents made by visitors, they buy a husband and with his assistance make their escape. Such elopements are often the subject of dramatic representation at Foo-chow-fu and are said to be not infrequent in real life. The frequency seems to be indicated by the fact that the phrase "to go on shore" has acquired by use the additional sense of turning from a life of vice to one of virtue and sobriety.[17]

In historic Fuzhou too, the sites of prostitution occupied marginal space, here on the river shore, and represented the tensions surrounding economic and sexual practices other to society.

Prostitution in contemporary China may be understood as an activity through which people contest gender values, engage in forms of social experimentation, and work within the high stakes economy generated under reform. Gail Hershatter (1997, p. 333), in her monumental study of prostitution, described how "The most striking feature of prostitution in late twentieth-century China was the proliferation of venues, prices, and migration patterns." Hershatter's discussion speaks to the diversity of participants in China's new prostitution, from migrant peasants to moonlighting university-educated professionals. Because working as a prostitute earns more money than the great bulk of jobs in the foreign export sector, many of the Chinese analyses of prostitution point to prostitutes' rational decision-making:

> Women often used the language of the market in speaking about their decision to take up sex work. Their capital, they said, was their youth and beauty, and, recognizing these as perishable assets, they intended to use them to accumulate more durable capital. Some planned to buy a taxi license, hire drivers, and secure a stable income that way; higher-class prostitutes set their sights on the purchase of a passport or a building. (Hershatter, 1997, p. 354)

From this discussion, women know they appropriate their own surplus value, rather than give it over to the capitalist employer. Of course women's decisions to take up prostitution reflect the problems of unemployment and especially diminished employment opportunities for women.

The state, though, has criminalized prostitution and has not addressed the economic conditions driving the problem. From 1981 to 1991 the official number of police detentions of people involved in prostitution (prostitutes, customers, and pimps) was 580,000, which must reflect only a fraction of total activity (Hershatter, 1997, pp. 334, 400). In the year 1996 alone, police arrested 420,000 prostitutes and their clients, a number officially estimated at 10 percent of the total (*Xinhua*, 1999b). The widespread resurgence of female prostitution in China under reform may represent the ultimate form of male bias in the development process.

Redirections

Industrialization is generally held up as the path to social and economic modernization, but where gender is concerned, the institutionalized cultures of development, and the historic models on which development is based, represent forms of male bias. Higher economic growth has been obtained in China under reform, but often at the expense of gender equity. Under reform, Chinese society has experienced the clashing forces of rapid economic restructuring, the arrival of globalized cultural and economic practices, and the resurgence of traditional cultural values, often in new forms. Reform has offered women new options, but women are also facing more discrimination in some sectors. From the resurgence of traditional social practices to new options for women to leave difficult rural circumstances by migrating to more prosperous regions, women face both the revival of traditional forms of patriarchal control and relatively uncharted opportunities.

The gendered conditions of economic development in south China represent the contradictions of reconstituted forms of patriarchal power and control, and new spaces for women's participation in society and economy. In 1998, though, the Ministry of Civil Affairs instituted new regulations controlling local social organizations, which required non-state women's organizations to come under the supervision of the Women's Federation, and forced some to close, while diminishing the formation of new ones (Howell, 2000, p. 361). Despite the state's attempts to shore up the role of the Women's Federation, increased mobility for women and the rise of regional economies have made the state's national organization an unrealistic forum for the representation of women's issues in contemporary Chinese society. The history of creating national symbolisms around women, from the dynastic era through the period of the communist planned economy, has encountered a disjuncture in contemporary times; the messages of the Party have lost relevance and internationalized cosmopolitan cultures about women's possibilities have become established in major cities, and widely accessible through diverse media. The mobile mainland woman in south China may have become an object of anxiety in Hong Kong and Taiwan, but migrant women are carving out new gendered spaces and alternative lifepaths that can only bring changes in society and economy. Women in the transboundary regional economy are experiencing new places of identity formation where they are able to live beyond the real boundaries of home towns and in some ways beyond the patriarchal strictures of the state itself.

NOTES

1 For example, in 1999 the first "people's condom" became available in China. Local authorities in Shanghai and Beijing announced the installation of condom vending machines in public bathrooms. Previously condoms had only been available through the state's standard birth control program, and as a result were not easily available to single people or promoted directly with men in general.

2 This chapter limits detailed discussion to three of these reforms, the one-child policy, the rural household responsibility system, and the open door policy and export-oriented industrialization regime.

3 Jinhua Emma Teng (1996), has suggested that Western theoretical meanings of gender, as expressed in Joan Scott's work, are too different from Chinese understandings of *xingbie* to serve as a primary mode of analysis for feminist issues in China, but the relevance of this distinction for humanist scholarship is not similarly applicable to critical economic analyses of patriarchal economic organization.

4 The United Nations has formulated the Gender-related Development Index (GDI) based on the major UN indicator of human development, the human development index (HDI). The GDI incorporates inequalities in living standards between men and women, based on categories of income, life expectancy, adult literacy, education, and other factors. Industrialized countries are the top ranking countries in the GDI, just as they are in the HDI. In the GDI list of 130 countries, China ranks 72; United Nations (1995).

5 Various authors, and especially Martin King Whyte (2000), have argued that there are insufficient data to accurately determine a baseline from which to measure the status of women in contemporary China.

6 Some of the terminology itself, especially the "big bang," invites a critical discursive analysis, for its obvious metaphorical potential to invoke a singular, momentous experience from which some substantial results would ensue and without responsible monitoring and follow-up. See Carol Cohn (1987) for the ways in which the use of sexual metaphors transfers male bias in the discursive sphere of strategic defense planning.

7 Henry Yeung has compared theoretical perspectives on the geography of business organizations and production systems and concluded that analysis of network relations in and between firms captures the cultural specificity of regional economic geographies. See the edited collection by Henry Yeung and Kris Olds (2000).

8 The assumption of trust relations derives from the discourse of scholarship on Chinese business networks and "oriental firms." For a critique of the related "Confucian values" thesis in the context of the Chinese family business enterprise see Susan Greenhalgh (1994).

9 In Elson's formulation, the recognition of male bias is not based on essentialist or biological categories, i.e. that all men would be biased against all women, or that all men are necessarily benefiting from reform, but on the development process tending to favor the accumulation of male-gendered power, privilege, and wealth. The fundamental contradiction about male bias is that it likely reduces overall productivity: denying women equal resource access and compensation lowers women's productivity and overall productivity by comparison to an economy free of gendered distortion in wages and capital distribution; see Elson (1991 pp. 6–7.).

10 Jasper Becker (1999) reports that "the one-child policy started as the brainchild of the Soviet-trained rocket scientist Song Jian, now a state controller in charge of science, who used the mathematical formulas used for calculating missile trajectories to make population projections." Apparently Song made a significant impression because it was the first time someone had used computer science technology to make a major state policy presentation. Criticisms of Maoist era policies charged that they had been unscientific, and so the emergence of scientific practice in itself in China under reform commanded notice.

11 Interview, Guangzhou, 1994. By 1998–9, discussion emerged about the end to the one-child policy, partly in response to the new social problems it has created. Different scenarios have emerged. At the national scale, the state may reconsider the one-child policy in 2010 when the population is predicted to reach a birth rate of 1 percent. A selective alternative to the one-child policy is "informed choice," which would allow people to have their first child without the official state permission required under the one-child policy; see Rosenthal (1998). Its proposed implementation is limited to major urban areas, and Shanghai and the leading cities of the Yangzi delta, where smaller families have become a matter of choice, are among the first places slated to implement the new policy. More concrete decisions will likely emerge with the results of the 2000 census, through which the state has sought to document previously unregistered births. Unregistered children born outside the state plan, and especially females, may number as high as tens of millions and cannot attend state schools; see Yabuki (1995, p. 15).

12 In addition, one of the state's leading measures against the illegal trade, the "six evils" campaign, grouped the illegal sale of women and children with a list of five other illegal activities, namely prostitution, gambling, producing and selling pornographic materials, and drug dealing. By comparison to

these crimes, the state has minimized the seriousness of trafficking in people. See the discussion by Sarah Biddulph and Sandy Cook (1999).

13 This is the most significant transformation in the gendered division of labor under reform. According to Francesca Bray (1997, p. 5), women in imperial China – unlike in Japan or Southeast Asia – did not generally engage in farming labor.

14 The suggested donations were 300 yuan (US$36) for one girl for one year, 20,000 yuan (US$2,500) for a full class of 50 girls above the fourth grade level, or 500,000 yuan (US$60,000) to be used more generally within a specific school.

15 No doubt the high percentages reflect in part the knowledge base of local prostitutes and their ability to avoid arrest. See also data reported by Gail Hershatter (1997, pp. 348–9).

16 FO 228/645 Amoy, 18 Nov. 1880, to Wade/from Giles.

17 FO 228/41 Fuzhou, 13 Dec. 1944, to Davis/from Lay.

7

Zone Fever

> Monopoly power over the use of land – implied by the very condition of landownership – can never be entirely stripped of its monopolistic aspects, because land is variegated in terms of its qualities of fertility, location, etc. Such monopoly power creates all kinds of opportunities for the appropriation of rent which do not arise in the case of other kinds of financial asset except under special circumstances. Monopoly control can arise in any sector, of course, but it is a chronic and unavoidable aspect which inevitably infects the circulation of interest-bearing capital through land purchase. The "insane forms" of speculation and the "height of distortion" achieved within the credit system stand, therefore, to be greatly magnified in the case of speculation in future rents. The integration of landownership within the circulation of interest-bearing capital may open up the land to the free flow of capital, but it also opens it up to the full play of the contradictions of capitalism. That is does so in a context characterized by appropriation and monopoly control guarantees that the problem of land speculation will acquire deep significance within the overall unstable dynamic of capitalism.
> (David Harvey, 1982, pp. 348–9)

In 1993, James Riady of the transnational conglomerate Lippo Group, headquartered in Indonesia, announced plans to invest US$10 billion in Fujian province, including development of a Mazu tourism theme park on Meizhou Island (Poh, 1993). Meizhou is part of Putian county, the legendary birthplace of Mazu, patron goddess of the south China coast. Thousands of people from Taiwan visit Mazu sites in Fujian every year, and the Riady family lineage is tied to Putian county. The Lippo Group identified the opportunity to develop Meizhou in advance of future predictions about direct travel between Taiwan and the mainland. Indeed, in January 2000 Taiwan moved to gradually relax restrictions on direct trade and transportation between Taiwan and the mainland with the opening of an indirect ferry passenger service from Taizhong to Meizhou via Hong Kong, followed by the "Mini-three-Links" policy, in the form of shipping and passenger service between Taiwan's offshore islands of Jinmen and Mazu, and Xiamen and Fuzhou, respectively. Taiwan's Mainland Affairs Council "stressed that the recent decision to allow the

Taichung–Meizhou line was instituted to ensure easier and safer transportation for Taiwan residents undertaking pilgrimages to Meizhou, the acknowledged birthplace of Matsu, the Chinese Goddess of the Sea who enjoys a wide following on Taiwan" (Marble and Wu, 2000). In south China the intersection of such otherwise apparently historically disparate events – shifting government policy across the Taiwan strait, industrial development associated with investment in a Mazu theme park, the transhistorical significance of the cult of Tianhou, and family origins in southern Fujian – invokes questions about transhistorical place relations intersecting with new cultural–economic formations. Such complexities of economic activity and their cultural contexts in the transboundary region recall Richard Peet's (1997, p. 38) observations, which underscore the difficulties of producing whole accounts of such complex geographies: "Thus economic forms are culturally created, economically maintained, interact with subjective processes of agents' identity formation, and to a degree are self-generating or at least self-maintaining. No wonder theorists have avoided the topic!"

Land development, at the nexus of issues concerning high growth development, transnational business networks, "megadevelopment" projects, human impacts on the environment, arable land conversion, and food security, is one of the most significant economic and natural resource issues China faces. This chapter examines the conditions of land development in south China under reform to reveal how rampant land development strained the national economy and bankrupted several major investment institutions in south China in the 1990s. Land development intensified in the early 1990s after Deng Xiaoping's 1992 southern tour, which coincided with a high point in property values in Hong Kong that sent Hong Kong investors across the border seeking development opportunities in China (Taylor, 1991; Sender, 1992, 1993). China's market in land use rights was pioneered in 1987 in Shenzhen, and the SEZs became the first sites of real estate development activity. The intersections of the state's continued promotion of rapid development and a geographical system of privileged development policies in the special zone model, with the evolution of a market in land use rights and an influx of real estate capital, made "real estate fever" and "zone fever" the mantras of rapid accumulation in the middle of the 1990s. During this period, the service sector industries of real estate and construction emerged as the most profitable economic sectors in south China, replacing manufacturing as favored investments for speculative development.[1]

While land development fever raged unchecked, an alternative anti-development discourse about China's "arable land problem" emerged in both domestic and international arenas. Excessive land development in the coastal zone and widespread conversion of arable land for industrial development engendered critical debate about impacts of reform on the agricultural resource base, and the country's future food security. In response, the state initiated campaigns against "irrational" land use conversions, situated in the context of concern for farmers and the material and ideological roles of agriculture at the foundation of Chinese society. The contradictions between policies exhorting continued rapid development and campaigns designed to conserve arable land were at least temporarily resolved by the events of the regional economic downturn: from Thailand to Japan, over-expansion in the real estate sector was a major factor of regional economic destabilization during the period 1997–9. In this context, China moved to control real estate speculation and implemented new controls on land development through revisions of the Land Administration Law. The state's initial campaign against irrational arable land conversions began in 1997, was extended in 1998, and took concrete form in the revised Land Administration Law, implemented in 1999. The new law has sought to protect arable land by regaining administrative controls over land development. Although the state's public discourse surrounding the implementation of the law focused on arable land conservation, the law has appeared designed to rein in real estate development as much as it serves to protect arable land. In the land law, the "arable land problem" has served as a partial discursive front for new controls over problems in the reform economy, which the state has kept minimally revealed.

The Proliferation of Special Development Zones

In the 1990s, *kaifaqu re* (development zone fever) and *fangdichan re* (real estate fever) swept the coastal zone and captured the economic imagination of local level administrations. Cities, counties, and towns all seemed to want their own special zones. After all, the state had sanctioned special zones as the geographical basis of the reform economy, the focus of industrial development, and centers of reform leadership. Zones had become emblematic of nationalist reform thinking. This focus on development zones, combined with their economic success, established special zones as models of industrial development, despite restrictions on excessive land conversions. The state's manner of continuing to sanction special zones, from the first four SEZs, to the declaration of provincial and SEZ status for Hainan Island in 1988, followed by the Pudong New Area in Shanghai in 1990, and then the extension of SEZ privileges to the China–Singapore Suzhou Industrial Park in 1994, established the land use pattern of using large-scale special zones for industrial development. The state allowed SEZs numerous advantages to promote rapid development, and then patterned other types of development zones after the success of the SEZ concept.

Through the 1980s state planners continued to promote a special zone strategy for major cities in a growing system of economic and technological development zones, high technology development zones or science parks, bonded zones or free trade zones, border region economic cooperative zones, and state tourist vacation zones (Gupta, 1996; Yang, 1997).[2] In 1995, there were 422 zones approved by the central government (Guo and Li, 1995). In addition to the establishment of various types of zones, many zones also enlarged in size through the second decade of reform, thereby extending special zone privileges into surrounding areas. The growth of the Xiamen SEZ illustrates the expansionary characteristics of the SEZ concept in practice. The Xiamen SEZ began in 1981 as an industrial district called Huli on the island of Xiamen. The Huli import and export processing district was originally a small area, just 2.5 km^2, but after Deng Xiaoping's 1984 tour of the SEZs, the state enlarged the SEZ to all of Xiamen island (Guo, 1995, pp. 6–11). In July 1987 the State Council strengthened measures to give preferential investment policies to Taiwan compatriots. Between 1989 and 1992, the State Council took further steps to promote Taiwan investment in the Xiamen area and opened three towns on the mainland across the causeway from Xiamen Island – Xinglin, Haichang, and Jimei – for the exclusive benefit of Taiwan investors. In a 1995 speech promoting reunification of the motherland, Jiang Zemin termed the Xiamen SEZ "the window and the bridge in relations across the Taiwan strait" (Guo, 1995, p. 1). In their political and economic functions, special zones became the focused sites of preferential development and geographical symbols of nationalist reform ideology (Crane, 1996).

The unplanned result of the SEZ model was widespread copying of the special zone concept. As a nationally legitimated strategy, implementation of SEZs became normative practice as places all over China at every level of administrative jurisdiction implemented their own special zones, often without required land use permissions from higher authorities. National estimates of the extent of special zone development indicate that the one time experiment transformed into an unplanned and uncontrolled land development regime. Available data on special zones differ and cannot be corroborated with precision, but existing figures are startling. An analysis by Anthony Yeh and Fulong Wu (1996, p. 345) enumerated 1,874 zones in 1990, and found that by 1992, the total extent of land area given over to zones had reached 15,000 km^2 – which exceeded the total preexisting urbanized area of the country.[3] He

Qinglian's (1997, p. 75) critique of China's reform economy cited National Construction Bureau statistics and repeated the 15,000 km^2 figure for total development zone area, but more than doubled the count of development zones generally, to over 6,000 by March 1993.[4] A Fujian province special investigation on the problems of conserving arable land cited 1993 Land Administration Bureau statistics to report that among 2,800 development zones nationwide, only 27 percent were national or provincial level zones (*BHGD*, 1997, p. 20). Based on these figures, three-quarters of industrial development zones were projects undertaken below the provincial level. Large numbers of low-level zones proliferated under the evolving market in land use rights even as they were often implemented without higher-level administrative approval. A 1993 report on the growth of local-level zones in Fujian province recorded 389 development zones at the township and village level, a figure 3.5 times higher than the count for the previous year (Zhao, 1994, p. 54). In the Su'nan area of Jiangsu, 275 out of a total 389, or 70 percent, of settlements at the village and township level had their own *xiaoqu* (small zones) by the middle of 1993 (Zhu and Sun, 1994, p. 266).

The SEZs were also the places for the first attempts at land marketization. The first transaction of a land use right occurred in Shenzhen in 1987, but the land market evolved haphazardly (J. Zhu, 1994). The municipality lacked the resources to prepare land and provide the infrastructure for industrial development, and became reliant on developers to engage in basic land grading, roadway construction, and more. In this way, developers gained control over the processes of land development, and profited substantially from land speculation in an environment of rapidly increasing rents. Xiamen began to engage in land lease sales in 1988, and sold the first two leases in the 480–830 yuan per m^2 range; four years later the average m^2 cost was 8,000 yuan (*Xinhua*, 1993b). Following on the Shenzhen and larger south China experience, land development became widely understood as a highly profitable venture in China at all scales. But rather than there being an evolution of a market in land use rights, land use rights were more often transferred by informal negotiation. A true market in land use leases never evolved (Huang and Yang, 1996; Wong and Zhao, 1999). In the negotiated price method, land lease costs are lower than would be expected at market rates. In addition, formal means of land apportionment more reliably send land lease revenue to the center, whereas informal negotiation of land leasing tends to profit the local state. Thus the anticipated market in land use rights did not evolve along the lines of market rationality under reform, and the local state guided reform toward its own financial ends.

"Model" development

The SEZ model of large-scale land development culminated in the China–Singapore Suzhou Industrial Park, China's largest scale joint venture development project (maps 7.1 and 7.2). The China–Singapore Suzhou Industrial Park (SIP) was planned for a 70 km^2 tract of prime agricultural land in the eastern suburbs of Suzhou and was granted SEZ privileges by the State Council in 1994. The SIP was also planned as an international model development project, specially sanctioned by Deng Xiaoping (Cartier, 1995). In 1978, on the eve of economic reform in China, Deng Xiaoping visited Singapore and discovered there the goals of his reform vision, "good public order" (*Selected Works of Deng Xiaoping*, 1984, p. 366) in polished capitalist modernity. China's leaders saw in Singapore the results of an economic development plan that reflected China's priorities in a combination of market economics, FDI-based export-oriented industrialization, and state limits on the evolution of civil society.[5] During his historic southern tour of the special economic zones in 1992, with Singapore apparently on his mind, Deng publicly remarked how China could learn from Singapore – and "surpass it" (*Selected Works of Deng Xiaoping*, 1984, p. 366). This was a key statement of consequent timing. In the second decade of reform, at a time of some economic stagnation and temporary economic uncertainty, Deng sanctioned further rapid growth and Singapore as the model of successful modernization.

Planning the SIP was conducted at the highest level between China and Singapore. From 1992, Senior Minister Lee of Singapore had been shuttling back and forth to China in negotiations over the SIP project.[6] These visits culminated in 1994 with the signing of a joint agreement between China and Singapore, in which the two countries pledged national support for the project. Chinese Premier Li Peng, Vice Premier Li Lanqing, Senior Minister Lee, and Singapore Prime Minister Goh Chok Tong met in Beijing to conclude the agreement, which lent the SIP project the ceremonious attention of high-level diplomatic negotiations. A project of this size could only be approved at the national level – the 70 km^2 site far exceeded the provincial limits of land use conversion approvals. Its scale and scope indicates close working connections between Suzhou city and Jiangsu province officials, the central government, and Singaporean authorities. The project was an object of pride for both countries, and especially for Singapore, whose government-linked urban and industrial planning bureau created an explicit model of the residential industrial estates in Singapore for the eastern flank of old Suzhou, one of China's most prominent and historically significant

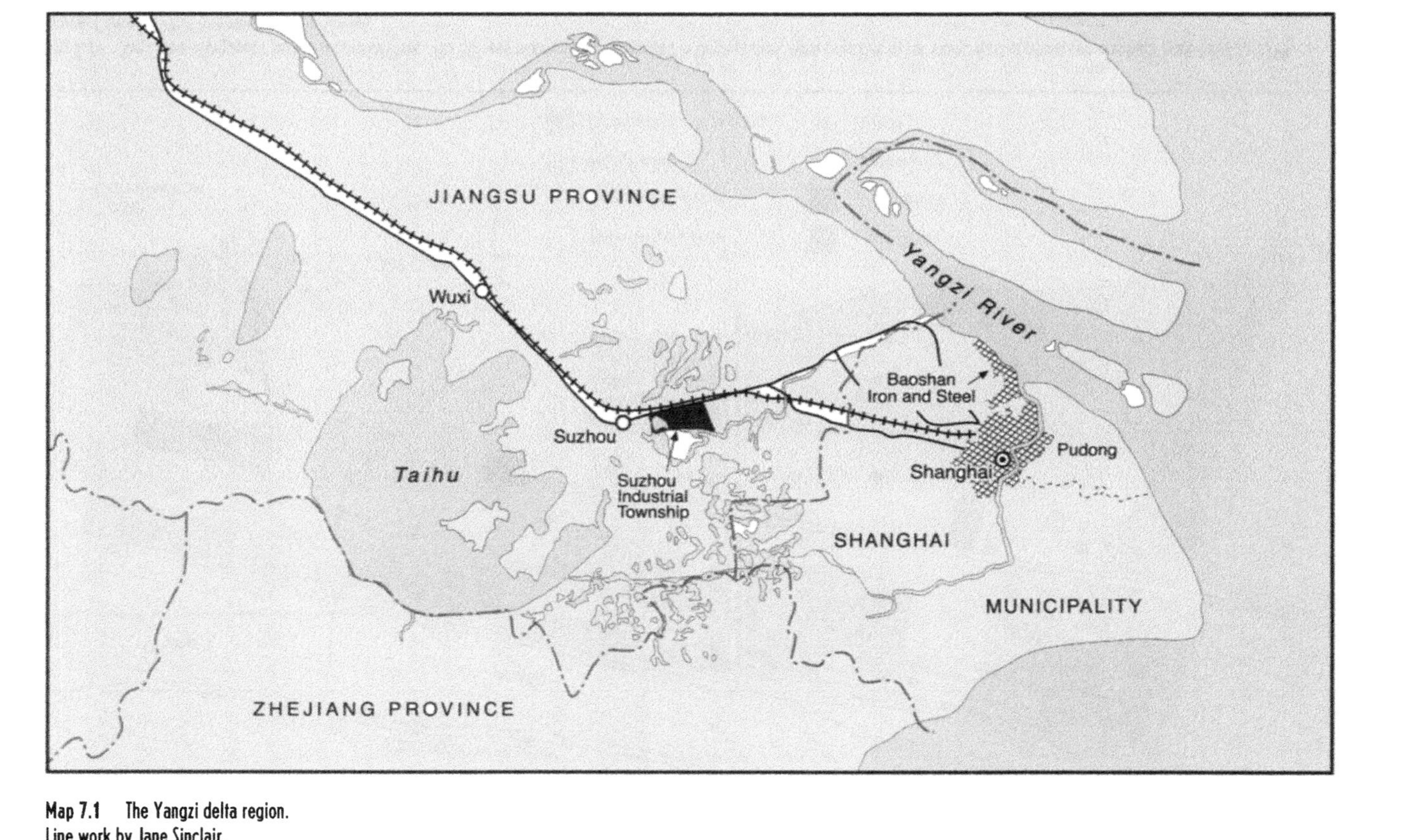

Map 7.1 The Yangzi delta region.
Line work by Jane Sinclair.

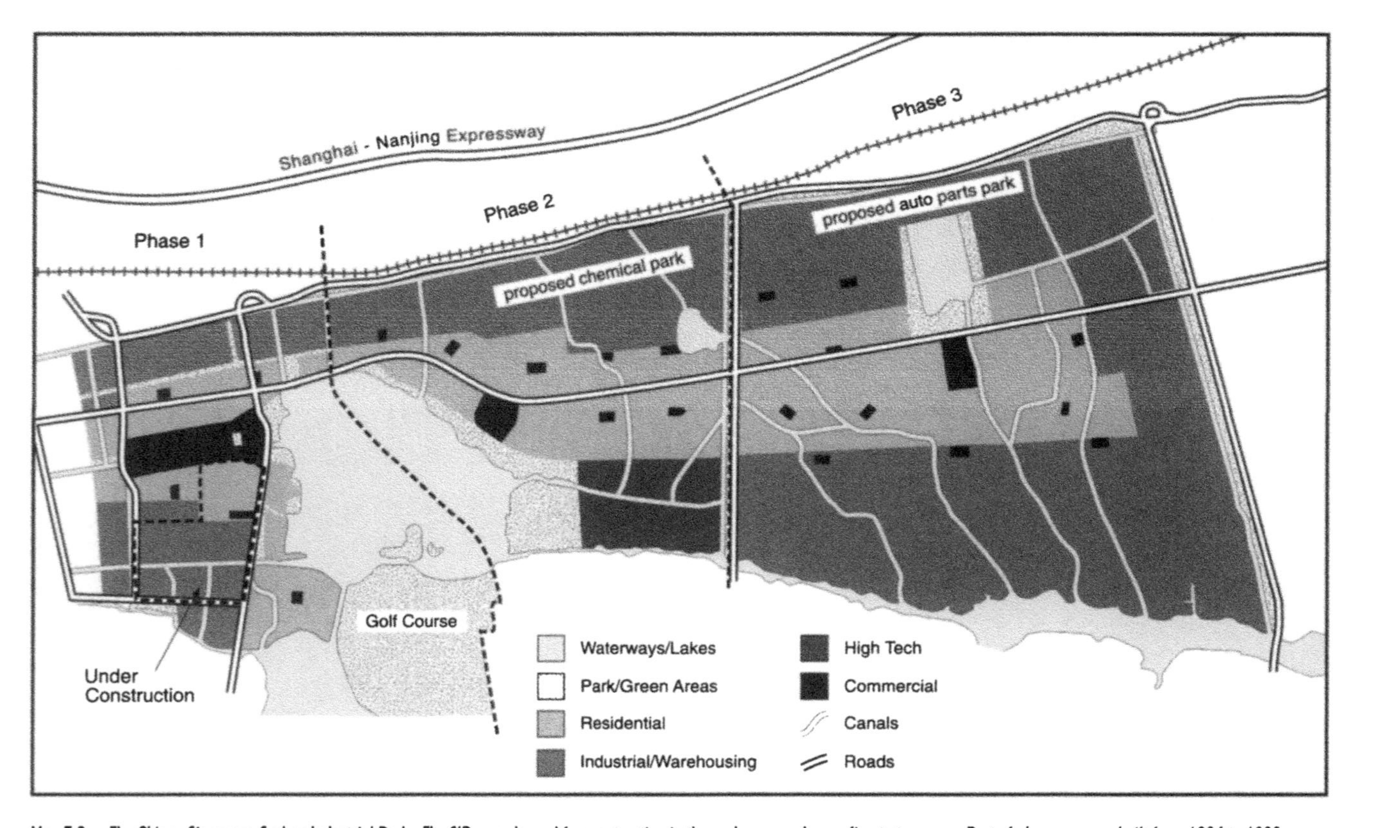

Map 7.2 The China–Singapore Suzhou Industrial Park. The SIP was planned for construction in three phases, each over five to ten years. Part of phase one was built from 1994 to 1999. Source: Cartier (1995); line work by Jane Sinclair.

cities. China's interest in installing a Singapore-style industrial park in the Su'nan region of Jiangsu province may have reflected its considerations of domestic regional development models. Among several domestic models, including the "Zhujiang model," "Minnan model," and so forth, the "Su'nan model" has received constant accolades from state planners for its high degree of collective ownership at the local level in township and village enterprises, minimal reliance on foreign capital and markets, and some of the highest overall growth rates in the country (Xu, et al., 1993; Zhu and Sun, 1994). But high growth in the delta could not depend on continued growth in township and village and enterprises, and after a decade of prioritizing Guangdong and Fujian for foreign-invested development, the state shifted priority to the Yangzi delta in the 1990s, first by establishing the Pudong New Area. In 1994 in Suzhou, the Su'nan and Singapore models together converged in a constellation of Chinese state planning goals.

Within months after the project broke ground, the SIP planning and administrative committee gained management autonomy in the park and received permission to approve investment projects valued up to US$50 million, which exceeded the level of provincial approvals (Lee, 1994). The SIP had more autonomy over negotiating and planning individual development projects than larger state territorial entities. The original investment consortium for the SIP from the Singapore side was the Singapore–Suzhou Township Development (SSTD), whose composition demonstrates transnational finance strategies in regional business networks, and alliances between the state and private capital interests. The SSTD held 65 percent of the project ownership, and in its initial configuration was made up of 19 members, including Singapore government-linked corporations and companies controlled by the Salim Group of Indonesia. The SSTD included Keppel, the Singapore shipping and engineering giant and lead member of the consortium, and United Industrial Corporation and Singapore Land, over both of which the Salim Group, the largest transnational corporation in Indonesia, has held a controlling stake. In the literature of Chinese overseas business networks, Salim Group is also the largest Chinese overseas business group in the world (Cai, 1995). In popular media and business lore, the founder of the Salim Group, Sudono Salim, whose *qiaoxiang* is Niuzhai (Cowshed) village in Putian county, Fujian, is "the richest Chinese in Southeast Asia" (*Asiaweek*, 1994). The combined participation of the Salim Group and Singapore corporations, led by Singapore state officials, lent special prominence to the international finance consortium, which contributed to national-level special zone privileges for the SIP in Suzhou.

Despite its considerable resources, national support, and expert planning, by 1999, after developing the first 8 km^2, the SIP faced various obstacles and the Singapore side of the consortium reconsidered its commitment to the project (Chan, 1999). The SSTD announced that it would remain involved in development of the site, but would scale back its participation and shift the majority of the financial responsibility to the Chinese side. The SIP likely will not be developed to the full extent of the original 70 km^2 plan – an enormous site, three times the size of the original city of Suzhou. Its ultimate fate remains unclear, while the timing of the restructuring coincided with the implementation of the revised land law.

The Land Conversion Moratorium

The problems generated by the spate of large-scale land development projects, especially after 1994, just after the 1992–4 peak years of arable land conversions, registered in serious national debate over land development. In May 1997 the State Council and the Central Committee of the Communist Party issued a joint notice promoting strict land use measures, including a one-year moratorium on arable land conversions (*Xinhua*, 1997a). A high-level briefing on diminishing agricultural land, based on comparative satellite imagery, likely prompted state leaders to take this unprecedented step.[7] In April 1998 the ban on non-agricultural use of farmland was extended for another year to 1999 (*China Daily*, 1998b). The state renewed the ban in response to continued difficulties with unapproved land use conversions, and moved to strengthen existing land use regulations by reviewing revisions of the Land Administration Law during the Ninth National People's Congress (*China Daily*, 1998c). Simultaneously, reorganization of the national government resulted in reshuffling of state ministries, which included the creation of the Ministry of Land and Resources from the combined offices of the former Ministry of Geology and Mineral Resources, the state Land Administration Bureau, the National Bureau of Oceanography, and the State Bureau of Surveying and Mapping. Thus, importantly, the former Land Administration Bureau was upgraded to ministerial status. The Ministry of Land and Resources will maintain a law enforcement supervision office, whose mission is to enforce the revised Land Administration Law. The revision of the land law was published in newspapers for public notice and response, which the state called "a leap forward in democratic legislation" (*Xinhua*, 1998d) because it marked the first time a law was drafted to incorporate public opinion. Clearly, the state was interested to generate public support for the law and its new system of controls.

The revised Land Administration Law, which took effect from January 1, 1999, covers land use rights, land use plans, and farmland protection (*Xinhua*, 1998a). The main goal of the revised law appears to be nothing less than the complete restructuring of the central focus of the land management regime: from facilitating land use transformations to controlling the purposes of land use. During the decade 1987–97, from the time of the first sale of land use rights to the moratorium period, land use policy structures, including the land law, focused on gaining rights to land use and how to achieve land use transformations. The central feature of the original system was a hierarchical approval system by administrative level, in which governments could approve land use transformations in a scaled system by size: the higher the administrative level, the larger the piece of land that level could approve for land use rights (see Keng, 1996). The structural focus of the new law is on the nature and purpose of land use by categorizing land in three main groups: agricultural, construction-related, and unused. This scheme shifts focus to current land use categories, and the maintenance of those categories, by emphasizing the use of land management plans. The revisions also signal an end to independent decision-making power over land use conversions at lower levels of administration by restoring higher-level checks on land use approval rights at the level of the provincial and central governments.

The revised Land Administration Law can be seen as the culmination of a series of policy measures and laws promulgated over the past decade to address chaotic land development and the arable land problem. From the late 1980s, when a market in land use rights first appeared in select urban areas, the land management regime has encompassed the conflicting needs of land development, land management, and arable land conservation. During this era, China's coastal zone experienced the highest economic growth rates in the world, the state introduced a complicated and variable socialist market economy in land rights, and the country faced problems of diminishing arable land. When the revised land law was published for public notice, Qu Geping, chairman of the Environmental and Resources Protection Committee of the NPC, characterized the effort as being to "draft the strictest land law in the world" (*China Business Information Network*, 1998). An editorial in *Renmin ribao* (*People's Daily*), which appeared with the onset of the original moratorium period, raised a broad range of concerns, and stated, "The current problem is, despite the grim situation about the protection of arable land, many people still lack a sense of crisis" (*Xinhua*, 1998c). It recommended solutions for increasing arable land, including reclaiming wasteland (which includes wetland, sand, eroded land, or alkaline soil) and abandoned land, and intensifying land use. Proposed land use intensification plans have involved settlement redistribution, such as a village and township restructuring plan for Jiangsu province in which more than 280,000 natural villages would be merged into over 50,000 villages of larger sizes to recoup arable land.

With the onset of the May 14, 1997 moratorium, the State Council called for a nationwide accounting of land use conversions. By end October 1997, all levels of government down to the township were required to have conducted an assessment of land use conversions under their jurisdictions since 1991. Governments were exhorted to: stop development zones unauthorized by the State Council or provincial government; reevaluate rational use of large-scale land areas taken for commercial real estate projects, especially golf courses and luxury villas; investigate possible illegal land transfers and problems of land speculation; remove brick kilns from arable land; and stop haphazard use of land by village and township enterprises, especially for temples and ancestral halls (*Xinhua*, 1997a). The report also sought to arrest provinces from selling land leases in military zones and large tracts of land for overseas investors. Entire islands, the state stressed, cannot be sold to foreign investors. The document did not name the island, but this stricture likely implicated the Lippo Group's "gigantic investment scheme" for Meizhou Island, centered on a "world class tourism resort" (*Xinhua*, 1995b) in association with a Mazu theme park.

Large-scale development projects in the south China coastal zone have resulted in widespread land use transformations and have contributed to the speed of rapid industrialization under reform. But in the context of this "industrial compression" (Wade, 1990, p. 42), the concentration of land use transformations in the coastal open areas raises a set of questions about intensive development in already densely populated areas, impacts on the natural environment, the problems of uneven development in China's larger space economy, and the real financial risks of widespread overextension in the real estate sector. Beginning in 1987, the state promoted a series of regulations and laws attempting to manage land use conversions. Yet analysis of the institutional contexts and realities of land use transformation demonstrates contradictory state policies in the evolving land use regime. The state's own definitions of the extent of the arable land problem complicated the issues.

Rapid Development and the "Arable Land Problem"

The moratorium on land use conversions and the revisions of the Land Administration Law appear to have responded to China's arable land problem. Through the middle of the 1990s, official reports and media analyses, both domestic and international, identified China's decreasing cultivated land per capita as a serious resource management problem. The publication of *Who Will Feed China?*, by Lester Brown (1995), especially galvanized popular concern in the USA. Brown argued that China's rapid industrialization and consequent conversion of arable land would generate demand for so much imported grain that China's future food consumption would lead to sharp rises in international grain prices, and destabilize the world food economy. Among a range of serious problems with the analysis, this neo-Malthusian "crisis" perspective turned out to be based in faulty though "official" data. The problems of measuring China's arable land are well known among China specialists.

Analysis of figures on arable land published in the *China Statistical Yearbook* has raised questions about the accuracy of government statistics and surveys (Smil, 1993, 1995, 1997; Heilig, 1997). Statistics from the early reform period measured China's arable land total at 95–96 Mha, but more recent estimates put the total in the range of 130–140 Mha (Ash and Edmonds, 1998). On this basis, China's arable land is 45 percent larger than the national total historically repeated by official statistical materials. Chinese scientists have also recorded and published these larger figures in the 130–140 Mha range (Sun, 1994; Wu and Guo, 1994). State policy changes in China's land taxation system, and the fact that the *mu* unit was not historically standardized, have contributed to the problems in statistical analysis. On this basis, some researchers take the position that China's arable land problem has been exaggerated, and that China's total cultivated land may be as large as 160 Mha (Smil, 1999).

While it is now generally recognized that official statistics on arable land losses differ and have been underreported during the reform era, all major sources confirm extensive transformations of arable land, especially in the coastal zone. Whether or not China has an arable land problem, the amount of arable land lost to construction has been significant. Recent research by Anthony Yeh and Xia Li (1997) utilizing remote sensing data from Dongguan county in Guangzhou concluded that the loss of agricultural land there has been underreported by as much as 61.3 percent. What can be agreed on is that extensive areas of vegetable land in the suburbs and urban periphery have succumbed to industrial development, and that the excesses of land use conversions during the 1980s and the first half of the 1990s have been politically and economically chaotic. From the widely publicized corrupt investment schemes of prominent officials to judgments by scholarly boards that Guangdong province has excess airport capacity as a result of five different cities rushing to build major airports (where the maximum distance between them is less than 250 km), it has become widely known that land and infrastructure development has not evolved along theoretical lines of market rationality under reform.[8]

These issues register acutely where historically low arable land per capita and widespread development have coincided in the coastal zone. During the first half of the 1990s per capita cultivated land in Fujian province dropped from 0.041 to 0.038 ha, making it the province with the overall lowest per capita arable land ratio (*Xinhua*, 1996). In 1998 the director of the Fujian Land Administration Bureau revealed the results of the mandated 1997 provincial surveys, which showed that in the previous seven years more than half of the 700,000 ha of land subject to industrial development had been arable land (*Xinhua*, 1998b). The Bureau reported handling 40,000 cases of illegal land use, but it rescinded land use rights for just 200 ha. The province responded by mandating that 80 percent of basic farmland would remain under "top protection." Even before the moratorium period, in 1995 the Fujian Governor's office had begun tying performance evaluations of lower level officials to farmland protection and reduced the amount of land lower administrative levels could approve for development (*Xinhua*, 1995a).

The land development problem in south China has particularly demonstrated scaled relations of economic activity, between local development sites and agents of land transfer, investors from Hong Kong and Taiwan in the transboundary region, international sources of FDI, and, from the perspective of Beijing, national fiscal conditions. Foreign direct investment from Hong Kong, Taiwan, and Chinese overseas sources has been the major driver of urban and industrial development in Fujian. Fujian's largest land use transformation is a 48 km^2 seaport industrial zone associated with the Salim Group, not far from Liem Sioe Liong's *qiaoxiang*, Niuzhai village. Local perspectives about this land use transaction have charged that officials allowed him 800 ha of prime farmland for "virtually nothing" (Chandra, 1994), which is an example of the common problem that land use rights have too frequently been given over by negotiation at below market rates (Ho, 1995). Elsewhere in Fujian, the city of Shishi, one of the country's top 100 county-level cities, had its 9.6 ha Zhenshi industrial development zone solely invested by a Hong Kong company (*Xinhua*, 1993a). Fujian successfully attracted many large-scale urban and industrial development projects especially during the period 1993–6. Yet even after the April 1997 moratorium announcement, policies encouraging coastal zone land development continued in Fujian. In May 1997 the province continued to promote foreign invested real estate projects, and announced new preferential measures for overseas funded infrastructure and industrial projects, including five-year exemptions from land use fees and "more preferential land-use prices . . . granted to those who develop wasteland, barren hillsides, and beaches," thereby apparently eliding the arable land problem (*Xinhua*, 1997b).

Policy shifts and arable land loss

China's statistical data on arable land have been problematic, and issues about the extent of the arable land problem have dominated debate, but the general statistical trends on land use change are worth retrieving. Selected data on decreases in arable land from two cities and two provinces demonstrate similar trends in land use change (table 7.1 and figure 7.1). The two cities selected, Shenzhen and Shanghai, represent different stages in the evolution of the land use and land marketization regimes. The two provinces selected, Fujian and Jiangsu, have both been subject to rapid industrialization and also have a mandate to conserve arable land because of their historic low per capita arable land conditions. Fujian has undergone substantial infrastructural development in the coastal zone and has the lowest arable land per capita ranking in the country. Jiangsu also has a low per capita arable ratio and had a particularly rich reform experience in the rural sector in the 1980s, in the successful growth of township and village enterprises. Accounting for localized differences, Shenzhen, Shanghai, Fujian, and Jiangsu all appear to have witnessed some similar rates in arable land losses. Shenzhen and Fujian, the selected datums associated with the first round of national SEZ policy, both experienced land losses in the early 1980s. After the announcement of the open coastal city and open coastal region policies in the middle of the 1980s, losses in Shanghai and Jiangsu increased substantially. Losses concomitantly dropped in Shenzhen, which could be expected with shifts in industrial development activities into the surrounding Zhujiang delta region. Losses were high again across the board in 1991, when increased capital flowed from Hong Kong into China, and after 1992, and when Deng Xiaoping promoted intensifying the pace of reform. Property prices doubled in Hong Kong between 1991 and 1994, which further propelled Hong Kong investment in south China (Sender, 1992). In general, the figures suggest that arable land losses are positively correlated with historic policy junctures in state promotion of land development in SEZs and open development regions. In these ways and more, the arable land problem is the land development problem.

The Geographical Contradictions

What are the internal contradictions in the land use policy regime in the 1980s and 1990s that promoted real estate development? The lack of coordination between state bureaus and ministries and their related policy regimes has been a well known institutional problem in China. This analysis instead focuses on the geographical nature of the problems that have

Table 7.1 Decreases in cultivated land under reform, selected coastal areas, 1000 ha

Year	*Shenzhen*	*Shanghai*	*Fujian*	*Jiangsu*	*Historical context*
1979		4.37	2.66	10.39	SEZ policy forumlated
1980	2.8	1.71	3.67	9.02	
1981	2.02	1.33	2	4.37	
1982	1.38	1.07	4.53	5.8	
1983	0.61	1.72	1.8	1.16	
1984	1.9	4.18	3.07	8.96	Coastal Cities opened
1985	4.1	6.18	18.4	17.06	Coastal Regions opened
1986	0.7	6.54	10.93	13.24	
1987	1.29	2.17	6.34	10.98	
1988	1.01	3.67	4.73	10.97	
1989	0.81	3.18	0.67	6.52	Tiananmen Incident
1990	0.16	0.82	2	4.46	
1991	3.56	2.24	2.06	7.89	Hong Kong property market high
1992	4.78	3.12	5.94	28.2	Deng Xiaoping's Southern Tour
1993	4.3	15.85	9.33	26.11	
1994	1.46	8.14	8.8	31.66	Farmland Protection Regulations
1995	0.19	3.89	6.41	15.69	Urban Real Estate Law
1996	0	2.71	7.79	12.87	
1997	0.87	0	7.67		Moratorium on arable land conversion

Sources: *Shenzhen tongji nianjian* (*Statistical Yearbook of Shenzhen*) (Beijing: Zhongguo tongji chubanshe, 1998), p. 181; *Shanghai tongji nianjian* (*Statistical Yearbook of Shanghai*) (Beijing: Zhongguo tongji chubanshe, 1997; 1998), p. 169; p. 163; *Fujian tongji nianjian* (*Statistical Yearbook of Fujian*) (Beijing: Zhongguo tongji chubanshe, 1996; 1998), p. 164; p. 134; *Jiangsu tongji nianjian* (*Statistical Yearbook of Jiangsu*) (Beijing: Zhongguo tongji chubanshe, 1998), p. 157.
Notes: Data calculated from year end total cultivated land. For 1996 Shenzhen reported increased in cultivated land of 0.72 over the previous year. Shanghai reported an increase in cultivated land of 10.8 in 1997. The 1998 Jiangsu yearbook did not report data for 1997.

emerged, in three categories: administrative hierarchy contradictions, land monitoring contradictions, and rural–urban contradictions. Spatial analysis of the policy regime reveals the problems in the evolving system of land use regulatory measures, and in the gaps between policy and practice both private and state agents of development located opportunities to make real estate development the most profitable sector of industrial expansion during the second decade of the reform era.

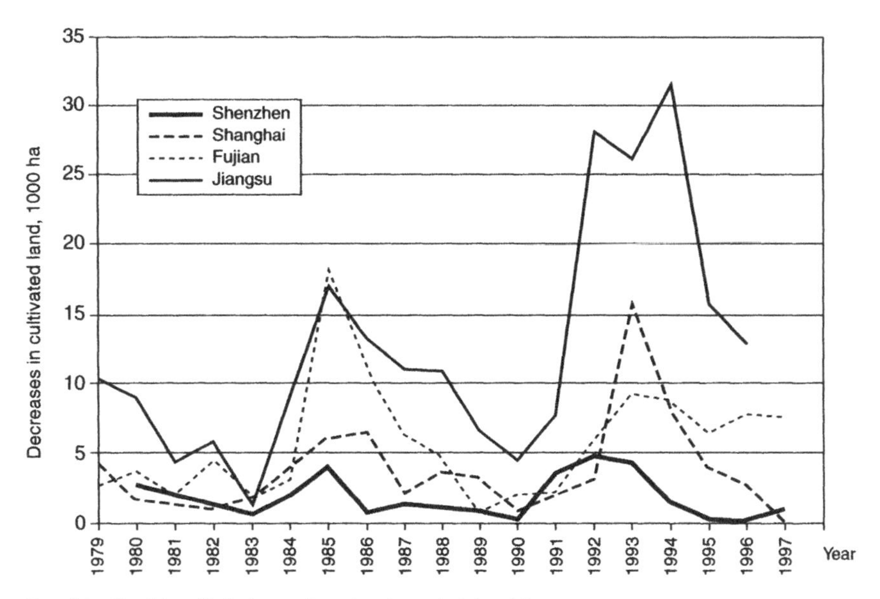

Figure 7.1 Trends in arable land conversion under reform, selected coastal areas.

The administrative hierarchy contradictions

The administrative hierarchy is China's scaled system of state organization, from the center to the provinces and to the local level in townships and villages. While land use policies were designed to administer land use in an orderly fashion by level of scale, the hierarchical nature of the administrative system contributed to problems in land use plan formulation, especially at the local level. The revised Land Administration Law and the stipulations of the 1997–8 moratorium on land conversions reiterated that jurisdictions down to the township level must prepare integrated land use plans highlighting both planned construction and the conservation of arable land. Yet in addition to the usual enforcement problems, plan preparation has not reliably taken place in an orderly fashion by administrative level. Lower level plans must be approved by higher levels of government, but not all provinces, for example, drafted regulations to enforce land use plans, which, in turn, compromised land use planning at all lower levels.

The system of land use approvals by administrative level, which prevailed during the decade up to the implementation of the new land law in 1999, led offices below the provincial level to devise ways of approving large-scale pieces of land for industrial development. One piecemeal solution has been known as "breaking the whole into parts."[9] In order to pursue relatively large-scale development projects at the local level, developers applied to the Land Administration Bureau for several different pieces of land up to the maximum size approved at that level, which would later be reunited into a larger land parcel. Since the majority of entrants in the land market have been government agencies and state-related enterprises, these arrangements between land administrators and land developers often represented common interests (Keng, 1996, p. 331). Counties preferred to be upgraded to city-level status and to approve their own projects and avoid petitioning higher-level authorities for large-scale land use conversion permissions. Approvals below the provincial level increased especially after the state relaxed the criteria for city-level designation in the administrative hierarchy. Between 1978 and 1994, the number of designated cities in Guangdong increased five times from 10 to 51. In Fujian and Jiangsu the number of designated cities more than tripled, from six in 1978 to 22, and 11 to 39, respectively. In Zhejiang, the number of cities increased 11 times, from three to 33 (see *XZQH*, 1978, 1995). In 1993 and 1994 alone, each year 53 counties were redesignated as cities, which is over twice the average rate of previous years (see Pu et al., 1995). As a result, many more cities were able to approve larger development projects on arable land, and 1992–4 were peak years of land development fever. The reclassification of counties to higher level city status was officially banned with the moratorium on arable land use conversions.

The role of the state in regulating land use conversions according to levels of administrative scale has also necessarily involved multiple offices at different administrative levels in planning and profiting from real estate development, especially the largest-scale projects. State land use requisitions include large-scale joint venture real estate projects, particularly for industry, commercial use, and housing. The existence of many government departments, agencies, and state-affiliated companies at each administrative level, and competing economic development interests among them, has made it unclear which office at which level should actually act as the bona fide land owner (Keng, 1996, pp. 328, 331). Under these circumstances, the higher level checks on large-scale projects have not reliably served a land conservation function but instead have given notice to new profit-sharing opportunities.

The land monitoring contradictions

Changes in the evolving land management system have given rise to historical trends to over- and underreport arable land, which exacerbates land management problems. Before the Land Administration Law took effect in 1987, there was a tendency to underreport arable land simply because it was taxed (G. Brown, 1995, p. 927). But the Administration also introduced fees and penalties associated with agricultural land conversions, and now the expectation is that arable land figures tend to be exaggerated. In some rural areas, rural land registration is at the root of the problem. Land may not be properly registered because the Land Administration Bureau claims that village collective boundaries are not clearly defined.[10] As a result, certificates of ownership are not issued to the collective and the potential for an evolving land management system breaks down at ultimately the lowest level of management – the land itself and the title which should define it.

In addition to the problems of accurately measuring arable land, the survey approach of the Land Administration Bureau has limited analysis of land loss. State statistical data have utilized five causal categories of arable land loss: expansion of forests, expansion of pastures, national infrastructure projects, township and village level infrastructure construction, and rural private construction. The difference between the five stated categories of arable land loss and the total arable land loss left unaccounted for the potentially largest category: joint venture development projects (G. Brown, 1995, p. 929). This gap in the data is another indicator of the lack of reporting on causes of arable land loss, especially in areas of high economic growth. By the middle of the 1990s, the state began to publish data on arable land converted for industrial development (Ho and Lin, 2001). The 1998 revisions of the land law tackle the monitoring problems in part by reducing land categorization to only three types: agriculture-related, construction-related, and unused.

In addition to the integrated land use plans required by the 1997–8 moratorium, administrative jurisdictions have also prepared sectoral plans – which have not been regularly articulated with integrated land use plans. For example, urban construction development plans of municipalities, planned by the Construction Bureau, and authorized by the Urban Planning Law, have tended to be in conflict with or have had priority over the agricultural priorities of land use plans and the basic farmland protection areas.[11] In this context, conflicts of interest between government officials charged with land conservation and approvals of land use conversions for construction create contradictory roles for the state. Implementation of land use management in Jiangsu demonstrated how some new land management officials at the township level were merged into offices of the township Construction Bureau (G. Brown, 1995, p. 923). While such mergers put officials from different departments in closer dialogue, the potential also exists to prioritize construction over land conservation. And like the strategy of "breaking the whole into parts," some localities have attempted to avoid development zone monitoring through discursive strategies, by simply not using the labels *xiaoqu* or *kaifaqu* in the names of their development zones, or by changing the names of existing development zones (*BHGD*, 1997, p. 20).

The rural–urban contradictions

Rural land in China is owned by the village collective and urban land is owned by the state. Land use rights may be obtained for both collective-owned and state-owned land, but national regulations for land use transfer rights have been only been formulated for urban land. Only urban land use rights can be sold and transferred for urban and industrial development. Rural land marked for industrial development must first be transferred to state ownership; then the state sells the development rights. Thus rural and urban land have experienced different tenure systems and different land market environments. Rural to urban land use conversions entail transformation fees, which, according to regulations, are supposed to be used for basic farmland expansion elsewhere, based on the logic that basic farm land must not decrease in overall extent. Certainly the contradictions in the evolving land use management system are well known in China, but the usual focus of criticism is the set of contradictory structures that characterize the urban land use rights market (Keng, 1996).[12] In the larger-scale view of the arable land debate, the fact that farmers are not sufficiently protected is one of the roots of the land management dilemma.

As anywhere, rural land costs are lower than urban land costs, but rural values have been especially low in China, where the comparative lack of an active market for rural land has depressed the value of rural land use rights compared to urban areas. Because the state can requisition rural land for development, some farmers have perceived that their land is owned by the state, or that their rights to land use, as administered through the village collective, are meaningless in the face of aggressive urban expansion by municipalities. Since compensation standards were set after the first version of the land law they were revised upward but did not keep pace with inflation and real estate values. Thus the state has been able to requisition rural land at set compensation prices which have been lower than the already low "market" (really negotiated) price. Farmers have been aware of these distinctions in land values, the consequent attractiveness of rural land for development, and unreliable opportunities to claim adequate compensation. As a result, farmers, in what might be called high development risk areas, have often viewed their responsibility land contracts as obligations without security of tenure. Further complicating the tenure problem is that contract length has varied considerably. Even though the state directed in 1995 that contracts be extended 30 years beyond the original 15-year term, in some low per capita arable land areas, including parts of Fujian, land use rights for agricultural land have been short term, for only one or two years. Contracts have been canceled outright when the state has requisitioned responsibility land holdings for industrial land use, and when land holding size must be readjusted based on change in family size. This latter condition has also promoted rural area son preference as married daughters continue to relocate to the husband's village. If contracts do not ultimately secure tenure, farmers are disinclined to make long-term investments to maintain and improve soil quality. Under this combination of circumstances, farmers near large cities have been officially encouraged to take the short-term view on arable land management. Thus the 1998 revision of the land law, in codifying a nationally uniform system of 30-year land use contracts for farmers, represents a potentially significant change for the rural sector.

Cities have a revenue-earning motivation in reclassifying rural land for industrial use. Land use transformations from rural to industrial use result in recategorization of the land to urban status, which incurs a higher tax rate – payable to the municipal government. Thus cities have considerable incentive to expand and bring suburban rural land under municipal jurisdiction. Rural land in the suburbs is also the logical location for large-scale industrial zone development. Fuzhou's mid-1990s urban area development plan projected expansion from 69 to 120 km^2.[13] Even in May 1997, coincident with the moratorium on arable land conversions, Xiamen tripled in size by absorbing Tong'an county, the province's largest county, which allowed it to extend urban land use transformation rights to an extensive area in the former urban hinterland. In Guangdong, Zhuhai's urban master plan projected especially ambitious urban expansion from 100 to 700 km^2; other cities and towns in the Zhujiang delta have attempted to follow Zhuhai's model (Wong and Zhao, 1999, p.121). The state, domestic companies, and developers with foreign capital have all looked to low cost agricultural land on the urban fringe as a prime site for new industrial development zones. Suzhou demonstrated the outstanding example of this logic with the plan for the 70 km^2 China–Singapore Suzhou Industrial Park.

Real Estate Fever and Financial Collapse

The institutionalization of the special zone system in China, from experiment to general strategy, was a gradual process that successfully promoted large-scale land development for both industrial and residential development. Since 1997, however, unapproved emulation of the zone concept by localities has earned land use plan penalties and zone closures. The widespread construction of development zones on rural lands of the urban periphery raised the stakes of the land use planning regime in China, where arable land, population, and processes of industrialization all concentrate and overlap in the plains and low-lying topography of the coastal zone.

In the early 1990s China's concern about land management centered on the arable land debate. By the late 1990s, the rhetoric of the arable land debate was maintained, while the larger problems of land management emerged in excessive and irrational land development. China's conditions mirror land development processes across the Asian region in the 1990s, an era that became distinguished by overextension in the real estate sector underwritten by domestic and international financial service industries. In China the state worked to dampen the feverish pitch of land development after the onset of the regional financial downturn. In early 1998, China's economists called for the closure and consolidation of a large number of local trust and investment companies, which have been particularly associated with funding real estate development (Johnson, 1998). The state also ordered all commercial banks to de-link from trust and investment companies.

Evidence for how closely the excesses of real estate development came to destabilizing the regional economy is found in the bleak picture of financial institutions in the late 1990s. In June 1998 the Hainan Development Bank, the main banker to the provincial government, was forced to close after the state ordered it to assume the liabilities of credit cooperatives on the island. The *Financial Times* of London reported that the causes of the financial disaster in Hainan could be traced to "1995 as a spectacular property bubble burst" (Kynge, 1998). Another analysis affirmed that "Shanghai is in the middle of the largest property collapse in the world and perhaps the most peculiar such deflation in history: side by side with double-digit growth are (according to some estimates) 70 percent vacancy rates, real estate prices that have slid 50 percent since 1995, and ironically, a housing shortage" (Ramo, 1998). After 1992, real estate speculation drove new housing prices to levels between 10 and 30 times household income (the cost of an affordable residence is three to six times household income; see Liu and Link, 1998).

In October 1998 the state announced that it would close the investment arm of the Guangdong provincial government, the Guangdong International Trust and Investment Corporation (GITIC), whose failure is the largest of any mainland company under reform. The state simultaneously demanded that 20 ministerial level units sever ties with trust and investment firms (*Ming pao*, 1998). The GITIC closure sent shock waves through the domestic economy and the international financial community: it effectively signaled an end to the era of the idea of a "miracle" economy in south China. A senior official reportedly characterized the scope of GITIC's problems as "no less serious than the financial storm breaking out in Thailand in October 1997 in terms of its size and implications. . . . it could endanger the country's whole financial system " (Wang X., 1999). At first the state announced that it would guarantee GITIC's international debts, but it later instead gave preference to individual investors. In January 1999 the People's Bank of China declared GITIC bankrupt, and under bankruptcy law foreign and domestic losses are treated equally. The state also declared bankrupt major GITIC subsidiaries, including Guangdong International Leasing Corporation, which in better days was characterized as south China's largest real estate development firm. Outside of GITIC's securities business, its most significant assets were in Guangdong real estate. Within a year of GITIC's closure, the provincial level international trust and investment firms, 239 across the country, were supposed to be restructured and consolidated. But this has yet to widely take place, likely because of the complexity of interlocking relationships between the non-bank financial institutions, local governments, and state-owned enterprises (Miller, 1999).

In the middle of 1999 the state announced that the People's Bank of China closed the Shanghai-based Pudong United Trust and Investment Corporation in the third quarter of 1998, in an apparent attempt to shield Shanghai from the economic pall brought over Guangdong by the closure of GITIC (Harding, 1999). This institution was heavily invested in property in Pudong, and ranked second in total assets after the Shanghai International Trust and Investment Corporation. In 2000, the Shanghai Agriculture, Industrial, and Commercial Group (SAIC), an investment company of the Shanghai government, was reportedly on the brink of failure (Miller, 2000a, b). SAIC was established in 1990 by the Shanghai Farm Administration Bureau, and maintained interests in more than 100 subsidiary companies, from supermarkets to securities firms. One of SAIC's reported problems was engaging in property development without obtaining land use rights: SAIC is believed to have developed property, presumably in part on agricultural land, without registering land use transformations. Based on this preliminary analysis, the newly revised Land Administration Law may help conserve arable land, and an equally if not more significant resource – national financial security.

NOTES

1 Through the 1970s and 1980s, export-oriented manufacturing was the leading investment sector across Asia. But by the early 1990s, profits from the secondary sector, manufacturing, relatively decreased, and investors switched to service sector industries, in this case primarily land development, and its associated industries in real estate, finance, engineering, and more. During this period both the state and large-scale private investors increasingly moved capital from manufacturing sectors into real estate and infrastructural development projects. See Robert Wade (1998) for an explanation of this capital circuit switching in world financial terms.
2 The many types and functions of industrial development zones below the national level are not well defined in the literature. The best comparative summary of development zones is S. P. Gupta (1996), a study sponsored by a Ford Foundation collaborative project between the Indian Council for Research on International Economic Relations, the Development Research Center in Beijing, and the Institute of Southeast Asian Studies in Singapore. See also Laurence C. Reardon (1996) for historical precedents of the development zone concept during the Maoist era.
3 The data cited by Yeh and Wu (1996, p. 345) do not specify precise types of zones or their administrative levels of approval. For example, the *Zhongguo jingji tequ kaifaqu nianjian* (*Yearbook of China's Special Economic and Development Zones*) for 1996 listed 34 nationally designated ETDZs, and so figures reported by Yeh and Wu include local and provincial level zones.
4 He Qinglian calls the zones *kaifaqu* (development zones) and so suggests all possible development zones rather than the category ETDZs.
5 The ability of Singapore government-linked corporations to organize and finance infrastructural "megaprojects" in China reflects several conditions of economic development in Singapore. By the middle of the 1980s, Singapore had successfully negotiated an export-oriented industrialization drive and became an important source of FDI. Singaporean government interests and private companies began to invest across the Asian region and in Western industrialized countries, maintaining favored investment destinations in Hong Kong, Malaysia, and Indonesia's offshore islands. The majority of Singapore's investments have been made in the Asian region, and over half of the total in other countries within ASEAN. By the early 1990s, Singapore's investment increasingly shifted north to China, so that by 1994 more than half of all FDI leaving Singapore went to China.
6 Singapore has also functioned as a diplomatic liaison of sorts between China and Taiwan. When China and Taiwan initiated political and economic discussions in 1993, Singapore was the site of the meetings.
7 Interview, Suzhou, 1998.
8 The former Beijing mayor, Chen Xitong, was tried and sentenced on corruption charges, including major real estate deals. China's largest-scale corruption scheme since 1949 involved a vast smuggling operation in Xiamen, and before it collapsed in 2000, its leader, Lai Changxing, had

begun excavation for an 88-story skyscraper in Xiamen, which would have been the city's tallest building. On the Guangdong infrastructure problem see Jiang Jingen (1998) and Shen Bin (1998).

9 This *chengyu* (maxim) is remembered in association with Mao Zedong, who used it to refer to a guerilla warfare strategy of dividing the army into small units.

10 Interview data, Xiamen, August 1995. Large-scale lineage feuds over village boundaries, which were not uncommon, especially in Fujian and parts of Guangdong, in the Ming and Qing dynasties, have been documented in rural Guangdong in the 1980s and 1990s; see I. Yuan (1996).

11 Interview, Xiamen, August 1995.

12 The pricing of land use rights, while not part of the geographical concerns of this chapter, has been one of the most serious issues plaguing the evolving land use market.

13 Interview data, Xiamen, August 1995.

8

Urban Triumphant

My human rights are constrained now because I cannot go to Hong Kong as often as I wish.
(Premier Zhu Rongji, 1998)

Turbulent times characterized China's twentieth-century urban scene. The end of the dynastic era, followed by the rise of Chinese modernities during the Republican period, played out distinctively in centers of urban culture, and nowhere more so than in Shanghai. The wars and rebellions that raged across China targeted cities as prizes of military power, and residents were widely subjected to shifting regimes of domestic warlord governments and foreign imperialism. Reconstruction of society during the Communist period depended on siphoning off the productive capital of cities and directing resources toward primary production economies in agriculture, and heavy industry in producer goods manufacturing. Urban built environments, in their transhistorically monumental and symbolic forms, representative of the world's longest urban tradition, were consequently allowed to decay, and, during the Cultural Revolution, actively destroyed. Debates about class formation and bourgeois elements took spatial terms over the ideological meaning of city walls, the defining element of the imperial Chinese city and symbolic boundary between urban and rural society. Most of them came down during the Maoist era, whereas in China under reform, massive wall fragments have become part of the touristed landscape, protected pieces of China's imperial past. One important spatial element of state policy generated during the Maoist era, the *hukou* system, has continued to exist under reform. However, it is breaking down under pressures of migration, and leading cities have adopted *hukou*-for-sale policies, which effectively continue to inscribe a geographical class system into the national landscape.

In China under reform the urban sphere has reclaimed its position of significance. With their special policies and special zones, cities have become the celebrated centers of reform achievements. Guangzhou transformed into the regional capital of a new south China, and Shanghai reemerged in the 1990s as a city reconstructed, studded with new architectural monuments to a transnational postmodern style, and a claimant to an evolving turn of the century cosmopolitan order. Even Deng Xiaoping recognized Hong Kong, once the colonial bastion, as a remarkable addition to state political economy, and repeated that he would like to see "many Hong Kongs" built across China. The return of the former Hong Kong colony to China in 1997 saw a week-long national holiday in Hong Kong and China, celebrated through myriad events, contests, "countdown to July 1" digital clocks, publications, and television specials. When Jiang Zemin arrived for the diplomatic ceremonies that returned Hong Kong's sovereignty to China, he had never been there before. In some ways, he was a stand-in for Deng who did not live to realize his wish to be in Hong Kong in 1997 to receive the restoration of national territory. In one day, Hong Kong officially became China's leading "world city" (Friedmann, 1986; Knox and Taylor, 1995) and international financial capital.

Hong Kong's entry into China's complement of major cities raised the stakes of urban and economic development country-wide, and on a larger scale across the Asian Pacific front. While a colonial territory, Hong Kong totally eclipsed Shanghai as the banking capital of China, and Hong Kong's financial industries have competed with those in Singapore for the number two position in Asia, after Tokyo. Compared to the dynamic transformations in Hong Kong, Guangzhou, and Shenzhen, political and economic elites in Shanghai chafed at the state controls through the first decade of reform that continued to compel the city to remit the majority of its revenue to Beijing and delay its own special development privileges (Ho and Tsui, 1996). The state sanctioned the Pudong New Area in Shanghai in 1990, which brought the largest SEZ to Shanghai (see map 7.1). Deng Xiaoping (*Selected Works of Deng Xiaoping*, 1994, p. 376) marked the occasion by stating that "his one major mistake was not to include Shanghai when we set up the four special economic zones," which offered a diplomatic gesture to an urban leadership unaccustomed to being sidelined while Guangdong and Fujian captured a decade's worth of foreign investment opportunities with little competition. By the end of the 1990s, however, the delay seemed to have mattered less as Shanghai and the Yangzi River delta region emerged, in Chinese terms, as "the dragon's head," the symbolic language of the leading economic region of the country (Sung, 1996; Foster et al., 1998). Now, assessments of Shanghai regularly measure its rapid transformation by comparison to Hong Kong, from the growth of Shanghai-based financial service industries to popular cultures of consumption and the city's new skyline, represented in a building boom that has propelled real estate prices to new highs in China second only to Hong Kong (Sender, 1994, p. 56).

This chapter examines urban and economic processes in Hong Kong and Shanghai against a broader backdrop of regional transformation generated by the sphere of transboundary processes across the East and Southeast Asian regions. The 1997 repatriation of Hong Kong to Chinese sovereignty propelled new questions about political economic relations across the region, especially between Hong Kong and Shanghai, Hong Kong, Taiwan, and China, and China and Singapore. In Hong Kong, the shift to SAR status compelled local concerns about the nature of Hong Kong identities, the future of the entrepôt economy, and cosmopolitan ideas about quality of life. In Shanghai, the rebuilding of the city broadly impacted both residential and institutional space, and the city became the leading center of new state policies to promote home ownership. Admiration for the "Hong Kong model" has registered in Shanghai along the Huangpu River shore of Pudong, where the city's new architectural statements concentrate to create a skyline of monumental proportions. In both cities, the urbanization of capital (Harvey, 1985) has registered dramatically in the built environment, as new and higher high rises prominently symbolize the significance of service industries in underwriting the continued expansion of the regional economy. Processes of globalization in China are making their distinctive marks in these cities, and propelling social and economic transformations that will be emulated across the country, if not the wider Asian region.

1997 and the Hong Kong Handover

The Hong Kong Special Administrative Region is the whole of the British colonial territory that grew as a result of British pressure to enlarge the colony over the second half of the nineteenth century (map 8.1a). When in 1843, as a result of the Treaty of Nanjing, the British received Hong Kong Island in perpetuity, it was little more than its geological configuration – an extinct volcanic plug, a large granitic rock in the South China Sea that China could afford to let slip from its southern territory without excessive loss of imperial face. It lacked fresh water and had little arable land. Its harbor was fine but its resource base was insufficient to sustain a settled population. The British pushed for enlargement, gaining Kowloon peninsula and Stonecutter's island via the Convention of Beijing, and ultimately the New Territories, the territorial extension whose terms of occupation led Hong Kong back to mainland rule. July 1, 1997 marked the end of a 99–year leasehold over the New Territories, the 919 km^2 extension of the Kowloon peninsula that provided the natural resource base otherwise missing in the early colony.

In the run-up to 1997, questions about the political and economic

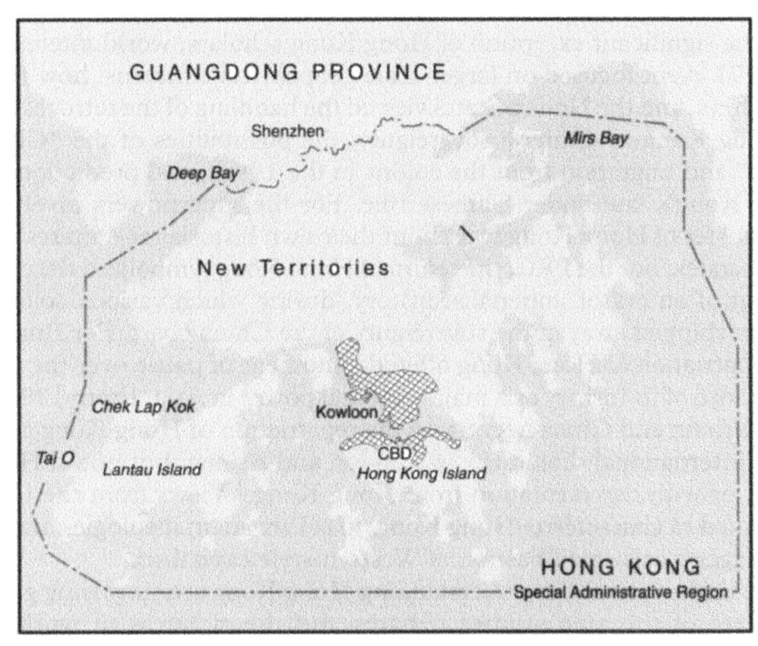

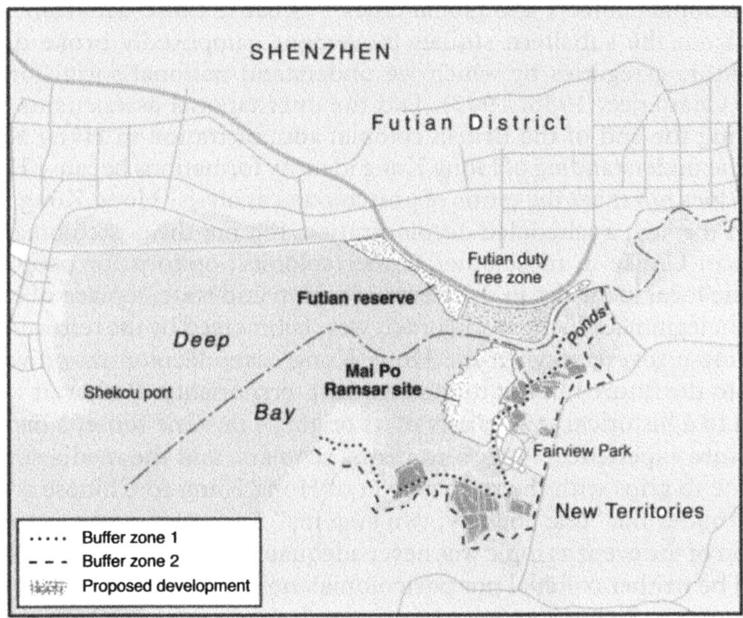

Map 8.1 (a) The Hong Kong SAR . (b) The Hong Kong–Shenzhen border area and Deep Bay .
Line work by Jane Sinclair.

status of Hong Kong received focused attention for the first time. Yet with the significant exception of Hong Kong scholars, world interest in the 1997 event focused on larger-scale geopolitical concerns: how Britain, China, and the United States viewed the handling of the retrocession of Hong Kong to Chinese sovereignty, the possibilities of the "China threat" and migration from the colony in the 1990s, and predictions of Hong Kong's fate under Chinese rule. For the superpowers involved, the transfer of Hong Kong was about their own histories as empires and nation states. For the PRC, the return of Hong Kong symbolized the overcoming of an era of imperialist history, during which various colonial powers chipped away at the sovereignty of the China coast. For Britain, the repatriation of Hong Kong offered a moment of pause over the ultimate close of its history as a major colonial power in Asia. From 1982 to 1984 Britain and China negotiated the repatriation of Hong Kong as an act of international diplomacy in London and Beijing, but not in Hong Kong, or with representation from Hong Kong.[1] Views from the USA continued to characterize Hong Kong, like Taiwan, in ideologies of mirror imagery, as a small bastion of Western-style capitalism.

The limitations about understanding Hong Kong are surprising given the state of the area studies debates and the richness of work on transnational subjects and global cities – because those debates, especially from the subaltern studies movement, supposedly broke down monolithic categories by which we understand national social formations (Chatterjee, 1986, 1993). Did the international assessments surrounding the end of the British colonial administration in Hong Kong deny the understanding of Hong Kong identity formations because Hong Kong does not meet the terms of postcolonial analysis? Hong Kong proceeded through a scheduled decolonization, but not the postcolonial experience. Unlike in most other former colonies, options for people to redefine local identities in the terms of nation and state, a place of independent legitimacy in the world order, were submerged by the retrocession to Chinese sovereignty. In the Hong Kong case, decolonization was a move to the future but not to independent territoriality. Rather, it was a return to a historical sovereignty of its origins – in some sense, a back to the future experience. As governments, scholars, and the media sought to come to grips with the retrocession of Hong Kong to Chinese sovereignty under the "one country, two systems" formulation, the utter seduction of the event as topic was never adequately conceived: Hong Kong would be neither colonial nor postcolonial, neither nation-state nor sovereign power. In that territorial never-land, the issues at stake in Hong Kong elided normative conceptualizations of national culture and territoriality. The slippages in meaning ask for explanations beyond the usual parameters of political geographies.

One country, all kinds of systems

In the context of the nation-state system, the "one country, two systems" formula appears not only out of step with prevailing norms, but a long-term impossibility, and created to satisfy the short-term territorial goals of an authoritarian state.[2] Against the backdrop of contemporary globalizing conditions and cosmopolitan possibilities, it may be read in less deviant terms, not just as Dengist pragmatism but even as a postmodern alternative to the problems of the modern state system. Analysts concerned with the future of Taiwan have suggested that China adopt a policy of "one country, four systems," in order to recognize the differences in political cultures between the mainland and Taiwan, in addition to Macao and Hong Kong. China's need to reinscribe territorial boundaries is driving the multiple systems model, but in a twist that prioritizes differences in political systems, suggestions for a "one-country, five systems" model include autonomy for Tibet. From Taiwan, former President Lee Teng-hui (1999) wrote, in his book *Taiwan's Viewpoint*, about China as a system of seven distinct economic regions. In this model, predictably, Taiwan is its own region. The other six are Tibet, Xinjiang, Mongolia, Northeastern China, Northern China, and Southern China, which recognizes the traditional divisions of Han China in northern and southern regions. President Lee defined the seven regions model as an administrative approach that would redistribute the power of the central government to the regions, but not create separate sovereign systems.[3] In response, the Chinese official press charged Lee with promoting Taiwan independence, elimination of the Chinese nation, and, based on joint efforts with the USA, splitting China apart (*China Daily*, 1999a). Clearly only the PRC may devise multiple systems, yet the discourses of multiple systems reflect the collisions between the historic norms of nation-state governance and the new realities of transboundary and transnational processes in south China. In reality, the "one country, two systems" formula may be best understood as a gradualist approach to territorial reunification (M. Chan, 1997).

Geographies of Globalization

Hong Kong and Shanghai are cities of globalizing processes, centers of trade, finance, capital flows, the circulation of information, the production of things, and high population mobility. Yet contemporary concerns about globalization seldom historicize the economic processes of such world city origins. We are concerned with flows of trade and capital but tend to ignore why it is interesting to understand where cargo ships sail; we write about mobility but do not consider the intermediary and facilitating spaces of population mobility like ports and airports in the globalization debates. A coded vocabulary of globalization, in "flows" and "local/global" events, stands in for material processes of transport and travel, their geographies, and the new cultural forms their collisions engender.[4] Globalization may have appeared as a relatively new concept in the late twentieth century, but its histories and geographies are traceable, along routes that have circulated through coastal settlements from the era of mercantile ports to contemporary centers of transnational mobility in world cities. Reasons why such divides are maintained arguably have to do in part with changing human perceptions of travel, technological changes in modes of travel, and economic restructuring of transport industries, in the shift from the industrial era to the period of late capitalism (Cartier, 1999b). In the mercantile and industrial eras, the port was the site of mobility in the maritime world city, for people and things, but in the period of late capitalism the technologies for moving travelers and cargoes have largely split between ports and airports. When in 2000 container cargo ships arrived at US West coast ports from Hong Kong with undocumented migrants eking out a passage inside canvas-topped shipping containers, the shock of discovery registered partly in the use of container ships as a mode of passenger travel.

In Hong Kong before 1997, every flight was an international flight and basic mobility in and out of the territory was a consummately transnational experience. Hong Kong architect Naonori Matsuda, observing the new Hong Kong airport megadevelopment at Chek Lap Kok, noted that the project was much more than a new mammoth international airport. It rather seemed that the theme of all Hong Kong in the middle of the 1990s was "city as airport" (Abbas, 1997). The airport's site location – a reclamation at the edge of Chek Lap Kok Island, sheared off and pushed into the sea to create its infrastructural base – required the whole scheme to be connected to the centers of settlement in Hong Kong by an elaborate transportation network of new bridges, roadways, and rapid transit links. The project was so extensive, and so commanding of resources and attention at all levels, that its processes of creation overtook other local issues in importance. When the airport opened and its computer systems failed, the Hong Kong government apologized in the international press, because they know that Chek Lap Kok is a central element in their contemporary stock of symbolic global capital. Chek Lap Kok, the high technology mega-airport, designed by Foster Associates, contains a 30,000 m^2 shopping center full of branches of Hong Kong restaurants and shops, to the degree that the experience of the airport mimics the consumption culture of the world city. Chek Lap

Kok opened with an annual passenger handling capacity of 35 million, and planned expansion for 87 million, at a future date when the airport serves the larger Zhujiang delta region or when mobility of the Hong Kong populace further intensifies. These service sector industries, including land and real estate development, transport, retail, and as underwritten by trade and financial services, have substantially driven Hong Kong's economy through the 1990s, and Chek Lap Kok airport, a constellation of service sector industries, was the largest development project of the late colonial period.

Historically, the existence of Hong Kong has been based on its seaport. During the 1990s, the port of Hong Kong ranked number one in the world by total cargo. Hong Kong's leading position in the world of port activity owes substantially to the fact that it functions as the ultimate port of departure for goods leaving south China for the world market (Sung, 1996). After 1997, the port of Hong Kong became the dominant ocean-going port in China as a whole. Although Shanghai, astride the Huangpu, maintains an image as a great functioning port city, it is really a river port and transshipment center for the Yangzi delta, and its international cargoes are transferred to ocean-going ships in Japan or Hong Kong. Shanghai transportation planners have devised various schemes in the attempt to transform Shanghai into an ocean-going port, but none are really practicable, and it is one significant way in which the city will never be able to challenge Hong Kong. Shanghai's new international airport, by contrast, is planned to become the major air transport hub of the East Asian region, and to capture Hong Kong's role as an airport hub for travel throughout China. The site for the Pudong International Airport was carved out of fishponds in reclaimed tidal marsh, and fishpond remnants front the terminal in a material fragmentation of nature–society relations in which the agricultural landscape has been transformed into an aesthetic design feature of the building. In China's era of "real estate fever," the development of ports and airports contributed substantially to local and regional expansion of services industries, and their consequent geographies now situate and gather the sets of transnational connections that make mobile the diverse people, products, and processes of the world city.

Shanghai and Hong Kong have been cities of migrants and high population mobility. Their populations distinctively enlarged and transformed in association with political upheavals and industrial growth during the nineteenth and twentieth centuries. Disruptions of the Taiping Rebellion led tens of thousands of people to flee Yangzi delta settlements for protection in Shanghai, where alliances between local and Western military powers protected the city. The Japanese occupation of the coast between 1937 and 1941 resulted in over three-quarters of a million people moving into the foreign concessions, bringing the total population of the concessions in Shanghai to over 2.4 million (Wakeman and Yeh, 1992, pp. 2–4). By 1911 over one-quarter of all foreign and Chinese-owned factories in the country were in Shanghai, and by 1949 the proportion had more than doubled to 60 percent, which absorbed the migrant labor force (Perry, 1993, p. 17). The war with Japan propelled hundreds of thousands of migrants into Hong Kong, followed by the flow of refugees after the Communist victory in 1949, and by 1950 Hong Kong's population reached two million. During the Maoist era, the strictures of the *hukou* system especially constrained population mobility, and *xiafang* (transfer to a lower level) policies decanted the urban population to the degree that Shanghai experienced a net loss of nearly 400,000 people (Deng et al., 1991, p. 15).[5] By the middle of the 1980s, the migration stream flowed back into Shanghai, and nearly a million migrants, both legal and "floating," entered the city (Wong, 1996, pp. 39–40); by 1993 the migrant flow was estimated at 2.5 million (Lee and Hook, 1998, p. 130). Between 1977 and 1982, in the first years of reform, almost half a million migrants entered Hong Kong from China (Skeldon, 1994, p. 23). Since then, both the colonial government of Hong Kong and the SAR have instituted policies to limit numbers of Chinese migrants moving to Hong Kong, and the "right of abode" in Hong Kong has become a charged political issue.[6] Emigration from Hong Kong surged in the late 1980s and through the 1990s as people sought alternative domiciles and citizenship in the run-up to 1997. Through the 1980s, an average of about 20,000 people left Hong Kong for international destinations each year, and in 1992, a high point of the emigration wave, 66,000 people emigrated (Skeldon, 1995, p. 57). During the 1990s, Shanghai had the highest proportion of total emigrants granted permission to leave the country, at 28 percent of the total (Miao, 1994, pp. 449–50). By 1999, the migration flow through Hong Kong reversed, as some people who left for international destinations in the early 1990s returned, especially from Canada. Throughout, Hong Kong and Shanghai have been the cities of China characterized by the highest degree of population mobility, cities whose populations in greatest numbers articulated with the cultural trends and economic orders of the global economy.

Connections between Hong Kong and Shanghai also grew through the twentieth century. The approaching Communist victory gave rise to capitalist flight in Shanghai, and significant numbers of Shanghai industrialists relocated to Hong Kong. Shanghai capital was especially concentrated in textile manufacture, which laid the basis for Hong Kong's textile and garment industry, and, ultimately Hong Kong's own industrialization and rise as a manufacturing power (Wong, 1988). The fortunes of the two cities thus became inextricably tied, and Hong Kong enjoyed cultural and economic transformations brought by Shanghai's economic elites. Shanghai film industry leaders also relocated to Hong Kong and contributed to the evolution of a vibrant regional school of filmmaking, Hong Kong cinema. Leo Lee (1999) has described how the "Shanghainization" of Hong Kong ensued through new cultural forms, which in part overwrote the British colonial qualities of the colony and made Hong Kong simultaneously a more Chinese and more cosmopolitan place. The presence of the elite Shanghai group in Hong Kong has also had longer-term political effects: when the process of selection for the first Governor of Hong Kong produced a list of three finalists, Tung Chee-hwa, Yang Ti Liang, and Peter Woo, all were natives of Shanghai. This result demonstrates the central state's preference for selecting leaders with non-local native place origins, and suggests a pattern of Shanghai ties among high profile positions of national leadership, including President Jiang Zemin and Premier Zhu Rongji, both of whom were posted to the capital from Shanghai.[7]

Cosmopolitics

In the international flux of ideas, the idea of cosmopolitanism, as a set of worldly social processes and values, has emerged as a humanist counterpart to globalization. Cosmopolitanism has a historical lineage in ethical and political philosophy, and has become central to debates about globalization across the disciplines. Two of the most commonly cited views about cosmopolitanism, in step with its historical lineage, emphasize universalizing qualities. David Held (1995) has explained cosmopolitanism as a new transscalar model of democracy for a globalizing world. Martha Nussbaum (1996, 1997) treats it as a set of values about a universal humanism. But as Harvey (2000) points out, these discussions simultaneously lack recognition of geographical realities, in uneven development, and decry the absence of knowledge about geographical differences that would foster a cosmopolitan ideal. Alternatively, Pheng Cheah and Bruce Robbins (1998) make the subject cosmopolitics to ask for greater accountability about the political contexts of the issues at stake. Still, as Scott Malcomson (1998, p. 238) points out, "the cosmopolitanism debate," on all sides, has avoided "actually existing cosmopolitanisms"; Malcomson too does not entangle himself with its geographies.

Adopting the cosmo*political* position signals a break with the elite qualities of the idea of cosmopolitanism, both contemporary and historical, because the contemporary global cosmopolitan is as reliably an informal laborer as a member of the international managerial or capitalist class.

Substantially increased labor migration for especially low wage service sector jobs – including Filipinas in domestic service, distinctively in Hong Kong and Singapore, and Bangladeshi construction workers, who helped build Hong Kong's new airport – has completely upended the question of who is cosmopolitan. The cosmopolitical suggests the contemporary issues at stake: understanding how the new global traveler/citizen lives through and forms ideas about processes that are shot through with politics, of identities, economies, and differences in cultures encountered in life paths of high mobility. Where the cosmopolitan ethic would appear to uphold liberal democratic values, cosmopolitical awareness understands how apparently "liberal" ethics are regularly uncomfortable with difference, especially in class and colored forms. In cosmopolitics, the theoretical universality of the cosmopolitan existence, in the historic Kantian ideal, embedded in the Enlightenment project, has yielded to the understanding of politically and geographically constituted differences in internationalized consciousness, which, in the area studies mode, presupposes no Western authority.[8]

If the cosmopolitical condition courts difference, then what conditions allow cultures of civility and tolerance to concentrate and flourish? In Shanghai – "an exceptional place where natives welcomed sojourners, while everywhere in the country the normal pattern was the reverse" – local society was historically characterized by "openness, amiability, tolerance, flexibility, and so on" (Lu, 1999, pp. 37, 36). The structures of state and society in the cosmopolitan environment are regularly flexible or institutionalized in ways that embrace diversity. In the mercantile period in Melaka, where diversity constituted and defined a settlement, the elite was neither large nor entrenched, and the settlement's population, traders and sailors of diverse origins, waxed and waned with lucrative trade opportunities or the shift of the monsoon. In James Scott's (1997, p. 7) view, "Melaka in 1500 *before* the Portuguese conquest was with its polyglot, trading population probably more diverse, open and cosmopolitan than its contemporary trading port, Venice." In the first half of the twentieth century that status in Asia belonged to Shanghai, a city so open to the world that during the Second World War the port could be entered without visas or immigration documents. Jewish refugees from Nazi Germany were barred entry from some countries but a community of some 20,000 Jews settled in Shanghai from 1939 to 1941 (Berenbaum, 1993, p. 59). The international community in Shanghai counted members of more than fifty nationalities. Shanghai's cosmopolitanism was about both a literal openness of minimal controls on population mobility, and an interest among its population in international events and diverse cultural, political, and economic forms. As Leo Lee (1999, p. 315) has written about Shanghai, "if cosmopolitanism means an abiding curiosity in 'looking out' – locating oneself as a cultural mediator at the intersection between China and other parts of the world – then Shanghai in the 1930s was the cosmopolitan city par excellence."

In the wake of the rhetoric of the Maoist era that sought to repress regional differences in support of the national socialist revolution, contemporary Chinese authors have been finding how ideas about regional culture have been maintained, and reemerged in new forms. In China under reform, an immensely popular book, Yang Dongping's (1994) *City Monsoon: The Cultural Spirit of Beijing and Shanghai*, provides a street-smart assessment of urban subcultures through a running account of comparisons between the characteristics of people from Beijing and Shanghai. In one comparison, which upholds Lee's perspective about Shanghai cosmopolitanism, Yang (1994, p. 462) has observed "those various 'crazes' in Bejing, from 'break-dancing' to *hulaquan* (hula hoops) will never drive Shanghainese crazy. Few things can get them excited except going abroad or investing in stocks." Critics have dismissed *City Monsoon* as stereotyping, but Yang's self-reflective perspective has proved disarming, as he writes, "If not scientific, at least it is interesting." In more systematic moments, Yang (p. 475) tallies aspects about standards of living, including how the city of Shanghai has the largest number of local clubs and special interest organizations, from "Jazz Band for the Aged" to support groups for people facing health problems and family hardships. The book's popularity, in multiple printings, speaks volumes to contemporary interest in place identity and the rise of new urban cultures under reform. If the cosmopolitan has a geography, cities of high mobility, whose populations have defined diversity and whose broadly scaled interactions have generated new cultural and economic forms, are reliable sites of cosmopolitical possibilities.

The idea of cosmopolitics tempers the often naively upbeat discourses of cosmopolitanism with a measure of recognition about class differences, and in this case, about the material geographies they produce. For example, writings about the cosmopolitan urban milieu often simply point to their distinctive built environments, in monumental architectural forms, as if representative of cosmopolitan characteristics. Accounts of pre-war Shanghai readily declare the city's stock of art deco buildings as the largest in the world, if not the finest. But a rhetorical style of acclaiming the built environment treats the material urban scene as if it were a straightforward signifier of cosmopolitan life. This could only be comopolitanism's elite representation. The processes of producing the built environment – the urbanization of capital – in the concentration of speculative investment in the building and rebuilding of high rent properties, are projects of political economic elites. The urban cores of Hong Kong and especially Shanghai have both been subject to considerable contemporary redevelopment, and both now claim some of the most prominent high rise buildings in the world. But rather than being simply symbolized in the monumental built environment, the case can be made, especially for Hong Kong, that urban cosmopolitical possibilities are emerging in the tensions between the control over land use and public space, and local resistances to the minimization of public space and public disclosure in the process of planning the built environment.

Property development in Hong Kong

With the onset of the reform era after 1978, Hong Kong entered a new era of entrepôt activity, as the single largest source of capital in China. Hong Kong became the place where political and economic elites have defined how capital is organized and deployed in China, and even what new development designs will be inscribed in China's landscapes and who will build them. This is the material reality of how "Shenzhen is Hong Kongized, Guangzhou is Shenzhenized, and the whole country is Guangdongized." Land development fever in China arguably had its origins in property development in Hong Kong. Like "zone fever," the rise of local Hong Kong firms in the property development sector was a relatively unforeseen result of larger-scale transformations, in which the position of Western producer services firms in Hong Kong relatively diminished during the period of transition toward Chinese sovereignty (Cartier, 1999a).

Writing from the perspective of the recessionary period in 1982–3 and 1985, which significantly depressed property prices in Hong Kong, Nigel Thrift (1986) predicted that producer service industries would substantially depart the colony in the run-up to 1997. Corporate producer services – banking, finance, insurance, law, real estate, property development, architecture, and engineering – generally provide services to other industries. Hong Kong's world city status rose with the expansion of its banking industry and related financial services, but insecurities about its future economic stability gave rise to ideas about substantial capital flight, relocation of corporation headquarters, and emigration of the transnational capitalist class. Yet just ten years later in 1992–3, Hong Kong property prices would be at an all time high. Only the post-handover recession, engendered by the larger regional financial crisis from 1997 to 1999, fundamentally stalled Hong Kong's economy, and so any immediate effects of the retrocession to Chinese sovereignty itself on the Hong Kong economy cannot be known. Hong Kong's service industries grew through the 1980s with the expansion of the Chinese economy and especially high growth in neighboring Guangdong province. The terms of the 1984 Joint Declaration guaranteed land lease arrangements in Hong Kong, and property prices continued to rise with renewed economic growth and a shortage of affordable housing. As Si-Ming Li (1990) has argued, Annex III of the Joint Declaration constrained the state's ability to dispose lands in undeveloped areas on the urban fringe, which prompted higher prices and land use intensity in the urban cores. Property prices doubled between 1991 and 1994, which coincided with new land lease sales opportunities in China at lower costs, and Hong Kong investors sought properties especially in Guangdong (Sender, 1992, 1993). In this regional atmosphere of "land development fever," new reclamation projects commenced on both sides of Victoria Harbour, and especially for transportation infrastructure associated with the airport facility at Chek Lap Kok. All told, land development hastened in Hong Kong during the 15 years running up to 1997.

Exhibitionary complex

In the Hong Kong central business district, a dense agglomeration of skyscrapers symbolizes the significance of producer services industries in local and regional economies, and the role of the state in facilitating high rise development. High rise Hong Kong would also appear to represent a certain cosmopolitan ethos, in dramatic building designs produced by leading transnational architecture firms. But the idea of the "exhibitionary complex" inverts the apparent public significance of the monumental built environment to focus on the buildings instead as spectacular representative forms of state power. The exhibitionary complex is the state's public side of the Janus face of power, by contrast to the relatively concealed disciplinary and carcereal elements. In Foucault's (1977) formulation of state disciplinary tactics, the individual is the object of the state's gaze and thus the subject of surveillance and control. The exhibitionary complex, by contrast, as Tony Bennett (1994, p. 126) has explained, "reversed the orientations of the disciplinary apparatuses in seeking to render the forces and principles of order visible to the populace . . . through the provision of object lessons in power – the power to command and arrange things and bodies for public display – they sought to allow the people . . . to become the subjects rather than the objects of knowledge." Through "object lessons in power", the populace will know and identify with the power of the state and its elite alliances. From this perspective, the Hong Kong government's primary "object lesson" is the spectacular imprimatur on rapid accumulation through high rise property development. In its instantiation of mega-scale office towers through the planning process, the state underwrites the accumulation goals of local property development firms. The monumental built environment it creates, in the state–property development alliance, speaks to the public: wealth, success, world city, leading center of the Asian regional and world economies. While Hong Kong's middle class and upper classes hold wealth in property, through real property and property stocks, the working and lower classes and some of the middle class, who form greater than 40 percent of the population, live in state housing, in high rise apartment blocks, which in their own way exhibit the state's power. At both ends of the economic spectrum, the Hong Kong populace is surrounded by the state's authority embedded in the high rise built environment.

In the 1990s, though, Hong Kong's property development elite experienced new challenges from the public sphere. Controversy over new and higher high rise buildings rose in response to an 88-story tower planned for Central district on Hong Kong Island, which would become the tallest building in Central and partially block the view of the Peak, the highest point on the island and backdrop to the city's international landmark viewscape. The International Finance Centre is designed to symbolize Hong Kong's position as a world financial capital, in an obvious state–property development alliance, in part by housing the leading financial institutions of the SAR, including the stock market and the monetary authority (figure 8.1). The developer of the project is the Hong Kong Mass Transit Railway Corporation (MTRC) and the architect is the US firm Cesar Pelli & Associates, which has received considerable notice in the region for the design of the Petronas Twin Towers in Kuala Lumpur. The institutional character of the MTRC defines the crux of the state–property development alliance in Hong Kong: brought into existence by a government legislative bill, the MTRC was the first public statutory corporation in Hong Kong and, unlike most state transportation interests, it was charged with operating at a profit and empowered to acquire and develop property (Harris, 1978, pp. 102–13). To finance the project, the MTRC formed a consortium with Sun Hung Kai Properties and Henderson Land, which are two of the four largest local property development companies in the SAR (*Building Journal Hong Kong*, 1992).[9]

The International Financial Centre tower raised the stakes of public acceptance for high rise buildings. Concerns over lack of public disclosure about the project began when the reclamation for the Hong Kong Station project, the city terminus for the airport rail link and base of the International Financial Centre project, simply began in 1993, unannounced by government. That year, the MTRC submitted a plan to the Planning Department for two 40–story buildings on the site. But in 1996 the government Planning Department approved a modified scheme for the 88–storey tower and an accompanying smaller tower in the same

Figure 8.1 Rendition of the future Hong Kong central business district.
Source: ARCHITECH Audio-Visual Ltd.

design, and admitted that it had waived existing height guidelines designed to preserve the view of the Peak (Wan, 1996, p. 3). In addition, the MTRC "refused to release illustrations of two towers it plans to build on the Central and West Kowloon reclamations because, it says, the public would not understand them." Local legislators responded critically and urged the MTRC to release the plans "in view of the public concern" (Tacey, 1996). The MTRC plans another tower for the West Kowloon reclamation, which, in pairing with the International Financial Center, will "create a gateway . . . helping to define Victoria Harbour" (Holland, 1996). In this context and under public scrutiny, the MTRC was compelled to make its case to the press. Instead of addressing the sheer height of the International Financial Centre tower, it promoted the importance of the project design as "suitable for a landmark building that would be a symbol of Hong Kong" (Holland, 1996). The MTRC further insisted that the higher tower would double the amount of open space at the site and even "'green' Central." The public relations discourse attempted to transform the meaning of the building, from a mega-development project blocking the viewscape into a public amenity. But as Alexander Cuthbert (1997, p. 302–3) has aptly noted in previous research on Hong Kong, so-called open space associated with tall buildings is generally not treated as freely usable public space but rather as an extension of the corporate space of the building. Local planning boards responded with a campaign to influence the developers to turn the 88th floor of the tower in to a "unique and cheap" viewing area where access would cost no more than HK$30. One promoter of the viewing area declared, "Hong Kong people have a right to go to the top of the building as they will have given up their view of the Peak" (Poole, 1997). Architects' descriptions of the project suggested a possible viewing area on the 88th floor, and so in their demands to have it, public representatives challenged the potentially disingenuous position of the state–property development alliance on providing public access.

New environmental movements

After extensive reclamations undertaken in the early 1990s for the new airport, the International Financial Centre, and related infrastructure, Hong Kong citizens called for a moderated approach to harborfront land reclamation. The Society for the Protection of the Harbour emerged as the first major grassroots organization to mandate a scheme of controls for land reclamation in central urban areas of Hong Kong. The Society for the Protection of the Harbour began as a private citizen initiative and enlarged to attract a diversified constituency and political support sufficient to generate legislation regulating reclamation projects. The day before the return of Hong Kong to Chinese sovereignty, the Hong Kong Legislative Council adopted the Protection of the Harbour Ordinance, which requires government to receive approval of the Legislative Council to engage in reclamations. The Society specified its opposition to the state's plans to reclaim 5 km^2 of the harbor, which would exceed the total area reclaimed over the past 150 years and render the harbor 800 meters wide at its narrowest point. As the petition argued, "in the history of mankind . . . there has never been a city which has converted its harbour into a river" (Petition of the Society for Protection of the Harbour, 1996, p. 2). Despite attempts to repeal it, the Harbour Ordinance has remained in force since it was adopted in June 1997. In 1998, in response to the regional economic downturn, property development projects stalled, and the immediacy of the reclamation debate subsided. Nevertheless, the Society for the Protection of the Harbor initiated a strong public discourse about local environmental quality and problems of unfettered development.

Beyond the urban cores, contemporary environmental concerns in Hong Kong have played out most distinctively in the New Territories at

Deep Bay. In the early 1990s, a coalition of supporters led by representatives of government and the Hong Kong office of the World Wide Fund for Nature successfully conserved a tidal wetland, the Mai Po Marshes, from development projects planned by leading Hong Kong developers, including Henderson Land and Li Ka-shing's Cheung Kong Group. The Mai Po Marshes site was successfully listed for protection under the Ramsar Convention on Wetlands[10] in 1995 and became Hong Kong's most celebrated environmental project. The Mai Po and Inner Deep Bay Ramsar site is an internationally important site for over 180 migratory bird species, and the protected area, the marsh and two buffer zones, encompasses the most extensive mangrove community in Hong Kong (World Wide Fund for Nature Hong Kong, 1996) (map 8.1b). With the repatriation of Hong Kong to Chinese sovereignty, China gained its only marine Ramsar site on the Asian-Pacific flyway.

The Ramsar site at the Mai Po Marshes is a small part of the larger Deep Bay ecosystem, which encompasses the Guangdong–Hong Kong border area and the Shenzhen SEZ. On the Shenzhen side of the bay, another pocket of mangrove shore has come under protection in Futian district of Shenzhen city. The Futian Nature Reserve was established in 1988 under the Chinese Ministry of Forestry reserve system. But nature conservation in China's system of environmental reserves has proceeded haphazardly, and many reserves have been subject to pressures of new economic activities in the climate of "development fever." Economic activities have encroached on the reserve at Futian, including the establishment of the Futian duty-free zone, one of the many types of special zones, and a road construction project. Futian Reserve might have succumbed to Shenzhen's rapid development, but the road work was stopped by positive state response to unprecedented legal action mounted by a coalition of local Shenzhen and Guangdong provincial authorities, local scholars, and Futian Reserve managers (Hua, 1994; Zhu, 1995). This test of China's relatively new environmental laws suggests that elements of an internationalized environmental regime are also taking hold in Guangdong, where policy has been "one step ahead" (Vogel, 1989) in China under reform.

The rise of environmental activism in Hong Kong and Shenzhen is a recent phenomenon, and similar to that in other world cities, where economic, political, and social concerns have coalesced in creating conditions for a new environmental outlook. In the Hong Kong business sector, the Private Sector Committee on the Environment formed in 1989 to promote local business interest in environmental quality. In government, values expressed in Hong Kong's Legislative Council (dismantled by Beijing after July 1997) began to shift when a minority of seats became popularly elected from 1991 to 1996 and several candidates ran in part on environmental quality platforms. WWF-Hong Kong and Friends of the Earth gained considerable notice in Hong Kong and Guangdong through the 1980s and 1990s after leading successful protests against private development projects in environmentally sensitive areas, and in mutual support of the Mai Po and Futian reserves. In Guangdong, Guangzhou "has been frequently cited as a model case of environmental management by law in China" (Lo, 1994, p. 39). The efforts of new local environmental organizations, new political parties, and the work of local offices of international environmental organizations, in combination with China's evolving regime of environmental law, signal the evolution of new social movement activity in Hong Kong and Guangdong. In the context of the typology of environmental movements offered by Manuel Castells (1997, p. 112), these movements are strategy-oriented and promoted by concerned citizens with definite ideas about long-term sustainability and the problems created by excesses of rapid land development and political economic power. These relatively new environmental priorities reflect high levels of economic development in Hong Kong, but also an internationalized outlook on environmental quality and environmental protection. This is especially the case in Shenzhen, where attempts to implement such world city-standard reforms have distinguished the municipal government's efforts to transform Shenzhen from largely a manufacturing zone into a diversified world city.

Rebuilding Shanghai

By the late 1980s, Shanghai was a city under a forest of construction cranes, widely observed to be the urban area most under construction in the world. Modernization of the industrial plant, such as the Baoshan Iron and Steel facility (map 7.1), proceeded in Shanghai during the Maoist era, but otherwise the physical infrastructure and built environment of the city lay in relative states of decay. Under reform, the built environment became subject to widespread rebuilding, especially for speculative development in office buildings and hotels, in addition to new public projects designed to reinstantiate Shanghai's world city status, such as the Shanghai city hall, museum, opera house, and subway. In the 1990s, introduction of the housing reforms and other related land use measures encouraged construction of new housing and propelled redevelopment of the residential landscape. In the housing policy arena, Shanghai replaced Guangzhou as the area "one step ahead," especially in the introduction of a mortgage loan system and other measures, including the government's housing fund, based on the Singapore model, designed to promote home ownership (Chiu, 1996; Wang and Murie, 1996).

The largest-scale transformation of the Shanghai municipality in the 1990s is its shift to urbanize the peninsula known as Pudong, the area east of the Huangpu river. The transformation of Pudong, at four-fifths the size of Singapore, from a largely agricultural area into an urbanized area of industry and housing, has created practically a whole new half of a central city for Shanghai. Along the Pudong shore Shanghai planners have concentrated the city's new financial center and the country's tallest skyscrapers to create a landscape of global proportions (figure 8.2). The first extraordinary skyscraper is the 88-story Jin Mao Building, designed by the Chicago office of Skidmore, Owings, Merrill. The building bears a postmodern design style along the lines of Chinese-pagoda-meets-millennial-Chrysler-Building (figure 8.3). When it was completed in 1998, the skyscraper became the tallest in the country. The engineering standards for the site, which required construction to withstand 200 k.p.h. typhoon winds as well as earthquakes, meant stablizing the underlying alluvial soil formation with 80-meter piles. The city's even taller blockbuster skyscraper will be the Shanghai World Financial Center, designed by the New York firm Kohn, Pederson, Fox (figure 8.4). More than any another other single structure in Shanghai, this building is designed to command prominence in the competitive landscape of international finance. It will be taller than Hong Kong's new International Finance Centre, and, arguably more futuristic in architectural style, symbolic of leading perspectives in urban and economic development. Contemporary global competition over tall building prominence has become so intense that the city of Chicago and Skidmore, Owings, Merrill have garnered support sufficient to raise the world's tallest skyscraper, at 108 stories, by 2004 in Chicago, the city that is the historic capital of skyscraper engineering (Skidmore et al., 1999). In response, the Shanghai World Financial Center may be built taller to ensure its first place position in the world list of tallest buildings (Mori Building, 1999).

The amount of and different types of capital (money, social, technological) devoted to high rise commercial development, as constellations of the activities of international service industries, contrasts sharply with the variable quality of new housing development in China. The development of a reliable supply of basic, affordable, and good quality housing has been hampered by oversupply of high end "villas" and low quality construction in moderately priced properties. In Shanghai, the vacancy rate in both commercial and residential properties was as high as 60 percent in the 1990s (Haila, 1999, p. 583). Housing reforms of 1995 and 1996 promoted increased development of basic housing and better terms for home mortgages, which made home purchase a more realistic

Figure 8.2 Rendition of the future Pudong New Area, Shanghai.
Source: Mori Building.

proposition. But the space economy of housing redevelopment has challenged stable household formation and the quality of daily life. Larger-scale planned restructuring of the urban space economy has reduced residential land use in the central cities and increased residential density in suburban areas. Municipalities generate considerable rents by transforming state housing to commercial uses, which has further compelled relocation of housing. This restructuring of land use significantly impacted central Shanghai, where at least 400,000 households, or roughly 1.6 million people, were forced to relocate outside the central city to make way for commercial redevelopment of the urban core (Li, 1997, pp. 201–2). Neighborhoods around Huaihai Road, the number two commercial district in the city after Nanjing Road, were especially impacted. Local residents organized street protests against mandated relocation, spurred by the lack of choice and public decision-making in the process. For many households, relocation to new suburban peripheries substantially increased the cost and duration of the journey to work.

Still, on a national scale, Shanghai and other major cities remain the most desirable places to live. Construction of new housing and privatiza-

Figure 8.3 The Jin Mao Building
Source: Skidmore, Owings & Merrill LLP; photograph by Hedrich Blessing.

tion of the housing market have proceeded most rapidly in Shanghai and Guangdong province, centered in Guangzhou, Shenzhen, and neighboring areas. Local officials have structured an economic relationship between urban desirability and the stalled housing market by offering *hukou* in exchange for local housing purchase. In 1994, Shanghai, which

Figure 8.4 The Shanghai World Financial Center.
Source: Skidmore, Owings, Merrill.

has been historically understood as having the strictest system of residence control, introduced the *lanyin hukou* policy or blue chop *hukou*, which offered local residency in exchange for local investment (one million yuan for a Chinese national) or for purchasing an expensive "foreigners-only" private apartment (Wong and Huen, 1998, pp. 978–

81). The system quickly became more flexible, and allowed anyone who purchased a regular housing unit in Pudong to become eligible for local *hukou*. The municipality continued to adjust the terms of investment in exchange, so that three *hukou* could be obtained by purchasing a flat in outlying districts, or one *hukou* for every 200 m^2 of office space. Thus the *hukou* system in China's leading cities became more flexible, but only on class terms, and further reinforced limitations on real and social mobility generated by uneven income distribution in society. Shenzhen introduced a similar blue chop *hukou* system in 1995. The *hukou*-for-sale scheme actually originated at lower scales in the settlement hierarchy, first at the township scale in Anhui and Henan provinces in the 1980s, and spread to other towns and counties (Wong and Huen, 1998, p. 977; Chan and Zhang, 1999, p. 836). In this way, from both ends of the urban hierarchy, the state has formally commodified legal mobility strategies. The activities of millions of people who have relocated informally should eventually challenge the integrity of the *hukou* system. In the meantime, legal *hukou* has become a fulcrum by which to leverage the success of the housing reforms.

New cosmopolitanisms?

In the global flux of ideas, cosmopolitanism has rapidly become an element in the contemporary discourse of regional government. Asian regional leaders have adapted the rhetorics of cosmopolitanism to local policy-making and new plans to promote economic growth. For China, contemporary concerns about globalization hinge on the potential domestic impacts of the WTO and other supra-state political and economic organizations, and achievement of a cosmopolitan revival in Shanghai and other major cities through rapid economic development and new cultures of consumption. In Taiwan, interest in new ideas about political globalization, in line with Held's discussion of a global cosmopolitan democracy, should hold special appeal in the face of the PRC's enduring modernist stand on state territoriality. In Hong Kong and Singapore, both governments are promoting continued growth based on the goal of achieving first-tier world city status and through a combination of service sector and information technology industries. Chief executive Tung Chee-hwa (*Xinhua*, 1999a) wants Hong Kong to become "a world-class, knowledge-based cosmopolitan city" vying for global significance with New York and London, and, ultimately, "the most cosmopolitan city in Asia." (It remains to be seen whether the government's newest mega-development project, Hong Kong Disneyland, set for construction on the edge of Lantau Island and a project in scope and scale comparable to the airport, will contribute to the production of a cultural milieu on a par with those of New York and London.) Prime Minister Goh Chok Tong, in his 1999 National Day speech, discussed a vision for Singapore as a "Renaissance city," in which economic growth based on high technology should propel new cultural vibrancy in arts and society, arenas which might be better developed in Singapore. The Prime Minister urged Singaporeans to adopt "a Silicon Valley state of mind" to promote creativity and draw quality industrial and human capital (*Business Times*, 1999). At the World Conference on Model Cities, the Prime Minister explained, "Our goal is to turn Singapore into a cosmopolis – an attractive, efficient and vibrant city exuding confidence and charm, a magnet hub of people, minds, talents, ideas and knowledge" (Tan, 1999). Goh Chok Tong's articulation of world city status has also depended on a characterization of Singaporeans as "cosmopolitans" and "heartlanders," in which the former are the members of the international managerial and capitalist classes, and the "heartlanders" are locally oriented, occupying low-skilled service sector positions in the Singaporean economy. By this definition, the Singaporean cosmopolitan is a member of the educational and economic elite and a loyal citizen who works in step with the nationalist vision. This characterization may suit the Kantian cosmopolitan ethic, but it is not the cosmopolitical condition that supports cultural difference. Such a dualistic categorization based on class ascriptions has engendered widespread criticism from Singaporeans.

The state discourses of cosmpolitanism in Hong Kong and Singapore are bound up in concerns about emigration of the transnational professional class. In both cities government programs seek to attract people associated with high-end service sector activity, while limiting numbers of people in low paid job sectors. Hundreds of thousands of construction workers, usually male, and domestic workers, overwhelmingly female, have migrated to these cities, but only on short-term contracts and without the opportunity to gain local citizenship. In Singapore the state has prioritized attracting "global talent" to "broaden the outlook of Singaporeans" (Chen et al., 1997), and the "need for a Chinese-proficient elite" (*The Straits Times*, 1997) to maintain local heritage, traditions of Chinese culture, and strong economic relationships with China. In these ways, regional ideas about cosmopolitanism are, foremost, elements of growth-oriented development strategies.

We may learn something different about cosmopolitical possibilities in the region from the late twentieth-century experience of what has been called the last remaining ethnic group in Hong Kong Chinese society, the Tanka (Cantonese for Dan). A community of Tanka people or "boat people" has continued to live in Hong Kong at Tai O on the western shore of Lantau Island (map 8.1a). In the early 1990s, Hong Kong government constructed high rise housing designated for the Tanka community, but upon completion, 80 percent of the population resisted relocation and remained in their "pile houses" along the water (Fung, 1998, p. 76). They refused to become "metropolitanized," and news accounts portrayed the population as set in their ways, backward, and disconnected from the world economy. A subsequent plan sought to make Tai O a tourist center, with a boat anchorage, plaza, and museum. Explanations for the condition of the Tanka have continued to rely on historic accounts in terms "other" to Han society, in their historic illiteracy, absence of ancestral records, and social prohibitions against taking part in official examinations, holding official titles, owning land, and marrying into landed society. Significant changes occurred in Tanka society and economy during the twentieth century, but ideas about their social marginality have been repeated by representatives of the press and the state. An impetus for the housing resettlement was a development plan for Tai O, and so the discourse of marginality revived in response to the new centrality of the site in government plans. The attempt to relocate the Tanka is not the first time government intervened in their settlement. In the 1970s the state enumeration scheme to manage informal housing, or "squatters," was extended to the Tanka houses, in which each dwelling was numbered and identified, room by room (Fung, 1998, p. 91). State management of informal housing proscribed additions, which constrained long-term suitability of the housing.

Thus marked and organized, the space of the Tanka community was brought under surveillance and control by the state. This narrative about the Tanka is a transhistorical social construction of marginality by the state and its elite representatives, whether imperial or colonial, in order to control land resources and maximize accumulation for elite political economic interests. Under the imperial gaze, the apparent geographical problem of the Tanka was their intimacy with the sea. The view of the modern state treats the Tanka resistance in terms approaching incivility, for their denial of interest in metropolitan housing. They refused to forgo their diurnal world of the sea for the concrete geometry of industrial housing, refused to trade their rough and concatenated spaces of community for the uniformity and individuation of the urban flat. In the case of the Tanka, the discourse of marginality (a fundamentally economic and class position) masks other histories and experiences, and contrasts with the state-sanctioned values of metropolitan society in Hong Kong. The Tanka position is neither marginal nor ideal, neither backward nor sublime, but is one which has understood how to maintain community and difference in the face of inimical power relations and their basis in the global economy. The Tanka are inheritors of the maritime culture in south China and continue to represent resistance to the norms of the state and landed society. Theirs is a cosmopolitical position, and one rehearsed in other places around the world in response to globalizing processes, one which is in fact central to the dynamic qualities of cosmopolitanism that Asia's leaders would claim to seek.

Globalizing cities

In the high drama of globalization, the protagonists of world cities often compare leading urban industrializing regions as competitors, for capital, corporate headquarters, human resources, tourists, and more. Among China's cities, Hong Kong's retrocession to Chinese sovereignty and Shanghai's reemergence under reform have made the two great port cities of the south China coast the would-be competitors for China's – and much of Asia's – economic leadership. This discourse of urban competitiveness is a popular way of organizing geographical knowledge, one which signifies conditions in common and ties between the two cities more than fundamental differences. The appearance of comparative scholarly studies about social conditions of urban life in Hong Kong and Shanghai underscores how people in different arenas are thinking about these ties (e.g. Yao, 1990; *GLTWYS*, 1998). In a commentary comparing the two cities, the Mayor of Shanghai, Xu Kuangdi, described Hong Kong as "very cosmopolitan" in blunt geographical terms of the regional finance economy: "Countries like Malaysia, Singapore, and Indonesia can borrow money from the American and Japanese banks in Hong Kong. You cannot do this in Shanghai" (Quak, 2000). On this basis, Hong Kong is China's international financial center while Shanghai's financial centrality will remain limited to China. Premier Zhu Rongji underscored the importance of this relationship when he verbally guaranteed Hong Kong's financial role in the region even if China had to slow development of financial services infrastructure in Shanghai (*The Straits Times*, 1999).

By several measures of Chinese nationalist outlook, the internationalism that makes Hong Kong and Shanghai cosmopolitan places also makes them relatively foreign cities on the Chinese domestic scene. Yet a cosmopolitanism on Chinese terms is what also establishes these cities as two of the most desirable destinations for citizens in China. The cultural transformations sweeping China under reform, in forms of popular and high culture, including television, the fine arts, architecture, literature, film, clothing, and debates over sex and gender, also originate and coalesce in these cities, and form a basis for the urban economy. The continued internationalization of these cities, whether through environmental movements, migration, travel, and tourism, or international trade and finance, makes them centers of political, economic, and cultural innovation as well as standard bearers of practice in these spheres. These processes are also the terms of societal formation invoked by the themes of "River Elegy": internationalization based on a maritime cultural economy. The contemporary urban conditions in Hong Kong and Shanghai need not invoke a political commentary on the historic condition of the Chinese state, though, since from the current northern vantage, geographical imaginaries of internationalization in south China are at the basis of the state's twenty-first-century economic project to transform China into an industrialized country and global economic power.

NOTES

1 See the work of Ming K. Chan (1996a, 1997) for analysis of the politics of repatriation and the drafting of the Hong Kong Basic Law.

2 Before it became a policy slogan for Hong Kong, the idea of "one country, two systems" had a short history of evolution in the late 1970s, in the first years of the reform era under Deng Xiaoping. Ming Chan (1997) explains how the policy was first evolved as an attempt to settle peacefully the issue of Taiwan reunification – a switch from the earlier Maoist era policy that Taiwan would be liberated by armed force. In January 1979 Deng Xiaoping, in a meeting with a US representative, explained how after peaceful reunification with the mainland Taiwan would maintain its autonomous status and keep its administrative power, military forces, economic and social systems, and way of life. This overture was detailed in the *santong* and *siliu*, or three links and four exchanges, which meant establishment of ties between the mainland and Taiwan in trade, transportation, and postal service, and exchanges involving economic, scientific, cultural, and sports activities. Chinese leaders saw this policy as fully consonant with larger goal of modernization, and which would be hastened by cooperation with Taiwan. A 1982 Chinese policy document on Hong Kong made use of the phrase for the first time: "the method of one country, two systems will be used to solve the question of sovereignty over Hong Kong and Macau."

3 Analysts have pointed out precedents for President Lee's approach, notably in the writing of Tokyo University professor Mineo Nakajima, who suggested that China should be divided into twelve different regions, in order to pose less threat to its neighbors.

4 This is to suggest that the vocabulary of "flows" stands in for and obscures material processes. Manuel Castells's 1989 writing about the "space of flows" in *The Informational City* arguably popularized the use of "flows." Some of Castells's (1989, p. 349) writings, especially out of context, would appear to deracinate local and regional meanings of place and resistances to globalizing processes: "The fundamental fact is that social meaning

evaporates from places, and therefore from society, and becomes diluted and diffused in the reconstructed logic of a space of flows whose profile, origins, and ultimate purposes are unknown." Such a perspective is characteristic of the early stages of the globalization debates.

5 *Xiafang* policies "sent down" members of the urban and educated classes to relocate to lower administrative levels and rural areas to engage in manual labor, sometimes for years; see Rensselaer Lee (1966) and Thomas Bernstein (1977).

6 The right of abode in Hong Kong became an especially contentious issue after 1997. The Hong Kong Basic Law held that all residents of China who had at least one parent in Hong Kong should have the right to reside in Hong Kong. Estimates of this population ranged from 310,000 to as high as 1.67 million, which caused alarm in Hong Kong. Such a large addition to the existing population of 6.8 million would strain existing resources; it also suggested, on the high estimate, that one out of five Hong Kong men in China had fathered a child out of wedlock. The Hong Kong Court of Final Appeals held up this provision, only to be disciplined by China's National People's Congress in 1999, which took the position that Beijing had the power to rule over decisions that affected both Hong Kong and China at large. The revised decision put conditions on the right of abode and reduced significantly the numbers of potential applicants, by measures such as proof of parental residence in Hong Kong at time of birth, or DNA testing and other means of verification for children born out of wedlock. See Kit Chun Lam and Pak Wai Liu (1998) for a detailed discussion of the issues.

7 See Peter Cheung (1996) for a discussion of the role of the Shanghai leadership in national politics.

8 This cosmopolitical alternative was not one that Kant articulated, particularly, as Harvey explains, if we consider Kant's geographies. Kant's dedication to geography was unparalleled among leading philosophers of his era, but unlike the globalization debates that have problematized the role of the nation-state, Kant's geography depended on the nation-state's geopolitical values and its "civilizing" opportunities. Kant envisioned cosmopolitan existence as an evolutionary process of nation-state development that would ultimately result in an international (literally inter-nation) society subject to international law, with the goal of achieving "perpetual peace." The evolutionary element of Kant's political geography was rooted in Enlightenment Europe, which, in its value system, also underwrote the global imperialist project. Kant's geography was at a loss to explain how people unattached to the nation-state system might achieve cosmopolitan existence. Indeed, Kant had problems with "different" people. Among a range of objections and superlatives, Kant wrote, "Humanity achieves its greatest perfection with the White race. The yellow Indians have somewhat less talent. The negroes are much inferior and some of the peoples of the Americans are well below them" (Kant, in Harvey, 2000, p. 533). One might excuse Kant as a product of his era, as many people are, but such work leaves no basis for an "enlightened" and contemporary geographical

understanding of cosmopolitan existence.

9 The two minor stakeholders in the development consortium are Hong Kong and China Gas, which is involved in energy infrastructure development in the Zhujiang delta, and Sun Chung Estate, a subsidiary of the Bank of China Group, which is the major holding group for several Chinese banks. The Hong Kong government privatized 20 percent of the MTRC in 2000.

10 The Ramsar Convention was the first mutlilateral environmental agreement aimed exclusively at wildlife habitat protection on a global scale, and the only convention dedicated to the conservation of particular ecosystems and associated species.

Epilogue

This preliminary assessment of the region on the south China coast questions ways of understanding regional formation and conceptualizes alternatives, in the context of debates in area studies and in geography, and issues at stake within the region itself in south China. It also provides a set of corrective views that substantially broaden understandings of the coastal region beyond recent questions about rapid economic growth and development. These views of the region are some among the diverse regional understandings about south China, which, in their multiplicity and overlap, work toward a coherent portrayal of regional meaning.

Rewriting the regional formation in south China is an important project in several respects. In the most general context, regions of diverse types and regionalist thinking have become central subjects in debates over the dynamics of globalization. Similarly, the spatialities of global processes are reliably taking shape in regional formations. In these critical ways, organizations of knowledge around regions address contemporary spatial transformations in society and economy. Globalization has also witnessed a proliferation of local and regional articulations of cultural forms, in the reality of local responses and resistances to globalizing forces. The emergence of regionalist thinking is one significant way in which the skeletal framework of the "local and the global" is being filled in with particular sites of transformation. In the context of south China itself, it is important to historicize contemporary economic portrayals of the region because it has been a transhistorical region of international economic activity. Part of what makes contemporary regional dynamism different at the national level is that it has been accompanied by regional identity reversal: south China early emerged as the leading region of reform and an alternative to the historically dominant north. Regions have also gained new significance in China under reform with decentralization of power to the provinces, and regionalism has become an important force of political economic power and identity formation. Just as important is how these "regions also imagine their own worlds" (Appadurai, 2000), and in relation to other regional formations. As Tim Oakes (2000) has assessed, provincial elites have promoted provincial identities in relation to ideas about national culture and the possibilities of attracting global capital. The restructuring of regional economies has transformed the national space economy, and propelled into flux how people understand their own regions and those of others.

In coastal south China windows on the regional past repeatedly reveal a zone of mercantile and maritime cultural economy, in coastal settlement and as the basis of wealth in society. The region found its settlement cores in port cities, but the focus of coastal life also regularly lay at the borders of the marine and riverine environments, where land reclamation was a common practice that demonstrated particularly intensive regional land use and the basis of an accumulation strategy. The south coast was also reliably China's region of foreign contact, the region of the first ports of call for nearly all sea-faring world visitors save those who arrived from northeast Asia. The maritime cultural economy and the mixing of worldviews carried by various agents of trade and exploration, from tribute bearers, mercantilists, pirates, and missionaries, to contemporary capitalists, have lent ideas about the wider world and showed the way to distant shores. The coastal south early became China's region of highest mobility, migration, and, ultimately, diaspora. Distinctive forms of social organization in support of all these practices have also emerged in the region, the paradigmatic variants of what we now call networked social relations. These general conditions, a mercantile and maritime cultural economy, intensive land use and land reclamation, mobility, migration, and diaspora, forms of social organization which supported those practices, and waves of foreign contact, have been the central geographical themes and processes of regional life in coastal south China. Discontinuous, layered, resurgent, these processes have constituted and sustained the region's transhistorical and transboundary cultural economy.

On the basis of a geographically specific cultural and economic history, the south China coast has emerged as a globally significant transboundary region, where historic regional characteristics have transformed and reemerged in new regional entities, unbounded and multiscalar, in the contemporary world order. How should we conceive a region whose formative cultural and economic processes transcend territoriality, and where, in "Greater China," the distanciation between the region of toponymy, the region on the ground, and the region of world imagination is so great that what constitutes the region cannot be set forth in mapped form, or a region whose greater than two millennia of history become a revisionist project when its once frontier position in the national political economic order is chaotically traded for a front seat on the roller coaster of millennial globalization?

I have approached these issues by pursuing alternative conceptualizations of regional formation as scaled, contextual geographies, in a transboundary cultural economic region. The idea of a regional cultural economy is a result of the cultural turn in economic geography, and a (long overdue) recognition of the reality that economic activity is culturally constituted and mediated, and has important symbolic dimensions. The cultural conditions of the economic region depend on understanding contextual geographies through place, and related concepts of embodiment, subjectivity, and identity, and human–environment relations, including history of control over natural resources, and environmental perception. Writing the transboundary cultural economy depends on narrative strategies that address the epistemological divisions opened up in the area studies debates – the questionable separation of historic and contemporary views on China, their parallel divisions into humanistic and social scientific modes of inquiry, and the limits of national bounded territory. This narrative style writes across these boundaries and transcends the conventions of administrative geographies to assess the transboundary and discontinuous region in historic perspective. The approach also draws on contemporary regionalist thinking, which has found alternatives to the conventions of area studies practice and traditional regional geography based in world regions, the nation-state, administrative geographies, and physical regions. Contemporary regional analysis is more interested in regional formation, transboundary regions, interregional relations, world city regions, and so-called marginal regions, approaches that foreground the dynamic and diverse processes of cultural economic activity.

Region and Globalization

Regional formations like south China, internationalized transboundary regions, are distinctive areas of globalizing processes. They are intermediary spatialities between the local and the global, even as their processes of formation gather within them constituent and distant places, and through flows and forces that characterize world economic activity. Regions are a mediating spatiality and a material space of globalizing processes, and, as I have earlier proposed, their slippages in meaning, like those of globalization, are embedded in their spatialities, in being simultaneously unbounded yet discrete and interconnected. Processes of globalization presuppose dynamic transboundary activity and so must unfix regional geographies, even as they also inscribe new material geographies in regional contexts.

The unfixity of the region is both its awkwardness and its elegance. The region as concept has been marginalized and disdained for its apparent absence of uniform theory, yet is emergent as the basis of social organization and economic activity in the contemporary world economy. We will visit this latter point further below. Debate over regional typologies and conditions of regional formation has been a long running exercise in geography and related fields, and regional approaches experienced mixed reviews over the twentieth century. In one important example, a field now known as traditional regional geography was once the core of the discipline. Prevailing critiques of traditional regional geography pointed to its descriptive nature, its identification of regions as if they are "natural" and "out there" rather than socially produced, and an absence of concern with causal processes (Pudup, 1988). The macroregion model, from the China area studies literature, is subject to similar assessments (Cartier, 2002). Traditional methods in regional geography were also concerned with mapping regions, and treating regional boundaries as containers of regional characteristics. In confronting these methdological problems, as Massey (1991, p. 27) once put it, "I remember some of my most painful times as a geographer have been spent unwillingly struggling to think how one could draw a boundary around somewhere like the 'east midlands'." Massey (1984) recognized that understanding causal forces of regional formation meant looking beyond the region to think about national and global processes. This point of conceptualization does not disregard the importance of the bounded region and regional boundaries, which have significant real effects, but recognizes theoretically that the act of drawing a boundary can be a highly arbitrary exercise, and that regional boundaries do not contain regional processes. Moreover, many regions are not formally bounded, and to force them into a mapped spatiality is to risk a certain feigned scientism and even violence, in real terms of political economic contention and war over land territory. This also does not mean that we should avoid mapping regions – because location and emplacement will always matter – but that if we require visual representations of regions, we need multiple two-dimensional maps that portray different regional possibilities, or dynamic visual maps, indeed moving, filmic maps, maps of virtual reality, that capture important qualities of mobility that are at the basis of dynamic regional formation. The spectral technologies of the emergent era should be able to synthesize visually such elements of regional formation.

Regions have become central subjects of analysis in political and economic accounts of world economic activity, and are beginning to emerge in humanist accounts of globalization, and as alternatives to nation-state spatialities. In theoretical research on economic organization, the role of regions has emerged from being considered a *result* of more fundamental economic processes, to being treated instead as the *basis* of economic organization and social life. The reconsideration of the fundamental role of regions is directly related to the reorganization of economic activity in the late twentieth century, in the transformation from the Fordist era of industrial production to contemporary conditions of flexible specialization. In *The Regional World*, Michael Storper (1997) has assembled current knowledge about regional economies and theorized the significance of industrial regions in the concept of "product-based technological learning." This concept combines understanding the dynamic realities of industrial production for world export in the context of local knowledge systems that propel continuous innovation of better and new products, and, in turn, the reorganization of production systems to create them. The implications of this argument are important in several respects. First, this perspective also holds that the formation and significance of economic regions is fundamentally linked to the conditions of late twentieth- and early twenty-first-century capitalism. These conditions, in increased internationalization of manufacturing production, the reorganization of firms, the speed of capital flows, and the new international division of labor, are also dependent in their movement on the technologies of globalization. More than spatialized economic events, this regional approach also foregrounds the social conditions of regional life. The knowledge component of product-based technological learning depends on understanding the conditions of located social interactions and human relationships, and that such learning spaces take place in regional formations. These relationships form the social basis of the region, bind places within the region, and transcend the region in ties to more distant places and regions in the world economy. Based on understanding these intertwined complexities, contemporary regionalization "might be more than merely another localization pattern: it might actually be central to the coordination of the most advanced forms of economic life today" (Storper, 1997, p. 4). The implications of this conclusion would suggest fundamental rethinking of several scholarly fields and arenas of public policy, including macroeconomic theory and the role of the state and structures of governance.

These reassessments have begun to take place. Scholars have described the "new regionalism" as the set of issues around "the 're-emergence' of the region as a unit of economic analysis and the territorial sphere most suited to the interaction of political, social and economic processes in an era of 'globalization'" (Tomaney and Ward, 2000). Widespread assumptions about the existence of a "new regionalism" are predictably emanating from Europe, but also have global dimensions in the recognition of regional economies. Several aspects of the new regionalism cross cut social, cultural, economic, political, and environmental spheres. These centrally include regional economic restructuring, regional consciousness and questions of regional identity, new forms of regional governance, and growing interest in environmental sustainability. The focus on regional identity is clearly related to the enhanced importance of regional economies and politics. The rising concern with environmental sustainability reflects how concern about international environmental quality is often debated at the regional scale of potential policy implementation. Several of these themes are characteristic of regional processes in south China, foremost, regional economic restructuring, new powers of regional governance at provincial and lower levels, the dynamic transformation of regional identities, and environmental movements.

The scholarship of the "new regionalism" finds its lineage in the analysis of economic regions, but what is interesting about the new regionalism is the degree to which it encompasses a range of concerns typically not associated with regional economies, from cultural questions about identity formation to environmental sustainability. We might reasonably add to this list of concerns the gendered characteristics of regional economies, which have distinctively appeared in the transboundary region in south China under reform. I have proposed that analysis of these diverse factors of regional formation is profitably pursued in scaled contexts through allied concepts of place, which have received considerable attention from theorists in human geography and related fields. Mine is certainly a preliminary statement, yet it argues for distinct and still related concepts of place and region that would overcome the problems of conflating these concepts, which appeared in debates over traditional regional geography and its legacy. Ron Johnston's (1991, p. 72) elaborations of this conflation explained a "preference for place (though region is occasionally used as a synonym!) because of its relative neutrality in the history of geography." From the current vantage, though, regionalism has come to the fore and will retain a distinct conceptual position, and so it is important to articulate the distinctiveness of these concepts.

Humanist considerations of regions and globalization have intervened in some of the more difficult elements of regional conceptualization. Focusing on *ideas* about regions leads to understanding the difficulties of attempting to arrive at general definitions about them. Leo Ching's (2000, p. 244) engagement with regionalist thinking argues that "any attempt to empirically ground and define regionalism would only confirm the changeable and indefinable nature of regionalism as an organizing concept." He agrees that "the regionalist imaginary is fundamentally complicit with the globalist project" (p. 237) in its challenges and alternatives to the nation-state, but that the problem of conceptualizing regions lies in their "intermediary relationality, in which the regional imaginary can neither conceive nor consolidate itself as a dominant discourse within the economic and political coordinates of the world system" (pp. 243-4). This is the problem of regional definition and also the significance of the regional spatiality, in its ability to occupy different scale positions, and sometimes in a temporary and mediatory status. Erica Schoenberger (1989, p. 134) observed this problem in other terms when she concluded, "in the final analysis, definitive theories of regional change elude us because the primary agents of change, capital and labour, are so extraordinarily creative, each in their own way." The regional spatiality, in its diverse scale positions, is more elusive and complex than the bounded nation-state form. Regional multiplicity is not a weakness of regional conceptualization, but rather its strength. As Ching (2000, p. 244) has aptly noted, "the effectiveness of regionalist imaginaries lies precisely in the coexistence and overlap of supposedly distinct regional designations." This is what we have seen in south China, where multiple and overlapping regional representations together constitute complex regional meaning.

Region, State, and Identity

Remaining discomforts about the unbounded character of regional processes must reflect the ideological underpinnings of the nation-state system. The forces of globalization have, to varying degrees, ideologically if not materially, undermined the state. They have also resulted in new spatialities, and at various scales. Among geographical forms, the region appears to have become the leading position and intervention in globalization's alternative spatial possibilities. Regionalization, like globalization, may be understood to hold the potential to undermine the priority of state-based territorialities, but may also serve as the basis for the state's interests to reposition itself in response to dynamic economic events. This is the case in the problem of "real estate fever" in south China under reform, in which the state responded to regional overextension in the real estate sector by repositioning land use policy and controls over land use. In regions, centralizing policies are both resisted and appropriated, and examples of how local officials overcame government restrictions on land use bear this out. What is especially critical about regional formation as a dynamic economic form is that it particularly emerges in times of economic restructuring. It must, as Etienne Balibar (1991, p. 89) has observed, because it is "quite impossible to 'deduce' the nation form from capitalist relations of production." Capital and labor are both fundamentally transnational in their operative capacities, and take multiscalar and regional forms of organization, even as national projects occasionally find interest to constrain their mobility. We should expect to see a wave of scholarly rediscovery of the region in the era of shift from industrial to late capitalism. In this regard, rather than the region versus the state position, what might be a more interesting proposition is a historical analysis of regional formation during major periods of economic restructuring.

By comparison to the nation-state, which, as a territorial model and set of political ideologies, entered modern history in the seventeenth century, regions of diverse types are arguably the most common and enduring forms of territorial spatiality. The nation-state project also brought the technologies of survey mapping for boundary-making, an activity which is closely associated with imperialism and capitalism. The boundary-making project, at the basis of conventional regional analysis, has really pressed regionality to conform to state-like forms of territorial expression. We have too commonly not questioned these conditions at the basis of conventional regional analysis. Transboundary and unbounded regions are antithetical to the state in that they do not uphold the nation-state ideal, and, critically, reveal it as a particular rather than as a general historical form. Regions are probably more important over the *longue durée* than nation-states; but like the fictional effects of nationalism, our perspectives on regions have been clouded by the past 300–400 years of the nation-state ideal.

Regional formations also offer an alternative to nationalism as a basis of identity formation. This is true in the supra-state region of the European Union, and to some degree in Southeast Asia, where ASEAN countries have maintained identity positions distinct from non-member countries. Regional identity formations are especially commonly associated with subnational regions at diverse scales, as in China, from south China in general, to the provinces, and more local contexts in the river deltas, cities, and towns. While it has been a common modernist outlook of the nation-state project to view regionally based identity positions as either parochial or threatening to national interests, what is more interesting in the contemporary period is how new regional identities intersect with national identities, other regional identities, homeland identities, and so forth.

Concerning identity formation among the Chinese overseas, the perspective on region (rather than nation) as homeland of origin works to both realistically and ideologically overcome the national imaginary of China as the primary pivot of historic and immigrant identity formation. In the idea of the region of homeland origin, Chinese people outside

China may both claim their roots and maintain positions of citizenship and nation orientation in their resident countries. They may also lay claim to diverse national identities, but it is important to bear in mind perspective on relations of scale: at a large scale, and far from one's homeland, it is commonplace to ascribe national-scale identity markers to travelers, migrants, and people in immigrant communities. Among those communities themselves, by contrast, people reliably differentiate based on subnational and local-scale factors of association, from common provinces to hometown and village origins, and also common dialect and its regional dimensions. Assessing identity through perspectives on place and region, in scaled formation, and in relation to national-scale interests, allows ways of seeing through to the realities of individual and group identity formation. The history of the diaspora in south China also suggests how increasing global population mobility is also arguably better understood in terms of regionally specific migrations.

Diasporic identities, formed along lifepaths of high mobility in the transboundary cultural economy, are comparatively complex and flexible identity formations. These kinds of identity formations have contrasted substantially with the common goals of nation-state formation and nationalism. Because the modernization project sought to command nation-state-based economies and nationalistic loyalties, it attached little importance to thinking about regions, and especially transboundary regions, coastal regions, and world city regions, all types of regions whose formative interactions may implicate diverse transnational connections. The modern nation-state and nationalism share intellectual history with the idea of cosmopolitanism and the rise of an international mercantile economy. In its philosophical origins, Kant "argued that international commerce was a historical condition of the cosmopolitical community" (Cheah, 1998, p. 23). For immediate purposes here, Kant offers early recognition of the relationship between an evolving mercantile economy and a distinctive type of community formation that would articulate meaning of global civil society and an international public sphere. Yi-fu Tuan (1996, p. 157) drew on similar ideas to describe how the merchant is "a cosmopolite, not only because he is at ease in the midst of strangers, but also because, potentially, he can operate on an extraregional and even global scale."

What are the geographies implied by this discussion? This is Melaka in the fifteenth century when the place, following Scott's view, was likely more cosmopolitan than Venice. In Venice, Melaka, the port cities of the south China coast, and other early centers of maritime culture and economy, cosmopolitan communities evolved before the word "international," of sixteenth-century origins, had entered the common lexicon of English-speaking peoples. The new relevance of cosmopolitanism in the contemporary era is its ability to conceptualize shared transnational human experiences and the formation of diverse supranational institutions and organizations. In this context, cosmopolitanism, in signifying societal condition in particular kinds of places and regions, might be rethought as pre-national and post-national, a would-be transhistorical concept if not for the intervening hegemony of nationalism as the primary force and concept of societal organization for the eighteenth, nineteenth, and twentieth centuries. In a related vein, Scott offers a view that virtually explains away the area studies debates – particularly the concern over the durability of paradigms based on nation-states and designated world regions – by recognizing transhistorical cosmopolitanism in Asia's complex cultural formations. "The fact, of course, was that all of these cultures, most especially those of Southeast Asia, were complex alloys at the center, constantly shifting over time, and tied to cultural peripheries and frontiers, which were themselves exotic amalgams. The 'new cosmopolitanism' linked to globalization, urbanization, and mass media is . . . a phenomenon alike in kind if unprecedented in scope and speed" (Scott, 1997, p. 7). The port cities of the south China coast were deeply tied to the cultural spheres Scott invokes. His view also suggests how complex transnational regional cultures have always been an integral aspect of societal formation in Asia, and, by extension, that concern around "boundary breakdown" is somewhat misplaced.

The concept of the region may be resituated in a similar logic. In the era of industrial capitalism and modernization based on the nation-state ideal, we might expect that regional conceptualizations would depend on the bounded territorial region. In an era of late capitalism, with its characteristic flexible organizations of production, forces of globalization have rendered many political boundaries more porous to flows of people, money, and things. Contemporary regionalism, then, is also a more flexible geography that captures characteristic processes of the era in transboundary, transnational, and world city–region geographies. We might even resituate the nation-state within a broad regional framework to underscore how it is one type (rather than the only type) of regional formation, in order to better understand how regions reflect and constitute the spatial processes of our times, and over the *longue durée*.

References

Abbas, Ackbar (1997) "Hong Kong: other histories, other politics." *Public Culture*, 9(3): 293–313.
Abdul Rahman, Tunku (1969) *May 13: Before and After*. Kuala Lumpur: Utusan Melayu Press.
Abraham, Itty and Ronald Kassimir (1997) "Internationalization of the social sciences and humanities: report on a ACLS/SSRC meeting, April 4–6, 1997." *Items* (SSRC), 51(2/3): 23–30.
ACWF (All-China Women's Federation) (1993) *The Impact of Economic Development on Rural Women in China*. Tokyo: The United Nations University.
Agnew, John A. (1989) "Introduction," pp. 1–8 in John A. Agnew and James S. Duncan, (eds), *The Power of Place: Bringing together Geographical and Sociological Imaginations*. Boston: Unwin Hyman.
—— (1993) "Representing space: space, scale, and culture in social science," pp. 251–71 in James Duncan and David Ley (eds), *Place/Culture/Representation*. London: Routledge.
—— (1994) "The territorial trap: the geographical assumptions of international relations theory." *Review of International Political Economy*, 1(1): 53–80.
—— (1999) "The new geopoltics of power," pp. 173–93 in Doreen Massey, John Allen, and Philip Sarre (eds), *Human Geography Today*. Cambridge: Polity Press.
Agnew, John A. and Duncan James S. (1989) "Introduction," pp. 1–8 in in J. A. Agnew and J. S. Duncan (eds), *The Power of Place: Bringing together Geographical and Sociological Imaginations*. Boston: Unwin Hyman.
Allen, John, Doreen Massey, and Allan Cochrane (1998) *Rethinking the Region*. New York and London: Routledge.
Anderson, Benedict (1983) *Imagined Communities: Reflections on the Origin and Spread of Nationalism*. London: Verso.
Appadurai, Arjun (1990) "The global cultural economy." *Public Culture*, 2(2): 1–11; 15–24.
—— (2000) "Grassroots globalization and the research imagination." *Public Culture*, 12(1): 1–19.

Ash, Robert F. and Richard Louis Edmonds (1998) "China's land resources, environment, and agricultural production." *The China Quarterly*, 156: 836–879.
Asiaweek, (1984). "Malaysia's Bukit China controversy." September 28, p. 94.
—— (1985) "A cloud over 'Chinese Hill'." January 11, p. 19.
Audemard, Louis (1960 and 1970) *Les Jonques Chinoises, volumes 3 and 9*. Rotterdam: Museum voor Land-en Volkenkunde en het Maritiem Museum "Prins Hendrik."
Balibar, Etienne (1991) "The national form: history and ideology," pp. 86–106 in Etienne Balibar and Immanuel Wallerstein, *Race, Nation, Class: Ambiguous Identities*. London and New York: Verso.
Barlow, Tani, E. (1994) "Politics and protocols of *funü*: (un)making national woman." pp. 339–59 in Christina K. Gilmartin, Gail Hershatter, Lisa Rofel, and Tyrene White (eds), *Engendering China: Women, Culture, and the State*. Cambridge, MA: Harvard University Press.
Barmé, Geremie (1999) *In the Red: On Contemporary Chinese Culture*. New York: Columbia University Press.
—— (2000) "The revolution of resistance," pp. 198–220 in Elizabeth Perry and Mark Selden (eds), *Chinese Society: Change, Conflict, and Resistance*. London and New York: Routledge.
Bauer, John, Wang Feng, Nancy E. Riley, and Zhao Xiaohua (1992) "Gender inequality in urban China: education and employment." *Modern China* 18(3): 333–70.
Beahan, Charlotte (1975) "Feminism and nationalism in the Chinese women's press, 1902–1911." *Modern China*, 1(4): 379–416.
Beard, William L. (1925) "Recent history," pp. 13–22 in *Fukien: A Study of a Province in China*. Shanghai: Presbyterian Mission Press.
Becker, Jasper (1999) "A failure to control the population: the mainland's one-child policy faces its moment of judgment with the approach of the new millennium." *South China Morning Post*, April 19, p. 17.
Bennett, Tony (1994) "The exhibitionary complex," pp. 123–54 in Nicholas B. Dirks, Geoff Eley, and Sherry B. Ortner (eds), *Culture/Power/History: A Reader in Contemporary Social Theory*. Princeton, NJ: Princeton University Press.
Berenbaum, Michael (1993) *The World Must Know: The History of the Holocaust as Told in the United States Holocaust Memorial Museum*. Washington, DC: United States Holocaust Memorial Museum Publication.
Bergère, Marie-Claire (1998) *Sun Yat-sen*. Stanford, CA: Stanford University Press.
Bernstein, Thomas (1977) *Up to the Mountains and Down to the Villages: The Transfer of Youth from Urban to Rural China*. New Haven, CT: Yale University Press.
BHGD (1997) *Baohu gengdi wenti zhuanti diaoyan zu* (Special investigative group on the problems of conserving arable land). "Luanzhan gengdi chengyin pouxi ji gengdi baohu tujing (Causes and analysis of indiscriminate occupation of arable land and means of arable land protection)." *Fujian tudi yanjiu* (Fujian land research), 2: 19–23.
Biddulph, Sarah and Sandy Cook (1999) "Kidnapping and selling women and

children." *Violence Against Women*, 5(12): 1437–68.
Bielenstein, Hans (1947) "The census of China during the period 2–742 AD," *Bulletin of the Museum of Far Eastern Antiquities*, 19: 125–63.
—— (1959) "The Chinese colonization of Fukien until the end of T'ang," pp. 98–122 in Søren Egerod and Else Glahn, (eds), *Studia Serica Bernard Karlgren Dedicata*. Copenhagen: Enjar Munksgaard.
Birch, Cyril (1976) "Introduction," pp. xiii–xix in K'ung Shang-jen, *The Peach Blossom Fan*. Berkeley: University of California Press.
Blaikie, Piers and Harold Brookfield (1987) *Land Degradation and Society*. London: Methuen.
Blunden, Caroline and Mark Elvin (1983) *The Cultural Atlas of China*. New York: Facts on File.
Blussé, Leonard (1990) "Minnan-jen or cosmopolitanism? The rise of Cheng Chih-lung alias Nicolas Iquan," pp. 245–96 in Eduard B. Vermeer, (ed.), *Development and Decline of Fukien Province in the Seventeenth and Eighteenth Centuries*. Leiden: E. J. Brill.
Boxer, Charle (1968) *Fidalgos in the Far East, 1550–1770*. Hong Kong: Oxford University Press.
Bourdieu, Pierre (1977) *Outline of a Theory of Practice*. Cambridge: Cambridge University Press.
Brandt, Conrad (1958) *Stalin's Failure in China, 1924–1927*. New York: Norton.
Bray, Francesca (1997) *Technology and Gender: Fabrics of Power in Late Imperial China*. Berkeley: University of California Press.
Brenner, Neil (1997) "Global, fragmented, hierarchical: Henri Lefebvre's geographies of globalization." *Public Culture*, 10(1): 135–67.
—— (1998) "Between fixity and motion: accumulation, territorial organization and the historical geography of spatial scales." *Environment and Planning D: Society and Space*, 16(4): 459–81.
—— (1999) "Beyond state-centrism? Space, territoriality, and geographical scale in globalization studies." *Theory and Society*, 28(1): 39–58.
—— (2000) "The urban question as a scale question: reflections on Henri Lefebvre, urban theory and the politics of scale." *International Journal of Urban and Regional Research*, 24(2): 361–78.
Brown C. C. (1952) "Sejarah Melayu, or 'Malay Annals': a translation of Raffles MS 18." *Journal of the Malayan Branch of the Royal Asiatic Society*, 25(2/3): 1–276.
Brown, George P. (1995) "Arable land loss in rural China: policy and implementation in Jiangsu province." *Asian Survey*, 35(10): 922–40.
Brown, Lester R. (1995) *Who Will Feed China? Wake-up Call for a Small Planet*. New York: W. W. Norton.
Bryant, Raymond L. and Sinéad Bailey (1997) *Third World Political Ecology*. London and New York: Routledge.
Building Journal Hong Kong (1992) "Hong Kong's largest landlords." January, p. 41.
Business Times (Singapore) (1999) "National Day rally." August 23, p. 2.
Cahill, James (1962) *The Art of Southern Sung China*. New York: The Asia Society.

Cai Renlong (1995) "Yindunixiya de huaren qiye jituan (Chinese enterprise groups in Indonesia)," pp. 26–88 in Zhu Muheng (ed.), *Dongnanya huaren qiye jituan yanjiu (A Study of Southeast Asian Chinese Enterprise Groups)*. Xiamen: Xiamen daxue chubanshe.

Campany, Robert Ford (1996) *Strange Writing: Anomaly Accounts in Early Medieval China*. Albany: State University of New York Press.

Carlson, Ellsworth C. (1974) *The Foochow Missionaries, 1847–1880*. Cambridge, MA: East Asian Research Center.

Carstens, Sharon (1988) "From myth to history: Yap Ah Loy and the heroic past of Chinese Malaysians." *Journal of Southeast Asian Studies*, 19(2): 185–207.

—— (1999) "Dancing lions and disappearing history: the national culture debates and Chinese Malaysian culture." *Crossroads: An Interdisciplinary Journal of Southeast Asian Studies*, 13(1): 11–64.

Cartier Carolyn L. (1993) "Creating historic open space in Melaka." *Geographical Review*, 83(4): 359–73.

—— (1995). "Singaporean investment in China: installing the Singapore model in Su'nan." *Chinese Environment and Development*, 6(1/2): 117–44.

—— (1996). "Conserving the built environment and generating heritage tourism in peninsular Malaysia." *Tourism Recreation Research*, 21(1): 45–53.

—— (1997) "The dead, space/place, and social activism: constructing the nationscape in historic Melaka." *Environment and Planning D: Society and Space*, 15(5): 555–86.

—— (1998) "Preserving *Bukit China*: a cultural politics of landscape interpretation in Melaka's Chinese cemetery," pp. 65–89 in Elizabeth Sinn (ed.), *The Last Half Century of the Chinese Overseas*. Hong Kong: University of Hong Kong Press.

—— (1999a) "The state, property development, and symbolic landscape in high rise Hong Kong." *Landscape Research*, 24(2): 185–208.

—— (1999b) "Cosmopolitics and the maritime world city." *The Geographical Review*, 89(4): 278–89.

—— (2001a) "'Zone fever', the arable land debate, and real estate speculation: China's evolving land use regime and its geographical contradictions." *Journal of Contemporary China*, 10(28), 445–69.

—— (2001b) "Regional geography in an era of theoretical invigoration: transnational arguments," paper presented at the annual meetings of the Association of American Geographers, New York, February 27 to March 4.

—— (2002) "Origins and evolution of a geographical idea: the macroregion in China." *Modern China*, 28(1).

Cartier, Carolyn and Jessica Rothenberg-Alami (1999) "Empowering the 'victim'? Gender, development and women in China under reform." *Journal of Geography*, 98(6): 283–94.

Casey, Edward S. (1996) "How to get from space to place in a fairly short stretch of time: phenomenological prolegomena," pp. 13–52 in Steven Feld and Keith H. Basso (eds), *Senses of Place*. Sante Fe, NM: School of American Research Press.

—— (1997) *The Fate of Place: A Philosophical History*. Berkeley: University of California Press.

Castells, Manuel (1989) *The Informational City: Information Technology, Economic Restructuring, and the Urban-Regional Process*. Oxford: Blackwell.

—— (1996) *The Rise of the Network Society. Volume 1 of The Information Age: Economy, Society and Culture*. Oxford: Blackwell.

—— (1997) *The Power of Identity. Volume 2 of The Information Age: Economy, Society and Culture*. Oxford: Blackwell.

Castells, Manuel and Peter Hall (1994) *Technopoles of the World: The Making of Twenty-first-century Industrial Complexes*. London and New York: Routledge.

Chan, Albert (1978) "Chinese–Philippine relations in the late sixteenth century and to 1603." *Philippine Studies*, 26: 51–82.

Chan, Anita (1997). "Regimented workers in China's free labor market." *China Perspectives*, 9(Jan./Feb.): 12–16.

—— (1998) "The conditions of Chinese workers in East Asian-funded enterprises: editor's introduction." *Chinese Sociology and Anthropology*, 30(4): 3–7.

Chan, Dennis (1999) "Suzhou handover in 2001." *The Straits Times* (Singapore), June 29, p. 1.

Chan Hwee Leng (1996) "The lust resort: sex tourism in Indonesia." Unpublished honors thesis (BA), Southeast Asian Studies Program, National University of Singapore.

Chan, Kam Wing and Li Zhang (1999) "The *hukou* system and rural–urban migration in China: processes and changes." *The China Quarterly*, 160: 818–55.

Chan, Ming K. (1996a) "Democracy derailed: realpolitik in the making of the Hong Kong Basic law, (1985–1990)," pp. 8–40 in M. K. Chan and G. A. Postiglione (eds), *The Hong Kong Reader: Passage to Chinese Sovereignty*. New York: Armonk.

—— (1996b) "A turning point in the modern Chinese revolution: the historical significance of the Canton decade, 1917–27," pp. 224–41 in Gail Hershatter, Emily Honig, Jonathan N. Lipman, and Randall Stross (eds), *Remapping China: Fissures in Historical Terrain*. Stanford, CA: Stanford University Press.

—— (1997) "The politics of Hong Kong's imperfect transition: dimensions of the China factor," in M. K. Chan (ed.), *The Challenges of Hong Kong's Reintegration with China*. Hong Kong: The Hong Kong University Press.

Chan, Wellington K. K. (1977) *Merchants, Mandarins, and Modern Enterprise in Late Ch'ing China*. Cambridge, MA: Harvard University Press.

Chandra, Rajiv (1994) "Prosperous sons return in search of more profit." *Inter Press Service*, April 8.

Chang, Chia-ao (1958) *The Inflationary Spiral: The Experience in China, 1939–1950*. Cambridge, MA: MIT Press.

Chang, Iris (1997) *The Rape of Nanking: The Forgotten Holocaust of World War II*. New York: Basic Books.

Chang, Peng (1957) "The distribution and relative strength of the provincial merchant groups in china, 1842–1911." Unpublished PhD dissertation, University of Washington.

Chang Pin-tsun (1990) "Maritime trade and local economy in late Ming Fukien," pp. 63–81 in Eduard B. Vermeer (ed.), *Development and Decline of Fukien Province in the Seventeenth and Eighteenth Centuries*. Leiden: E. J. Brill.

Chang, Sen-dou (1963) "The historical trend of Chinese urbanization." *Annals of the Association of American Geographers*, 53(2): 109–43.
Chatterjee, Partha (1986) *Nationalist Thought and the Colonial World: A Derivative Discourse?* London: Zed Books.
—— (1993) *The Nation and Its Fragments: Colonial and Postcolonial Histories*. Princeton, NJ: Princeton University Press.
Chau Ju-Kua (1911) *Chau Ju-Kua: His Work on the Chinese and Arab Trade in the Twelfth and Thirteenth Centuries*. St Petersburg: Printing Office of the Imperial Academy of Sciences.
Cheah Boon Kheng (1979) *The Masked Comrades: A Study of the Communist United Front in Malaysia, 1945–48*. Singapore: Times Books International.
Cheah, Pheng (1998) "The cosmopolitical – today," pp. 20–41 in Pheng Cheah and Bruce Robbins (eds), *Cosmopolitics: Thinking and Feeling beyond the Nation*. Minneapolis: University of Minnesota Press.
Cheah, Pheng and Bruce Robbins (eds) (1998) *Cosmopolitics: Thinking and Feeling beyond the Nation*. Minneapolis: University of Minnesota Press.
Chen Baiming (1997) "Comment on *Who Will Feed China?*" *Chinese Geographical Science*, 7: 11–18.
Chen Kao et al. (1997) "Reeling in talent from all over the world." *The Straits Times* (Singapore), August 26, p. 29.
Chen Guoqiang (1995) "Gulangyu de jianzhu fengge (Gulangyu archectural styles)." *Gulangyu wenshi ziliao* (*Gulangyu Historical Materials*), 1: 179–83.
Ch'en Ta (1939) *Emigrant Communities in South China: A Study of Overseas Migration and Its Influence on Standards of Living and Social Change*. Shanghai: Kelly and Walsh.
Ch'en, Ta-tuan (1968) "Investiture of Liu-ch'iu kings in the Ch'ing period," pp. 135–164 in John K. Fairbank (ed.), *The Chinese World Order*. Cambridge, MA: Harvard University Press.
Chen, Yung-fa (1986) *Making Revolution: The Communist Movement in Eastern and Central China, 1937–1945*. Berkeley: University of California Press.
Chen, Shih-hsiang and Harold Acton (1976) *The Peach Blossom Fan*. Berkeley: University of California Press.
Cheung, Peter T. Y. (1996) "The political context of Shanghai's economic development," pp. 49–92 in Y. M. Yeung and Sung Yun-wing (eds), *Shanghai: Transformation and Modernization under China's Open Policy*. Hong Kong: Chinese University Press.
Chen, Xiangming (1995) "The evolution of free economic zones and the recent development of cross-national growth zones." *International Journal of Urban and Regional Research*, 19(4): 593–621.
Cheng, Chu-yuan (1993–4) "Concept and practice of a 'Greater Chinese Economic Market'." *Chinese Economic Studies*, 26(6): 5–12.
Cheng Hoon Teng Temple (1949) *Cheng Hoon Teng Temple Incorporation Ordinance* (supplement to the Federation of Malaya Government Gazette of September 28, 1949, no. 20, vol. II, notification federal no. 3247, Kuala Lumpur).
Cheng Lim-Keak (1985) *Social Change and the Chinese in Singapore: A Socio-Economic Geography with Special Reference to Bang Structure*. Singapore: Singapore University Press.

Cheng, Tiejun and Mark Selden (1994) "The origins and social consequences of China's *hukou* system." *The China Quarterly*, 139: 644–68.

Chi, Ch'ao-ting (1963) *Key Economic Areas in Chinese History, as Revealed in the Development of Public Works for Water Control*. London: G. Allen & Unwin.

Chiang Soon (1984) "Bukit China geomancy, 'foong swee'." Melaka: unpublished report.

China Business Information Network (1998) "China NPC to revise environmental laws: Qu Geping." May 28. http://www.lexis-nexis.com/universe

China Continuation Committee (1922) *The Christian Occupation of China*. Shanghai: China Continuation Committee.

China Daily (Beijing) (1998a) "Yellow river weeps for water." February 11. http://www.lexis-nexis.com/universe

—— (1998b) "China policy protects resources; freeze on land use continues." April 15. http://www.lexis-nexis.com/universe

—— (1998c) "NPC opens new session; farmland protection examined." September 7. http://www.lexis-nexis.com/universe

—— (1999a) "China – Lee Teng-hui's book exposes his true bias." May 31. http://www.lexis-nexis.com/universe

—— (1999b) "China – training brings women farmers prosperity." June 19. http://www.lexis-nexis.com/universe

China, Imperial Maritime Customs (1880) *Reports on Trade at the Treaty Ports for the Years 1865–1881*. 16 vols. Shanghai: Statistical Department of the Inspectorate General of Customs.

—— (1893) *Decennial Report on the Trade, Navigation, Industries, etc. of the Ports Open to Foreign Commerce in China, 1882–1891*. 2 vols. Shanghai: Statistical Department of the Inspectorate General of Customs.

Ching, Leo (2000) "Globalizing the regional, regionalizing the global: mass culture and Asianism in the age of late capital." *Public Culture*, 12(1): 237–57.

Chiu, Rebecca L. H. (1996) "Housing," pp. 341–74 in Y. M. Yeung and Sung Yun-wing, (eds), *Shanghai: Transformation and Modernization under China's Open Policy*. Hong Kong: The Chinese University Press.

Cho, George (1990) *The Malaysian Economy: Spatial Perspectives*. London and New York: Routledge.

Christerson, Brad, and Constance Lever-Tracy (1997) "The Third China? Emerging industrial districts in rural China." *International Journal of Urban and Regional Research*, 21(4): 569–79.

Clad, James (1984). "No place to rest." *Far Eastern Economic Review*, August 23, p. 13.

Clammer, John (1979) *The Ambiguity of Identity: Ethnicity and Change among the Straits Chinese Community of Malaysia and Singapore*. Singapore: Institute of Southeast Asian Studies, occasional paper no. 54.

—— (1983) "The Straits Chinese of Melaka," pp. 156–73 in Kernial Singh Sandhu and Paul Wheatley (eds), *Melaka: The Transformation of a Malay Capital, volume 2*. Kuala Lumpur: Oxford University Press.

Clark, Hugh (1991) *Community, Trade, and Networks: Southern Fujian Province from the Third to the Thirteenth Century*. Cambridge: Cambridge University Press.

Clifford, James (1997) *Routes: Travel and Translation in the Late Twentieth Century*. Cambridge, MA: Harvard University Press.
Cochrane, Thomas John (1913) *Survey of the Missionary Occupation of China*. Shanghai: The Christian Literature Society for China.
Cohen, Paul A. (1984) *Discovering History in China: American Historical Writing on the Recent Chinese Past*. New York: Columbia University Press.
Cohn, Carol (1987) "Sex and death in the rational world of defense intellectuals." *Signs*, 12(4): 687–718.
Collis, Maurice (1946) *Foreign Mud: Being an Account of the Opium Imbroglio at Canton in the 1830s and the Anglo-Chinese War that Followed*. London: Faber and Faber.
Comaroff, Jean and John L. Comaroff (2000) "Millennial capitalism: first thoughts on a second coming." *Public Culture*, 12(2): 291–343.
The Conservation Atlas of China (1990) Compiled by Changchun Institute of Geography, Chinese Academy of Sciences. Beijing: Science Press.
Corbin, Alain (1994) *The Lure of the Sea: The Discovery of the Seaside in the Western World, 1750–1840*. London: Penguin Books.
Cordier, Henri (1902) "Les marchands hanistes de Canton." *T'oung Pao*, 3: 281–315.
Cornue, Virginia (1999) "Practicing NGOness and relating women's space publicly: the women's hotline and the state," pp. 68–91 in Mayfair Mei-hui Yang, (ed.), *Spaces of Their Own: Women's Public Sphere in Transnational China*. Minneapolis: University of Minnesota Press.
Cosgrove, Denis (1993) "Landscapes and myths, gods and humans," pp. 281–306 in Barbara Bender (ed.), *Landscape: Politics and Perspectives*. Providence, RI, and Oxford: Berg Publishers.
Coughlin, Richard J. (1960) *Double Identity: The Chinese in Modern Thailand*. Hong Kong: Hong Kong University Press.
Crane, George T. (1996) "'Special things in special ways': national identity and China's special economic zones," pp. 148–68 in Jonathan Unger (ed.), *Chinese Nationalism*. Armonk, NY: M. E. Sharpe.
Croizer, Ralph (1977) *Koxinga and Chinese Nationalism: History, Myth, and the Hero*. Cambridge, MA: East Asian Research Center, Harvard University.
Croll, Elizabeth (1978) *Feminism and Socialism in China*. London: Routledge & Keegan Paul.
—— (1983) *Chinese Women since Mao*. London: Zed Books.
—— (1994) *From Heaven to Earth: Images and Experiences of Development in China*. London and New York: Routledge.
—— (1995) *Changing Identities of Chinese Women: Rhetoric, Experience and Self-perception in Twentieth-century China*. Hong Kong: Hong Kong University Press.
Cronon, William (ed.) (1996) *Uncommon Ground: Rethinking the Human Place in Nature*. New York: W. W. Norton.
Crouch, Harold (1992) "Authoritarian trends, the UMNO split and the limits to state power," pp. 21–43 in Joel S. Kahn and Francis Loh Kok Wah (eds), *Fragmented Vision: Culture and Politics in Contemporary Malaysia*. Honolulu: University of Hawaii Press.
Cumings, Bruce (1997a) "Boundary displacement: area studies and interna-

tional studies during and after the Cold War." *Bulletin of Concerned Asia Scholars*, 29(1): 6–26.
—— (1997b) "Futures of Asian studies." *Asian Studies Newsletter*, 42: 8.
Cushman, Jennifer Wayne (1993) *Fields from the Sea: Chinese Junk Trade with Siam during the Late Eighteenth and Early Nineteenth Centuries*. Ithaca, NY: Southeast Asia Program, Cornell University.
Cuthbert, Alexander R. (1997) "Ambiguous space, ambiguous rights–corporate power and social control in Hong Kong." *Cities*, 14: 294–311.
Dai Yifeng (1988) "Jindai Fujian huaqiao churuguo guimo jiqi fazhan bianhua (The scale of emigration, the returns from emigration, and the changes in development among Overseas Chinese in Fujian province in modern times)." *Huaqiao huaren lishi yanjiu* (*Overseas Chinese Historical Studies*), 2: 95–106.
—— (1996) "Overseas migration and the economic modernization of Xiamen City during the twentieth century," pp. 159–68 in Leo Douw and Peter Post (eds), *South China: State, Culture, and Social Change during the 20th Century*. Amsterdam: Royal Netherlands Academy of Arts and Sciences; New York: Oxford University Press.
Davin, Delia (1991) "Women, work and property in the Chinese peasant household of the 1980s," pp. 29–50 in Diane Elson (ed.), *Male Bias in the Development Process*. Manchester: Manchester University Press.
—— (1996a) "Gender and rural–urban migration in China," *Gender and Development*, 4(1): 24–30.
—— (1996b) "Migration and rural women in China: a look at the gendered impact of large scale migration." *Journal of International Development*, 8(5): 655–65.
—— (1999) "Gender and migration in China," pp. 23–240 in Flemming Christiansen and Junzuo Zhang (eds), *Village Inc.: Chinese Rural Society in the 1990s*. Surrey: Curzon Press.
Davis, Richard L. (1996) *Wind Against the Mountain: The Crisis of Politics and Culture in Thirteenth-century China*. Cambridge, MA: Harvard University Press.
Dean, Kenneth (1993) *Daoist Ritual and Popular Cults of Southeast China*. Princeton, NJ: Princeton University Press.
Deleuze, Gilles and Felix Guattari (1987) *A Thousand Plateaus: Capitalism and Schizophrenia*. Minneapolis: University of Minnesota Press.
Delman, Jørgen, Clemens Stubbe Østergaard, and Flemming Christiansen (eds) (1990) *Remaking Peasant China: Problems of Rural Development and Institutions at the Start of the 1990s*. Aarhus, Denmark: Aarhus University Press.
Dirlik, Arif (1989) *The Origins of Chinese Communism*. New York: Oxford University Press.
—— (1991) *Anarchism in the Chinese Revolution*. Berkeley: University of California Press.
——(1994) *After the Revolution: Waking to Global Capitalism*. Hanover, NH: University Press of New England.
—— (1995) "Confucius in the borderlands: global capitalism and the reinvention of Confucianism." *Boundary 2*, 22(3): 229–73.
—— (1996) "Reversals, ironies, hegemonies: notes on the contemporary historiography of modern China." *Modern China*, 22(3): 243–84.

Dong, Xin (1996) *Nüxing weifa fanzui jieshi* (*An Analysis of Female Criminal Offenses*). Chongqing: Chongqing chubanshe.
Deng Weizhi, Wu Xiuyi and Hu Shensheng (eds) (1991) *Shanghai shehui fazhan sishi nian* (*Forty Years of Social Development in Shanghai*). Shanghai: Zhishi chubanshe.
Driver, Felix (1992) "Geography's empire: histories of geographical knowledge." *Environment and Planning D: Society and Space*, 10(1): 23–40.
Duara, Prasenjit (1995) *Rescuing History from the Nation: Questioning Narratives of Modern China*. Chicago: University of Chicago Press.
—— (1998) "The regime of authenticity: timelessness, gender, and national history in modern China." *History and Theory*, 37(3): 287–308.
Duncan, James and David Ley (1993) "Introduction: representing the place of culture," pp. 1–24 in *Place/Culture/Representation*. London and New York: Routledge.
Dunch, Ryan (1996) "Piety, patriotism, progress: Chinese Protestants in Fuzhou society and the making of a modern China, 1857–1927." Unpublished PhD dissertation, Yale University.
Dutton, Michael (1988) "Policing the Chinese household: a comparison of modern and ancient forms." *Economy and Society*, 17(2): 195–224.
Duyvendak, J. J. L. (1939) "The true dates of the Chinese maritime expeditions in the early fifteenth century." *T'oung Pao*, 38: 341–412.
Eberhard, Wolfram (1968) *The Local Cultures of South and East China*. Leiden: E. J. Brill.
—— (1982) *China's Minorities: Yesterday and Today*. Belmont, CA: Wadsworth.
Ebrey, Patricia Buckley and James L. Watson (eds) (1986) *Kinship Organization in Late Imperial China, 1000–1940*. Berkeley: University of California Press.
Elson, Diane (ed.) (1991) *Male Bias in the Development Process*. Manchester and New York: Manchester University Press.
Elvin, Mark (1984) "Female virtue and the state in China." *Past and Present*, 104: 111–52.
Entrikin, J. Nicholas (1991) *The Betweenness of Place: Towards a Geography of Modernity*. Baltimore: Johns Hopkins University Press.
Esherick, Joseph (1972) "The apologetics of imperialism." *Bulletin of Concerned Asia Scholars*, (4)4: 9–16.
Esherick, Joseph W. and Mary Backus Rankin (1990) *Chinese Local Elites and Patterns of Dominance*. Berkeley: University of California Press.
Evans, Harriet (1995) "Defining difference: the 'scientific' construction of sexuality and gender in the People's Republic of China." *Signs* 20(2): 357–94.
—— (1997) *Women and Sexuality in China: Female Sexuality and Gender since 1949*. New York: Continuum.
Fairbank, John K. (1953) *Trade and Diplomacy on the China Coast: The Opening of the Treaty Ports, 1842–1854*. Cambridge, MA: Harvard University Press.
Fairbank, John K. (ed.) (1968) *The Chinese World Order: Traditional China's Foreign Relations*. Cambridge, MA: Harvard University Press.
Fan, C. Cindy (1996) "Economic opportunities and internal migration: a case study of Guangdong province, China." *Professional Geographer*, 48(1): 28–45.
—— (2000) "Migration and gender in China," pp. 423–54 in Lau Chung-ming

and Jianfa Shen, *China Review 2000*. Hong Kong: Chinese University Press.
Fan, C. Cindy and Youqin Huang (1998) "Waves of rural brides: female marriage migration in China." *Annals of the Association of American Geographers*, 88(2): 227–51.
Fang Xiongpu and Xie Chengjia (eds) (1993) *Huaqiao huaren gaikuang* (*General Conditions of the Chinese Overseas*). Beijing: Beijing huaqiao chubanshe.
Faure, David and Helen Siu (eds) (1996) *Down to Earth: The Territorial Bond in South China*. Stanford, MA: Stanford University Press.
Faure, David (1986) *The Structure of Chinese Rural Society: Lineage and Village in the Eastern New Territories, Hong Kong*. Hong Kong: Oxford University Press.
—— (1989) "The lineage as cultural invention: the case of the Pearl river delta." *Modern China*, 15(1): 4–36.
Featherstone, Mike (ed.) (1990) *Global Culture: Nationalism, Globalization and Modernity*. Special issue of *Theory, Culture and Society*. London and Thousand Oaks, CA: Sage Publications.
Feng, Wang (2000). "Gendered migration and the migration of genders in contemporary China," pp. 231–245 in Barbara Entwisle and Gail E. Henderson (eds), *Re-drawing Boundaries: Work, Households, and Gender in China*. Berkeley: University of California Press
Feuchtwang Stephen (1974) *An Anthropological Analysis of Chinese Geomancy*. Vithagna: Vientiane.
—— (1992) *The Imperial Metaphor: Popular Religion in China*. London and New York: Routledge.
Feuerwerker, Albert (1958) *China's Early Industrialization: Sheng Hsuan-Huai (1844–1916) and Mandarin Enterprise*. Cambridge, MA: Harvard University Press.
—— (1969) *The Chinese Economy, ca. 1870–1911*. Ann Arbor: University of Michigan, Michigan papers in Chinese studies no. 5.
—— (1976) *The Foreign Establishment in China in the Early Twentieth Century*. Ann Arbor: University of Michigan, Michigan papers in Chinese Studies, no. 29.
Fewsmith, Joseph (1985) *Party, State, and Local Elites in Republican China: Merchant Organizations and Politics in Shanghai, 1890–1930*. Honolulu: University of Hawaii Press.
Fitzgerald, C. P. (1972) *The Southern Expansion of the Chinese People*. New York: Praeger.
Fitzgerald, Stephen (1972) *China and the Overseas Chinese: A Study of Peking's Changing Policy, 1949–1970*. Cambridge: Cambridge University Press.
FO 17 (1815–1905) China, General Correspondence, British Foreign Office, Public Record Office, Kew, England.
FO 228 (1842–1930) Amoy, Fuzhou, and Ningbo. Embassy and Consular Archives, British Foreign Office, Public Record Office, Kew, England.
FO 663 (1843–67) Amoy. Embassy and Consular Archives, Local, British Foreign Office, Public Record Office, Kew, England.
FO 802 (1815–90) China, Registers and Indexes to General Correspondence, British Foreign Office, Public Record Office, Kew, England.
Forward, Roy (1991) "Letter from Shanghai," pp. 187–203 in Jonathan Unger

(ed.), *The Pro-democracy Protests in China: Reports from the Provinces*. Armonk, NY: M. E. Sharpe

Foster, Harold D., David Chuenyan Lai, and Naisheng Zhou (eds) (1998) *The Dragon's Head: China's Emerging Megacity*. Victoria, BC: Western Geographical Press.

Foucault, Michel (1970) *The Order of Things: An Archaeology of the Human Sciences*. New York: Pantheon.

—— (1977) *Discipline and Punish: The Birth of the Prison*. New York: Pantheon Books.

Fox, Richard G. (1990) "Introduction," pp. 1–14 in Richard G. Fox (ed.), *Nationalist Ideologies and the Production of National Cultures*. Washington, DC: The American Anthropological Association.

Franck, Harry A. (1925) *Roving through Southern China*. New York and London: The Century Co.

Frank, André Gunder (1998) *Reorient: Global Economy in the Asian Age*. Berkeley: University of California Press.

Freedman, Maurice (1958) *Lineage Organization in Southeast China*. London: University of London.

—— (1966) *Chinese Lineage and Society: Fukien and Kwangtung*. London: University of London.

Friedman, Edward (1993) "China's north–south split and the forces of disintegration." *Current History*, 92(575): 270–4.

—— (1994) "Reconstructing China's national identity: a southern alternative to Mao-era anti-imperialist nationalism." *Journal of Asian Studies*, 53(1): 67–91.

Friedmann, John (1986) "The world city hypothesis." *Development and Change*, 17(1): 69–83.

Fu Yiling (1956) *Ming Qing shidai shangren ji shangye ziben* (*Merchants and Commercial Capital during the Ming and Qing Dynasties*). Shanghai: Shanghai renmin chubanshe.

—— (1957) *Mingdai Jiangnan shimin jingji shitan* (*Exploration of the Economy of Urban Residents in the Jiangnan Region during the Ming Period*). Shanghai: Shanghai renmin chubanshe.

—— (1961) *Ming Qing nongcun shehui jingji* (*Rural Society and Economy during the Ming and Qing Dynasties*). Beijing: Sanlian.

Fu Yiling and Chen Zhiping (1987) "Shangpin jingji dui mingdai fengjian jieji jiegou de chongji jiqi yaozhe" (Impact of the commodity economy on the structure of the feudal economy and its early demise), pp. 1–17 in Fu Yiling and Yang Guochen (eds), *Mingqing Fujian shehui yu xiangcun jingji* (*Fujian Society and Rural Economy during the Ming and Qing Dynasties*). Xiamen: Xiamen daxue chubanshe.

Fujian tongji nianjian (*Fujian Statistical Yearbook*) (1998, 1999) Beijing: Zhongguo tongji chubanshe.

Fung, Hung Ho (1998) "Thousand-year oppression and thousand-year resistance: the Tanka fisherfolk in Tai O before and after colonialism." *Chinese Sociology and Anthropology*, 30(3): 75–99.

Gardella, Robert (1990) "The Min-pei tea trade during the late Ch'ien-lung and Chia-ch'ing eras: foreign commerce and the mid-Ch'ing Fukien highlands," pp. 317–47 in Eduard B. Vermeer (ed.), *Development and Decline of Fukien Province in the Seventeenth and Eighteenth Centuries*. Leiden: E. J. Brill.

—— (1994) *Harvesting Mountains: Fujian and the China Tea Trade, 1757–1937*. Berkeley: University of California Press.

Gates, Hill (1989) "The commoditization of Chinese women." *Signs*, 14(4): 799–832.

Gernet, Jacques (1995) *Buddhism in Chinese Society: An Economic History from the Fifth to the Tenth Centuries*. New York: Columbia University Press.

Gibson-Graham, J. K. (1996) *The End of Capitalism (as We Knew It): A Feminist Critique of Political Economy*. Cambridge, MA, and Oxford: Blackwell Publishers.

Giddens, Anthony (1984) *The Constitution of Society: Outline of the Theory of Structuration*. Berkeley: University of California Press.

Gilmartin, Christina (1993) "Gender in the formation of a Communist body politic," *Modern China*, 19(3): 299–329.

Gilmartin, Christina K., Gail Hershatter, Lisa Rofel, and Tyrene White (eds) (1994) *Engendering China: Women, Culture, and the State*. Cambridge, MA: Harvard University Press.

GLTWYS Guoli Taiwan yishu jiaoyu guan (Taiwan National Arts Teaching Institute) (1998) *Lianggan xinsheng: dangdai huayu* (New voices: contemporary art dialogue between Taipei, Hong Kong, and Shanghai). Taipei: Guoli Taiwan yishu jiaoyuguan.

Godley, Michael R. (1981) *The Mandarin-capitalists from Nanyang: Overseas Chinese Enterprise in the Modernization of China, 1893–1911*. Cambridge: Cambridge University Press.

—— (1989) "The sojourners: returned Overseas Chinese in the People's Republic of China." *Pacific Affairs*, 62(3): 330–52.

Goodman, Bryna (1995a) "The locality as microcosm of the nation? Native place networks and early urban nationalism in China." *Modern China*, 21(4): 387–419.

—— (1995b) *Native Place, City and Nation: Regional Networks and Identities in Shanghai, 1853–1937*. Berkeley: University of California Press.

Goodman, David S. G. (ed.) (1989) *China's Regional Development*. London and New York: Routledge.

—— (1997) *China's Provinces in Reform: Class, Community, and Political Culture*. London and New York: Routledge.

Goodman, David S. G. and Gerald Segal (eds) (1994) *China Deconstructs: Politics, Trade and Regionalism*. London and New York: Routledge.

Greenberg, Michael (1951) *British Trade and the Opening of China, 1800–42*. Cambridge: Cambridge University Press.

Greenhalgh, Susan (1994) "De-orientalizing the Chinese family firm." *American Ethnologist*, 21(4): 746–75.

de Groot J. J. M. (1897) *The Religious System of China, volume 3, book 1*. EJ Brill: Leiden.

Guo Kesha and Li Haijian (1995) "Zhongguo duiwai kaifang diqu chayi yanjiu (A study of regional variations in China's opening to the outside)." *Zhongguo*

gongye jingji (China's Industrial Economy), 8: 61–8.
Guo Zhemin (1995) *Xiamen jingji tequ jianshe yu fazhan yanjiu* (*A Study of the Construction and Development of the Xiamen Special Economic Zone*), Xiamen: Xiamen daxue chubanshe.
Gupta, S. P. (ed.) (1996) *China's Economic Reforms: The Role of Special Economic Zones and Economic and Technological Development Zones*. New Delhi: Allied Publishers.
Habermas, Jürgen (1989) *The Structural Transformation of the Public Sphere: An Inquiry into a Category of Bourgeois Society*. Cambridge, MA: MIT Press.
Haila, Anne (1999) "Why is Shanghai building a giant speculative property bubble?" *International Journal of Urban and Regional Research*, 23(3): 583–8.
Hainan tongji nianjian (*Hainan Statistical Yearbook*) (1999) Beijing: Zhongguo tongji chubanshe.
Hamilton, Gary G. (1977) "Nineteenth century Chinese merchant associations: conspiracy or combination? The case of the Swatow opium guild." *Ch'ing-shih wen-t'i*, 3(8): 50–71.
Hamilton, Gary (ed.) (1991) *Business Networks and Economic Development in East and Southeast Asia*. Hong Kong: Centre of Asian Studies, University of Hong Kong.
Hanan, Patrick (ed.) (1990) *Silent Operas*. Hong Kong: Chinese University of Hong Kong.
Hao, Yen-p'ing (1986) *The Commercial Revolution in Nineteenth-century China: The Rise of Sino-Western Mercantile Capitalism*. Berkeley: University of California Press.
Harding, Harry (1993) "The concept of "Greater China": themes, variations and reservations." *The China Quarterly*, 136: 660–86.
—— (1994) "Taiwan and Greater China," pp. 235–75 in Robert G. Sutter and William R. Johnson (eds), *Taiwan in World Affairs*. Boulder, CO: Westview.
Harding, James (1999) "China shuts finance company." *Financial Times* (London), July 8, Companies and Finance sec., p. 27.
Harris, Peter (1978) *Hong Kong: A Study in Bureaucratic Politics*. Hong Kong: Heinemann Asia.
Harvey, David (1982) *The Limits to Capital*. Chicago: University of Chicago Press.
—— (1985) *The Urbanization of Capital: Studies in the History and Theory of Capitalist Urbanization*. Baltimore: Johns Hopkins University Press.
—— (1989) *The Condition of Postmodernity*. Oxford: Blackwell.
—— (1996) *Justice, Nature, and the Geography of Difference*. Oxford: Blackwell.
—— (2000) "Cosmopolitanism and the banality of geographical evils." *Public Culture*, 12(2): 529–64.
Hawkes, David (1985) *The Songs of the South*. New York: Viking.
He Qinglian (1997) *Zhongguo de Xianjing* (*China's Pitfall*): *Primary Capital Accumulation in China* (in Chinese; Chinese title, English subtitle). Hong Kong and Flushing, NY: Mirror Books.
Heijdra, Martin (1995) "A preliminary note on cultural geography and Ming history." *Ming Studies*, 34(July): 30–60.

Heilig, Gerhard K. (1997) "Anthropogenic factors in land-use change in China." *Population and Development Review*," 23(1): 139–68.
Held, David (1995) *Democracy and the Global Order: From the Modern State to Cosmopolitan Governance*. Stanford, CA: Stanford University Press.
Henderson-Sellers, Ann and Peter J. Robinson (1986) *Contemporary Climatology*. New York: John Wiley and Sons.
Heng Pek Koon (1988) *Chinese Politics in Malaysia: A History of the Malaysian Chinese Association*. Singapore: Oxford University Press.
Herdt, Gilbert (1994) "Introduction: third sexes and third genders," pp. 21–81 in Gilbert Herdt (ed.), *Third Sex, Third Gender: Beyond Sexual Dimorphism in Culture and History*. New York: Zone Books.
Herrman, Albert (1966) *An Historical Atlas of China*. Chicago: Aldine Publishing Company.
Hershatter, Gail E. (1997) *Dangerous Pleasures: Prostitution and Modernity in Twentieth-Century Shanghai*. Berkeley: University of California Press.
Hertslet, Godfrey E. P. (1908) *Hertslet's China Treaties, volume 1*, 3rd edn. London: Harrison and Sons.
Hevia, James (1995) *Cherishing Men from Afar: Qing Guest Ritual and the Macartney Embassy of 1793*. Durham, NC: Duke University Press.
Hicks, George L. (ed.) (1993) *Overseas Chinese Remittances from Southeast Asia, 1910–1940*. Singapore: Select Books.
Hinsch, Bret (1990) *Passions of the Cut Sleeve: The Male Homosexual Tradition in China*. Berkeley: University of California Press.
Hirschon, Renee (ed.) (1984) *Women and Property – Women as Property*. London: Croom Helm; New York: St Martin's Press.
Ho, David Y. W. (1995) "Editor's notes," pp. 222–5 in *The Encyclopedia of Chinese Law, volume 2, June 1993–Dec. 1994*. Hong Kong: Asia Law and Practice.
Ho Lok-sang and Tsui Kai-yuen (1996) "Fiscal relations between Shanghai and the central government," pp. 153–70 in Y. M. Yeung and Sung Yunwing (eds), *Shanghai: Transformation and Modernization under China's Open Policy*. Hong Kong: Chinese University Press.
Ho, Ping-ti (1955) "The introduction of American food plants into China." *American Anthropologist*, 57(2): 191–201.
—— (1956) "Early-ripening rice in Chinese history." *The Economic History Review*, 9(2): 200–18.
—— (1959) *Studies on the Population of China, 1368–1953*. Cambridge, MA: Harvard University Press.
—— (1962) *The Ladder of Success in Imperial China*. New York: Columbia University Press.
—— (1966) *Chung-kuo hui kuan shih lun (A historical survey of Landsmannschaften in China)*. Taiwan: Taiwan hsueh sheng shu chu.
Ho, Samuel P. S. and George C. S. Lin (2001) "China's evolving land system." Discussion paper, Centre for Chinese Research, University of British Columbia Centre for Asian Research, Vancouver.
Ho, Virgil K. Y. (1998) "Whose bodies? Taming contemporary prostitutes' bodies in official Chinese rhetoric." *China Information*, 13(2/3): 14–35.

Hobsbawm Eric J. (1990) *Nations and Nationalism since 1780: Programme, Myth, Reality*. Cambridge: Cambridge University Press.
Holland, Fiona (1996) "Tower will 'green' Central: 88-story office building planned to provide more open space." *South China Morning Post*, November 26, p. 8.
Honig, Emily (1992) *Creating Chinese Ethnicity: Subei People in Shanghai, 1850–1980*. New Haven, CT: Yale University Press.
Hou Xueyu (1988) *Zhongguo ziran shengtai quhua yu da nongye fazhan zhanlue (China's Biome Divisions and Large-scale Agricultural Development Strategies)*. Beijing: Kexue chubanshe.
Howell, Jude (1993) *China Opens Its Doors: The Politics of Economic Transition*. Hemel Hempstead: Harvester Wheatsheaf; Boulder, CO: Lynne Reinner.
—— (2000) "Organising around women and labour in China: uneasey shadows, uncomfortable alliances." *Communist and Post-Communist Studies*, 33(3): 355–77.
Howitt, Richard (1993) "'A world in a grain of sand': towards a reconceptualisation of geographical scale." *Australian Geographer*, 24(1): 33–42.
—— (1998) "Scale as relation: musical metaphors of geographical scale." *Area*, 30(1): 49–58.
Hsieh, A. C. K. and Jonathan Spence (1981) "Suicide and the family in pre-modern Chinese society," pp. 29–47 in Arthur Kleinman and Tsung-yi Lin (eds), *Normal and Abnormal Behavior in Chinese Culture*. Dordrecht, Holland: D. Reidel.
Hsing, You-tien (1998) *Making Capitalism in China: The Taiwan Connection*. New York: Oxford University Press.
Hua Fen (1994) "Jiujiu hongshulin (Please save the red forest)." *Shenzhen tequ bao (Shenzhen Special Zone Daily)*, August 31, p. 1.
Huang Guosheng (1999) "Qingchu sisheng haiguan yashu shizhi xintan (Preliminary discussion of the maritime customs offices in the four coastal provinces in the early Qing period)." *Fujian shifan daxue xuebao (Journal of Fujian Normal University)*, 2: 1–7.
Huang, Philip C. C. (1985) *The Peasant Economy and Social Change in North China*. Stanford, CA: Stanford University Press.
Huang Xiyi (1999) "Divided gender, divided women: state policy and the labour market," pp. 90–107 in Jackie West, Zhao Minghua, Chang Xiangqun, and Cheng Yuan (eds), *Women of China: Economic and Social Transformation*. London: Macmillan Press; New York: St Martin's Press.
Huang, Yanzhong and Dali L. Yang (1996) "The political dynamics of regulatory change: Speculation and regulation in the real estate sector." *Journal of Contemporary China*, 5(12): 171–85.
Hucker, Charles O. (1978) *The Ming Dynasty, Its Origins and Evolving Institutions*. Ann Arbor: Center for Chinese Studies, University of Michigan.
Human Rights Watch (1998) *Indonesia: The Damaging Debate on Rapes of Ethnic Chinese Women*. New York: Human Rights Watch.
I Yuan, (1996) "Center and periphery: cultural identity and localism of the southern Chinese peasantry." *Issues and Studies*, 32(6): 1–36.

Jacka, Tamara, (1997) *Women's Work in Rural China: Change and Continuity in an Era of Reform*. Cambridge: Cambridge University Press.

Jackson, Peter and Jan Penrose (1993) "Placing 'race' and nation," pp. 1–23 in P. Jackson and J. Penrose (eds), *Constructions of Race, Place and Nation*. London: UCL Press.

Jameson, Fredric (1987) *Houxiandai zhuyi yu wenhua lilun* (Postmodernism and Cultural Theory). Xi'an: Shanxi shifan daxue chubanshe.

—— (1998a) "Preface," pp. xi–xvii in Fredric Jameson and Masao Miyoshi, eds., *The Cultures of Globalization*. Durham, NC: Duke University Press.

—— 1998b. "Notes on globalization as a philosophical issue," pp. 54–77 in Fredric Jameson and Masao Miyoshi (eds), *The Cultures of Globalization*. Durham: Duke University Press.

Jameson, Fredric and Masao Miyoshi (eds) (1998) *The Cultures of Globalization*. Durham, NC: Duke University Press.

Jesudason, James V. (1989) *Ethnicity and the Economy: The State, Chinese Business, and Multinationals in Malaysia*. Singapore: Oxford University Press.

Jiang Bingzhe (1989) "Hui'an diqu changzhu niangjia hunsu de lishi kaocha (A historical investigation of extended natolocal residence marriage customs in the Hui'an region)." *Zhongguo shehui kexue* (*Chinese Social Science*), 3: 193–203.

Jiang Jingen (1998) "State urged to improve construction planning." *China Daily*, July 24, p. 2.

Johansson, Sten and Ola Nygren (1991) "The missing girls of China: a new demographic account." *Population and Development Review*, 17(1): 35–51.

Johnson, Graeme (1998) "Changing structures, changing contexts: research implications of a transformed context," pp. 3–22 in Sidney C. H. Cheung (ed.), *On the South China Track: Perspectives on Anthropological Research and Teaching*. Hong Kong: The Chinese University of Hong Kong.

Johnson, Ian (1998) "Beijing plans massive closure of trust firms." *Asian Wall Street Journal* (Hong Kong), January 7, pp. 1, 4.

Johnston, R. J. (1991) *A Question of Place: Exploring the Practice of Human Geography*. Oxford: Blackwell.

Jordan, Amos A. and Jane Khanna (1995) "Economic interdependence and challenges to the nation-state: the emergence of natural economic territories in the Asia Pacific." *Journal of International Affairs*, 48(2): 433–62.

Judd, Ellen R. (1994) *Gender and Power in Rural North China*. Stanford, CA: Stanford University Press.

Kamm, Henry (1980) "Singapore worries about its values as it seeks even greater prosperity." *The New York Times*, September 23, sec. A, p. 8.

Kancil, Sang (1984) "Preserving Malacca's historical heritage." *The Star* (Petaling Jaya), September 29.

Kapferer, Bruce (1988) *Legends of People, Myths of State*. Washington, DC: Smithsonian Institution Press.

Kelkar, Govind and Wang Yunxian (1997) "Farmers, women, and economic reform in China." *Bulletin of Concerned Asia Scholars*, 29(4): 69–77.

Kellogg, Claude R. (1925) "Fauna and flora," pp. 34–44 in *Fukien: A Study of a Province in China*. Anti-Cobweb Club, Shanghai: Presbyterian Mission Press.

Keng, C. W. Kenneth (1996) "China's land disposition system." *Journal of Contemporary China*, 5(13): 325–45.
Kessler, Lawrence D. (1976) *K'ang-Hsi and the Consolidation of Ch'ing Rule, 1661–1684*. Chicago: University of Chicago Press.
Khan Azizur Rahman and Carl Riskin (1999) "Income and inequality in China: composition, distribution and growth of household income." *The China Quarterly*, 154: 221–53.
Kirkby, R. J. R. (1985) *Urbanisation in China Town and Country in a Developing Economy, 1949–2000 AD*. London: Croom Helm.
Knox, Paul and Peter Taylor (eds) (1995) *World Cities in a World System*. Cambridge: Cambridge University Press.
Kohl, David G. (1984) *Chinese Architecture in the Straits Settlements and Western Malaya: Temples, Kongsis and Houses*. Kuala Lumpur: Heinemann.
Kraus, Richard Curt (1989) *Pianos and Politics in China: Middle-Class Ambitions and the Struggle over Western Music*. New York: Oxford University Press.
—— (2000) "Public monuments and private pleasures in the parks of Nanjing: a tango in the ruins of the Ming emperor's palace," pp. 287–311 in Deborah Davis (ed.), *The Consumer Revolution in Urban China*. Berkeley: University of California Press.
Krugman, Paul (1994) "The myth of Asia's miracle." *Foreign Affairs*, 73(6): 62–79.
Kua Kia Soong (1984) "Bukit China belongs to the people." *The Star* (Petaling Jaya, Malaysia), December 22, p. 24.
—— (1990) *Malaysian Cultural Policy and Democracy*. Kuala Lumpur: The Malaysian Chinese Resource and Research Center.
Kuriyama, Shigehisa (1994) "The imagination of the winds and the development of the Chinese conception of the body," pp. 23–41 in Angela Zito and Tani E. Barlow, (eds), *Body, Subject and Power in China*. Chicago: Chicago University Press.
Kynge, James (1998) "Hainan bank closure sounds warning bell." *Financial Times* (London), July 23, Asia-Pacific sec., p. 6.
Lam, Kit Chun and Pak Wai Liu (1998) *Immigration and the Economy of Hong Kong*. Hong Kong: City University of Hong Kong Press.
Language Atlas of China (1988). Edited by S. A. Wurm et al. Hong Kong: Longman Group.
Lary, Diana (1974) *Region and Nation: The Kwangsi Clique in Chinese Politics, 1925–1937*. London and New York: Cambridge University Press.
—— (1996) "The tomb of the King of Nanyue – the contemporary agenda of history." *Modern China*, 22(1): 3–27.
Latour, Bruno (1993) *We Have Never Been Modern*. Cambridge, MA: Harvard University Press.
Laufer, Berthold (1908) "The relations of the Chinese to the Philippine islands." *Smithsonian Miscellaneous Collections*, 50: 248–84.
Lee, Ching Kwan (1995) "Engendering the worlds of labor: women workers, labor markets, and production politics in the south China economic miracle." *American Sociological Review*, 60(3): 378–97.
—— (1998) *Gender and the South China Miracle: Two Worlds of Factory Women*.

Berkeley: University of California Press.
Lee, Leo Ou-fan (1999) *Shanghai Modern: The Flowering of a New Urban Culture in China, 1930–45*. Cambridge, MA: Harvard University Press.
Lee, Rensselaer W. III (1966) "The *hsia fang* system: marxism and modernisation." *The China Quarterly*, 28: 40–62.
Lee, Schultz (1994) "Beijing gives new powers to Suzhou township management." *Business Times* (Singapore), November 21, p. 3.
Lee Teng-hui (1999) *T'ai-wan ti chu chang* (*Taiwan's Viewpoint*). T'ai-pei: Yüan liu ch'u pan shih yeh ku fen yu hsien kung ssu.
Lee, T. Y. (1991) *The Growth Triangle: The Johor–Singapore–Riau Experience*. Singapore: Institute of Southeast Asian Studies.
Lee Wing On and Brian Hook (1998) "Human resources," pp. 119–46 in Brian Hook (ed.), *Shanghai and the Yangtze Delta: A City Reborn*. Hong Kong: Oxford University Press.
Lefebvre, Henri (1991) *The Production of Space*. Oxford: Blackwell.
Lencek, Lena and Gideon Bosker (1998) *The Beach: The History of Paradise on Earth*. New York: Viking Penguin.
Leong, Sow-Theng (1997) *Migration and Ethnicity in Chinese History: Hakkas, Pengmin, and Their Neighbors*. Stanford, CA: Stanford University Press.
Li, Cheng (1997) *Discovering China: Dynamics and Dilemmas of Reform*. Lanham, MD: Rowman and Littlefield.
Li Chengrui and Zhang Zhuoyuan (1984) "An outline of economic development (1977–1980)," pp. 3–69 in Yu Guangyuan (ed.), *China's Socialist Modernization*. Beijing: Foreign Languages Press.
Li Fuxi (1984) "Cong shijie gumo tan dao San Bao Shan (From the world's ancient graves to Bukit China)," pp. 192–3 in *Lishi de zuyin: San Bao Shan ziliao xuanji* (*Sound of Historical Footsteps: Selected Materials about Bukit China*). Kuala Lumpur: Chinese Materials Research Center.
Li Guoqi (1982) *Zhongguo xiandaihua de quyu yanjiu – Minzhetai diqu, 1860–1916* (*Modernization in China: A Regional Study of Social, Political, and Economic Change in Fujian, Zhejiang, and Taiwan, 1860–1916*). Taipei: Zhongyang yanjiuyuan, jindaishi yanjiusuo.
Li Jinming, (1990) "Qingdai qianchao Zhongguo yu Dongnanya de da mi maoyi (Rice trade between China and Southeast Asia during the early Qing dynasty)." *Nanyang wenti yanjiu* (*Southeast Asia Research*), 4: 96–104.
Li, Si-Ming (1990) "The Sino-British Joint Declaration, 1997, and the land market of Hong Kong." *Review of Urban and Regional Development Studies*, 29: 84–101.
Li Xueqin (1991) "Chu bronzes and Chu culture," pp. 1–22 in T. Lawton (ed.), *New Perspectives on Chu Culture during the Eastern Zhou Period*. Princeton, NJ: Princeton University Press.
Lim, Shirley Goek-Lin (1994) *Monsoon History: Selected Poems*. London: Skoob.
—— (1996) *Among the White Moon Faces: An Asian-American Memoir of Homeland*. New York: The Feminist Press at the City University of New York.
Lin, Alfred H. Y. (1997) *The Rural Economy of Guangdong, 1870–1937: A Study of the Agrarian Crisis and Its Origins in Southernmost China*. New York: St. Martin's.

Lin Chuancang (1936) *Fuzhou Xiamen dijia yanjiu* (*A Study on Land Prices in Fuzhou and Xiamen*). Shanghai: Zhongguo dizheng yanjiu suo.

Lin, George C. S. (1997) *Red Capitalism in South China: Growth and Development in the Pearl River Delta*. Vancouver: UBC Press.

Lin Jinzhi (1988) *Jindai huaqiao touzi guonei qiye gailun* (*Introduction to Overseas Chinese Investment in Domestic Enterprises in Modern Times*). Xiamen: Xiamen daxue chubanshe.

Lin Jinzhi and Zhuang Weiji (1985) *Jindai huaqiao touzi guonei qiye shi ziliao xuanji: Fujian juan* (*Selected Materials on Modern Overseas Chinese Investment in Domestic Industry: Fujian Volume*). Fuzhou: Fujian renmin chubanshe.

Lin, Jing (1995) "Women and rural development in China." *Canadian Journal of Development Studies*, 16(2): 229–40.

Liu Binhong (1997) "A summary of issues in women's employement." *Chinese Sociology and Anthropology*, 29(3): 7–51.

Liu Binyan and Perry Link (1998) "A great leap backward." *The New York Review of Books*, October 5, pp. 19–23.

Liu Meng and Cecilia Lai Wan Chan (2000) "Deprivation of status and resources: a case study of violent marriage in the rural mainland." *Journal of Women and Gender Studies*, 11(April): 79–97.

Liu, Hong (1998) "Old linkages, new networks: the globalization of Overseas Chinese voluntary associations and its implications. *The China Quarterly*, 155: 582–609.

Liu Kang (1998) "Is there an alternative to (capitalist) globalization? The debate about modernity in China," pp. 164–90 in Fredric Jameson and Masao Miyoshi (eds), *The Cultures of Globalization*. Durham, NC: Duke University Press.

Liu Weijun and Ray Zhang (1997) "Chinese women face discrimination in job search." *China Daily*, December 22. http://www.cnd.org/CND-Global/CND-Global.97.4th/CND-Global.97–12–21.html

Liu Zhiwei (1995) "Lineage on the sands: the case of Shawan," pp. 22–43 in Helen F. Siu and David Faure (eds), *Down to Earth: The Territorial Bond in South China*. Stanford, CA: Stanford University Press.

Lo, Carlos Wing-Hung (1994) "Environmental management by law in China: the Guangzhou experience." *Journal of Contemporary China*, 6: 39–58.

Lu, Hanchao (1999) *Beyond the Neon Lights: Everyday Shanghai in the Early Twentieth Century*. Berkeley: University of California Press.

Luo, Qi and Christopher Howe (1993) "The case of Taiwanese investment in Xiamen." *The China Quarterly*, 136: 746–69.

Ma, Laurence J. C. (1971) *Commercial Development and Urban Change in Sung China (960–1279)*. Ann Arbor: Department of Geography, University of Michigan.

—— (2002) "Space, place, and transnationalism in the Chinese diaspora," in Laurence J. C. Ma and Carolyn L. Cartier (eds), *A Geography of the Chinese Diaspora*. Lanham, MD: Rowman and Littlefield.

Ma Yuanxi (1999) "Self-reinstating and coming to 'conscious aloneness'," pp. 77–100 in Sharon K. Hom (ed.), *Chinese Women Traversing Diaspora: Memoirs, Essays, Poetry*. New York and London: Garland.

Mackie, J. A. C. (1976) "Anti-Chinese outbreaks in Indonesia, 1959–1968," pp. 77–138 in J. A. C. Mackie (ed.), *The Chinese in Indonesia: Five Essays*. Melbourne: Nelson.

Mackie, J. A. C. and Charles Coppell (1976) "A preliminary survey," pp. 1–18 in J. A. C. Mackie (ed.), *The Chinese in Indonesia: Five Essays*. Melbourne: Nelson.

McDowell, Linda (1996) "Spatializing feminism: geographic perspectives," pp. 28–44 in Nancy Duncan (ed.), *Bodyspace: Destabilizing Geographies of Gender and Sexuality*.

—— (1999) *Gender, Identity and Place: Understanding Feminist Geographies*. Minneapolis: University of Minnesota Press.

Malcomson, Scott L. (1998) "The varieties of cosmopolitan experience," pp. 233–45 in Pheng Cheah and Bruce Robbins (eds), *Cosmopolitics: Thinking and Feeling beyond the Nation*. Minneapolis: University of Minnesota Press.

Mallee, Hein (1996) "Reform of the *hukou* system: introduction." *Chinese Sociology and Anthropology*, 29(1): 3–14.

Malpas, J. E. (1999) *Place and Experience: A Philosophical Perspective*. Cambridge: Cambridge University Press.

Mann, Susan (1972) "Finance in Ningbo: The 'Ch'ien Chuang,' 1750–1840," pp. 47–78 in W. E. Willmott (ed.), *Economic Organization in Chinese Society*. Stanford, CA: Stanford University Press.

—— (1987) *Local Merchants and the Chinese Bureaucracy, 1750–1950*. Stanford, CA: Stanford University Press.

Mao Zedong (1954) *The Chinese Revolution and the Chinese Communist Party*. Beijing: Foreign Languages Press.

Marble, Andrew and Wu Yiyi (2000) "'Three links' policy to be loosened only gradually: Taiwan official." *China News Digest*, January 26. http://www.cnd.org/CND-Global/

Marks, Robert (1997) *Tigers, Rice, Silk, and Silt: Environment and Economy in Late Imperial South China*. Cambridge: Cambridge University Press.

Marmé, Michael (1993) "Heaven on earth: the rise of Suzhou, 1127–1550," pp. 17–46 in Linda Cooke Johnson (ed.), *Cities of Jiangnan in Late Imperial China*. Albany: State University of New York Press.

Massey, Doreen B. (1984) *Spatial Divisions of Labor: Social Structures and the Geography of Production*. New York: Routledge.

—— (1991) "A global sense of place." *Marxism Today*, 35(June): 24–9.

—— (1993) "Power geometry and a progressive sense of place," pp. 59–69 in John Bird, Barry Curtis, Tim Putnam, George Robertson, and Lisa Tickner (eds), *Mapping the Futures: Local Cultures, Global Change*. London and New York: Routledge.

—— (1994) *Space, Place, and Gender*. Minneapolis: University of Minnesota Press.

—— (1995) "The conceptualization of place," pp. 45–86 in Doreen Massey and Pat Jess (eds), *A Place in the World? Places, Cultures and Globalization*. Buckingham: The Open University Press.

Meacham, William (1983) "Origins and development of the Yüeh coastal neolithic: a microcosm of culture change on the mainland of East Asia," pp.

147–75 in David N. Keightley (ed.), *The Origins of Chinese Civilization*. Berkeley: University of California Press.

Means, Gordon (1970) *Malaysian Politics*. London: University of London Press.

—— (1991) *Malaysian Politics: The Second Generation*. Singapore and New York: Oxford University Press.

Means Laurel (1994) "Introduction," pp. xii–xxii in Shirley Lim, *Monsoon History: Selected Poems*. London: Skoob.

Meng, Xin and Paul Miller (1995) "Occupational segregation and its impact on gender wage discrimination in China's rural industrial sector." *Oxford Economic Papers*, 47(1): 136–55.

Merrifield, Andrew (1993) "Place and space: a Lefebvrian reconciliation." *Transactions of the Institute of British Geographers*, 18(4), 516–31.

Metzger, Thomas (1970) "The state and commerce in imperial China." *Journal of Asian and African Studies*, 6: 23–46.

MHHW (Melaka Chinese Movement to Protect Bukit China) (1984) *Bukit China: The Road to Survival*. Melaka*: Mahua hanwei San Bao Shan xingdong* (English title, Chinese text).

Miao Jianhua (1994) "International migration in China: a survey of emigrants from Shanghai." *Asian and Pacific Migration Journal*, 3(2/3): 445–63.

Miller, Matthew (1999) "Debt-chaos legacy of Gitic collapse." *South China Morning Post*, December 17, Business Post sec., p. 14.

—— (2000a) "Shanghai group's woes echo Gitic." *South China Morning Post*, April 21, Business Post sec., p. 1.

—— (2000b) "SAIC scandal confirms continuing sorry state of commerce." *South China Morning Post*, April 28, Business Post sec., p. 1.

Ming pao (Hong Kong) (1998) "Ministerial units delinked with trust and investment firms," October 19, trans. in FBIS-CHI-98-292.

Mitamura, Taisuke (1963) *Chinese Eunuchs: The Structure of Intimate Politics*. Rutland, VT, and Tokyo: Charles E. Tuttle Company.

—— (1970) *Chinese Eunuchs: The Structure of Intimate Politics*. Rutland, VT and Tokyo: Charles E. Tuttle.

Miyoshi, Masao (1998) "'Globalization', culture, and the university," pp. 247–70 in Fredric Jameson and Masao Miyoshi, (eds), *The Cultures of Globalization*. Durham, NC: Duke University Press.

Mori Building (1999) "Shanghai World Financial Centre." http://www.worldstallest.com

Morse, Hosea Ballou (1910) *The International Relations of the Chinese Empire. Volume 1, The Period of Conflict, 1834–1860*. London: Longmans, Green.

—— (1926) *The Chronicles of the East India Company Trading to China, 1635–1834, volume 1*. Cambridge, MA: Harvard University Press; Oxford: Clarendon Press.

Mote, Frederick W. (1995) "Urban history in later imperial China." *Ming Studies*, 34(July): 61–76.

Murphey, Rhoads (1953) *Shanghai: Key to Modern China*. Cambridge, MA: Harvard University Press.

—— (1970) *The Treaty Ports and China's Modernization: What Went Wrong?* Ann Arbor: Center for Chinese Studies, University of Michigan.

—— (1980) *The Fading of the Maoist Vision: City and Country in China's Development*. New York: Methuen.

Murray, Christopher J. L. and Alan D. Lopez, (1996) *Global Health Statistics: A Compendium of Incidence, Prevalence, and Mortality Estimates for over 200 Conditions*. Cambridge, MA: Harvard University Press.

Murray, Dian H. (1981) "One woman's rise to power: Cheng I's wife and the pirates," pp. 147–61 in Richard W. Guisso and Stanley Johannesen (eds), *Women in China: Current Directions in Historical Scholarship*. Youngstown, NY: Philo Press.

—— (1987) *Pirates of the South China Coast, 1790–1810*. Stanford, CA: Stanford University Press.

—— (1993) "Migration, protection, and racketeering: the spread of the *Tiandihui* within China," pp. 177–89 in D. Ownby and M. S. Heidhues (eds), *Secret Societies Reconsidered: Perspectives on the Social History of Modern South China and Southeast Asia*. Armonk, NY: M. E. Sharpe.

Musk, Leslie F. (1988) *Weather Systems*. Cambridge and New York: Cambridge University Press.

Myers, Ramon (1991) "How did the modern Chinese economy develop? A review article." *Journal of Asian Studies*, 50(3): 604–28.

Naisbitt, John (1996) *Megatrends Asia: Eight Asian Megatrends that Are Reshaping Our World*. New York: Simon & Schuster.

Nan Qi (1966) "Taiwan Zheng shi wushang zhi yanjiu (A study of the five merchant firms established by the Zhengs of Taiwan)." *Taiwan yanjiu congkan (Research Series on Taiwan)*, 90: 43–51.

Naquin, Susan and Evelyn Rawski (1987) *Chinese Society in the Eighteenth Century*. New Haven, CT: Yale University Press.

Naughton, Barry (1995) *Growing out of the Plan: Chinese Economic Reform, 1978–1993*. Cambridge: Cambridge University Press.

Needham, Joseph (1956–62) *Science and Civilization in China*, volumes 2, 3, 4: 1. Cambridge: Cambridge University Press.

Ng Chin-keong (1972) "A study on the peasant society of south Fukien, 1506–1644." *Nanyang University Journal*, 6: 189–212.

—— (1983) *Trade and Society: The Amoy Network on the China Coast, 1638–1735*. Singapore: Singapore University Press.

Ngai, Pun (1999) "Becoming *Dagongmei* (Working Girls): the politics of identity and difference in reform China." *The China Journal*, 42(July): 1–19.

Nonini, Donald (1997) "Shifting identities, positioned imaginaries: transnational traversals and reversals by Malaysian Chinese," pp. 203–27 in Aihwa Ong and Donald Nonini (eds), *Ungrounded Empires: The Cultural Politics of Modern Chinese Transnationalism*. London and New York: Routledge.

Norman, Jerry (1988) *Chinese*. Cambridge: Cambridge University Press.

Nussbaum, Martha (1996) *For Love of Country: Debating the Limits of Patriotism*. Boston: Beacon Press.

—— (1997) "Kant and stoic cosmopolitanism." *Journal of Political Philosophy* 5(1): 1–25.

Oakes, Tim (2000) "China's provincial identities: reviving regionalism and reinventing 'Chineseness'. *Journal of Asian Studies*, 59(3), 667–92.

Ocko, Jonathan K. (1991) "Women, property, and law in the People's Republic of China," pp. 313–46 in Ruby S. Watson and Patricia Buckley Ebrey (eds), *Marriage and Inequality in Chinese Society*. Berkeley: University of California Press.

Oi, Jean C. (1996) "Transitional economy," pp. 170–87 in Andrew G. Walder (ed.), *China's Transitional Economy*. Oxford: Oxford University Press.

—— (1999) *Rural China Takes Off: Institutional Foundations of Economic Reform*. Berkeley: University of California Press.

Olsen, John (1992) "Archaeology in China today." *China Exchange News*, 20(2): 3–6.

Ong, Aihwa, (1997) "Chinese modernities: narratives of nation and of capitalism," pp. 171–202 in Aihwa Ong and Donald Nonini (eds), *Ungrounded Empires: The Cultural Politics of Modern Chinese Transnationalism*. London and New York: Routledge.

—— (1999) *Flexible Citizenship: The Cultural Logics of Transnationality*. Durham, NC: Duke University Press.

Ong, Aihwa and Donald Nonini (1997) "Introduction: Chinese transnationalism as an alternative modernity," pp. 3–33 in Aihwa Ong and Donald Nonini (eds), *Ungrounded Empires: The Cultural Politics of Modern Chinese Transnationalism*. London and New York: Routledge.

Oo, Yu Hock (1991) *Ethnic Chameleon: Multiracial Politics in Malaysia*. Petaling Jaya, Malaysia: Palanduk.

Ortner, Sherry B. (1999) "Thick resistance: death and the cultural construction of agency in Himalayan mountaineering," pp. 136–164 in S. B. Ortner (ed.), *The Fate of "Culture": Geertz and Beyond*. Berkeley: University of California Press.

Panagariya, Arvind (1993) "Unravelling the mysteries of China's foreign trade regime." *The World Economy*, 16(1): 51–8.

Pandiyan M. Veera (1984) "23 nabbed at Bukit China." *The Star* (Petaling Jaya, Malaysia), August 20, p. 1.

Parsonage, J. (1992) "Southeast Asia's growth triangle: a subregional response to a global transformation." *International Journal of Urban and Regional Research*, 16(2): 307–17.

Peet, Richard (1997) "The cultural production of economic forms," pp. 37–46 in Roger Lee and Jane Wills (eds), *Geographies of Economies*. London: Arnold.

—— (2000) "Culture, imaginary and rationality in regional economic development." *Environment and Planning A*, 32(7): 1215–34.

Peet, Richard and Michael Watts (1996) "Liberation ecology: development, sustainability, and environment in an age of market triumphalism," pp. 1–45 in Richard Peet and Michael Watts (eds), *Liberation Ecologies: Environment, Development, Social Movements*. London and New York: Routledge.

Peng Shifan (1990) *Bai Yue minzu yanjiu* (*A Study of the Bai Yue Peoples*). Nanchang: Jiangxi jiaoyu chubanshe.

Peng Xizhe (1989) "Major determinants of China's fertility transition." *China Quarterly*, 117: 1–37.

Penrose, Jan (1993) "Reification in the name of change: the impact of nationalism on social constructions of nation, people and place in Scotland and the

United Kingdom," pp. 27–49 in Peter Jackson and J. Penrose (eds), *Constructions of Race, Place and Nation*. London: UCL Press.

Perkins, Dwight (1991) "The lasting effects of China's economic reforms, 1979–1989," pp. 341–63 in Kenneth Lieberthal, Joyce Kallgren, Roderick MacFarquhar, and Frederic Wakeman, Jr. (eds), *Perspectives on Modern China: Four Anniversaries*. Armonk, NY: M. E. Sharpe.

Perry, Elizabeth J. (1993) *Shanghai on Strike: The Politics of Chinese Labor*, Stanford, CA: Stanford University Press.

Petition of the Society for Protection of the Harbour (1996), to His Excellency the Governor-in-Council The Right Honourable Chris Patten, in the matter of the Territorial Development Strategy Review 1996 and in the matter of the Protection of the Harbour Bill 1996.

Philips, Michael R., Huaqing Liu and Yanping Zhang (1999) "Suicide and social change in China." *Culture, Medicine, and Psychiatry*, 23(1): 25–50.

Pincus, Steve (1995) "'Coffee politicians does create': coffeehouses and restoration political culture." *Journal of Modern History*, 67(4): 807–34.

Piore, Michael J. and Charles F. Sabel (1984) *The Second Industrial Divide: Possibilities for Prosperity*. New York: Basic Books.

Poh, Quah Choon (1993) "Lippo chief plans US$ 10b development in Fujian." *Business Times* (Singapore), August 7, Weekend edition.

Poole, Oliver (1997) "Drive to view city from new high-rise." *South China Morning Post*, July 13, p. 3.

Poston, Dudley and Mei-Yu Yu (1990) "The distribution of the Overseas Chinese in the contemporary world." *International Migration Review*, 24(3): 480–508.

Potter, Jack M. (1968) *Capitalism and the Chinese Peasant: Social and Economic Change in a Hong Kong Village*. Berkeley: University of California Press.

Pred, Allan (1984) "Place as historically contingent process: structuration and the time-geography of becoming places." *Annals of the Association of American Geographers*, 74(2): 279–97.

—— (1986) *Place, Practice, and Structure: Social and Spatial Transformation in Southern Sweden, 1750–1850*. Totowa, NJ: Barnes & Noble.

Pritchard, C. (1996) "Suicide in the People's Republic of China categorized by age and gender: evidence of the influence of culture on suicide." *Acta Psychiatrica Scandinavica*, 93(5): 362–67.

Pu Shanxin, Chen Deyu, and Zhou Yi (1995) *Zhongguo xingzhengqu hua gailun* (*An Overview of Changes in China's Administrative Divisions*). Beijing: Zhishi chubanshe.

Pudup, Mary Beth (1988) "Arguments within regional geography." *Progress in Human Geography*, 12(3): 369–90.

Purcell, Victor (1947) "Chinese settlement in Melaka." *Journal of the Malayan Branch of the Royal Asiatic Society*, 20(1): 115–25.

—— (1965) *The Chinese in Southeast Asia, 2nd edn*. London: Oxford University Press.

Pye, Lucian W. (1956) *Guerilla Communism in Malaysia: Its Social and Political Meaning*. Princeton, NJ: Princeton University Press.

Qian Dongxiang and Tan Songshou (1995) *Zhongguo lishi dituji* (*Historical Atlas*

of China). Taipei: Tianwei wenhua tushu youxian gongsi.
Qin Xiangyin and Ni Jianzhong (1993) *Nanbei chunqiu: Zhongguo huibuhui fenlie?* (*The History of the North and the South: Will China Disintegrate?*). Beijing: Renmin Zhongguo chubanshe.
Qin Yunming (1937) "Zheng He qi ci xia xiyang nianyue kaozheng (Analysis of Zheng He's seven trips to the Western Ocean)." *Fujian wenhua* (*Fujian Culture*), 26: 1–48.
Qiu Dezai (1979) *Taiwan miao shen zhuan* (*Temple Deities in Taiwan*). Douliu: Xintong shuju.
Qiu Zhang and Lin Cuifen (1994) *Yiguo liangqi: liang'an hunyin baipishu* (*One Country, Two Wives: White Paper on Cross-strait Marriages*). Taibei: Jingmei chubanshe.
Quak Hiang Whai (2000) "Mayor plays down rivalry between Shanghai, HK." *Business Times* (Singapore), December 5, p. 17.
Quanzhou fu zhi (*Quanzhou Prefecture Gazeteer*) (1612) Reprint, Taipei (1987).
Rafael, Vincente L. (1994) "The cultures of area studies in the United States." *Social Text* 41: 91–111.
Rai, Shirin M. (1994a) "Gender issues in China: a survey." *China Report*, 30(4): 407–20.
—— 1994b. "Modernisation and gender: education and employment in post-Mao China," *Gender and Education*, 6(2): 119–29.
Rai Shirin and Zhang Junzuo (1994) "'Competing and learning': women and the state in contemporary rural mainland China." *Issues and Studies*, 30(3): 51–66.
Ramo, Joshua Cooper (1998) "The Shanghai bubble." *Foreign Policy*, 111: 64–75.
Rankin, Mary Baskus (1986) *Elite Activism and Political Transformation in China: Zhejiang Province, 1865–1911*. Stanford, CA: Stanford University Press.
Ranson, Brian H. A. (ed.) (1997) *Guild-hall and Government: An Exploration of Power, Control and Resistance in Britain and China. Volume One, A Preliminary Study of the Social Organisation of Guilds in China*. Hong Kong: David C. Lam Institute for East–West Studies, Hong Kong Baptist University.
Rawski, Evelyn S. (1972) *Agricultural Change and the Peasant Economy of South China*. Cambridge, MA: Harvard University Press.
—— (1988a) "A historian's approach to Chinese death ritual," pp. 20–36 in James L. Watson and Evelyn S. Rawski (eds), *Death Ritual in Late Imperial and Modern China*. Berkeley: University of California Press.
—— (1988b) "The imperial way of death: Ming and Ch'ing emperors and death ritual," pp. 228–353 in James L. Watson and Evelyn S. Rawski (eds), *Death Ritual in Late Imperial and Modern China*. Berkeley: University of California Press.
Rawski, Thomas (1989) *Economic Growth in Prewar China*. Berkeley: University of California Press.
Readings, Bill (1996) *The University in Ruins*. Cambridge, MA: Harvard University Press.
Reardon, Laurence C. (1996) "The rise and decline of China's export processing zones." *Journal of Contemporary China*, 5(13): 281–303.
Reid, Anthony (1988) *Southeast Asia in the Age of Commerce, 1450–1680. Volume*

1: The Lands below the Winds. New Haven, CT: Yale University Press.
Remer, C. F. (1968) *Foreign Investments in China*. New York: Howard Fertig.
Rigg, Jonathan (1997) *Southeast Asia: The Human Landscape of Modernization and Development*. London and New York: Routledge.
Robinson, H. (1967) *Monsoon Asia: A Geographical Survey*. New York: Praeger.
Rofel, Lisa (1999) "Museum as women's space: displays of gender in post-Mao China," pp. 116–131 in Mayfair Mei-hui Yang (ed.), *Spaces of Their Own: Women's Public Sphere in Transnational China*. Minneapolis: University of Minnesota Press.
Rose, Gillian (1993) *Feminism and Geography: The Limits of Geographical Knowledge*. Minneapolis: University of Minnesota Press.
Rosen, Stanley and Gary Zou, eds. (1991/1992) "The Chinese television documentary 'River Elegy' (Part I)." *Chinese Sociology and Anthropology*, 24(2): 3–90.
—— (1992a). "The Chinese television documentary 'River Elegy' (Part II)." *Chinese Sociology and Anthropology*, 24(4): 3–91.
—— (1992b) "The Chinese television documentary 'River Elegy' (Part III)." *Chinese Sociology and Anthropology*, 25(1): 3–91.
Rosenthal, Elisabeth (1998) "For one-child policy, China rethinks iron hand." *The New York Times*, November 1, sec. 1, p. 1.
Rostow, W. W. (1960) *The Stages of Economic Growth, a Non-communist Manifesto*. Cambridge: Cambridge University Press.
Rowe, William T. (1984) *Hankow: Commerce and Society in a Chinese City, 1796–1889*. Stanford, CA: Stanford University Press.
Ruo Shi and Feng Yuan, (1987) "Nüxing lixiang di fansi yu qiuzheng (Seeking and reflecting on women's ideals)," *Funü zuzhi yu huodong: yinshua baokan, ziliao* (*Women's Organizations and Activities: Published Press Materials Reader*), 2. Beijing: Renmin daxue; see Barlow (1994, n. 6 and 25).
Said, Edward W. (1978) *Orientalism*. New York: Pantheon.
Sakai, Robert K. (1968) "The Ryukyu (Liu-ch'iu) islands as a fief of Satsuma," pp. 112–34 in John K. Fairbank (ed.), *The Chinese World Order*. Cambridge, MA: Harvard University Press.
Sandhu, Kernial Singh (1983) "Chinese colonization in Melaka," pp 93–136 in K. S. Sandhu and Paul Wheatley (eds), *Melaka: The Transformation of a Malay Capital, volume 2*. Kuala Lumpur: Oxford University Press.
Sandhu, Kernial Singh and Paul Wheatley (1983) "Preface," pp. v–vii in K. S. Sandhu and P. Wheatley (eds), *Melaka: The Transformation of a Malay Capital, volume 1*. Kuala Lumpur: Oxford University Press.
Sanger, David E. and Don Van Natta, Jr. (1998) "'China area' tied to 'illegal' gifts: Senate Republicans can't link the contributions to Beijing." *The New York Times*, February 11, sec. A, p. 1.
Scalapino, Robert, A. (1991) "The United States and Asia: future prospects." *Foreign Affairs*, 70(5): 19–40.
Schafer, Edward H. (1954) *The Empire of Min*. Rutland, VT: C. E. Tuttle Co.
—— (1967) *The Vermilion Bird: T'ang Images of the South*. Berkeley: University of California Press.
—— (1969) *Shore of Pearls*. Berkeley: University of California Press.

Schaller, George B. (1993) *The Last Panda*. Chicago: Chicago University Press.

Schama, Simon (1995) *Landscape and Memory*. New York: A.A. Knopf.

Schein, Louisa (1996) "The other goes to market: the state, the nation, and unruliness in contemporary China." *Identities*, 2(3): 197–222.

Schiffrin, Harold Z. (1968) *Sun Yat-sen and the Origins of the Chinese Revolution*. Berkeley: University of California Press.

—— (1980) *Sun Yat-sen, Reluctant Revolutionary*. Boston: Little, Brown.

Schoenberger, Erica (1989) "New models of regional change," pp. 115–41 in Richard Peet and Nigel Thrift (eds), *New Models in Geography: The Political Economy Perspective, volume 1*. London and Boston: Unwin Human.

Schoppa, R. Keith (1982) *Chinese Elites and Political Change: Zhejiang Province in the Early Twentieth Century*. Cambridge, MA: Harvard University Press.

Schurz, William Lytle (1939) *The Manila Galleon*. New York: Dutton.

Scott, Allen J. (1988) *New Industrial Spaces: Flexible Production Organization and Regional Development in North America and Western Europe*. London: Pion.

Scott, James C. (1997) "Futures of Asian studies." *Asian Studies Newsletter*, 42: 7.

Scott, Joan Wallach (1999) *Gender and the Politics of History*, rev. edition. New York: Columbia University Press.

Seagrave, Sterling (1995) *Lords of the Rim: The Invisible Empire of the Overseas Chinese*. New York: Putnam's Sons.

Selected Works of Deng Xiaoping (1975–1982) (1984) Beijing: Foreign Languages Press.

Selected Works of Deng Xiaoping, volume 3 (1982–1992) (1994) Beijing: Foreign Languages Press.

Selvarani, P. (1984) "Bukit China: fighting an uphill battle." *Malaysian Business*, August 16, pp. 35–6.

Sender, Henny (1992) "And now for China: Hong Kong's speculators move north of the border." *Far Eastern Economic Review*, December 3, pp. 59–60.

—— (1993) "Just like Hong Kong: Canton prospers by welcoming property investment." *Far Eastern Economic Review*, August 12, pp. 71–2.

—— (1994) "Passion for profit." *Far Eastern Economic Review*, June 23, pp. 54–6.

Shambaugh, David (1993) "Introduction: the emergence of 'Greater China'." *The China Quarterly*, 136: 653–9.

Shen Bin (1998) "Port projects run out of control." *China Daily* (Beijing), Business Weekly, May 18, p. 1.

Shen Shiji and Wan Guochen (1997) "10 million children drop out of schools due to financial problems." *China News Digest*, April 25. http://www.cnd.org/CND-Global/

Shi, Shu-mei (1998) "Gender and a new geopolitics of desire: the seduction of mainland women in Taiwan and Hong Kong media." *Signs* 23(2), 287–319.

Shiba, Yoshinobu (1977) "Ningbo and its hinterland," pp. 391–439 in G. William Skinner (ed.), *The City in Late Imperial China*. Stanford, CA: Stanford University Press.

Shirk, Susan L. (1993) *The Political Logic of Economic Reform in China*. Berkeley: University of California Press.

—— (1994) *How China Opened Its Door: The Political Success of the PRC's Foreign Trade and Investment Reforms*. Washington, DC: The Brookings Institution.
Sidhu, M. S. (1976) "Chinese dominance of West Malaysian towns, 1921–1970." *Geography*, 61(1): 17–33.
Siu, Helen F. (1989) *Agents and Victims in South China: Accomplices in Rural Revolution*. New Haven, CT: Yale University Press.
—— (1990a) "Recycling tradition: culture, history, and political economy in the chrysanthemum festivals of South China." *Comparative Studies in Society and History*, 32(4): 765–94.
—— (1990b) "Where were the women? Rethinking marriage resistance regional culture in South China." *Late Imperial China*, 11(2): 32–62.
—— (1993) "Cultural identity and the politics of difference in south China." *Daedalus*, 122(2): 19–43.
Skeldon, Ronald (1994) "Hong Kong in an international migration system," pp. 21–51 in R. Skeldon (ed.), *Reluctant Exiles? Migration from Hong Kong and the New Overseas Chinese*. Armonk, NY: M. E. Sharpe.
—— (1995) "Emigration from Hong Kong, 1945–1994: the demographic lead-up to 1997," pp. 51–77 in R. Skeldon (ed.), *Emigration from Hong Kong: Tendencies and Impacts*. Hong Kong: The Chinese University Press.
Skidmore, Owings, and Merrill (1999) "Projects: 7 South Dearborn, Chicago, Illinois." http://www.som.com/html/7_south_dearborn.html
Skinner, G. William (1957) *Chinese Society in Thailand: An Analytical History*. Ithaca, NY: Cornell University Press.
—— (1964) "Marketing and social structure in rural China (part I)." *Journal of Asian Studies*, 24(1): 3–44.
—— (1965a) "Marketing and social structure in rural China (part II)." *Journal of Asian Studies*, 24(2): 195–228.
—— (1965b) Marketing and social structure in rural China (part III)." *Journal of Asian Studies*, 24(3): 363–400.
—— (1977a) "Cities and the hierarchy of local systems," pp. 275–351 in G. William Skinner (ed.), *The City in Late Imperial China*. Stanford, CA: Stanford University Press.
—— (1977b) "Regional urbanization in nineteenth-century China," pp. 211–49 in G. William Skinner (ed.), *The City in Late Imperial China*. Stanford, CA: Stanford University Press.
—— (1985) "Presidential address: the structure of Chinese history." *Journal of Asian Studies*, 44(2): 271–92.
Smil, Vaclav (1993) *China's Environmental Crisis: An Inquiry into the Limits of National Development*. Armonk, NY: M. E. Sharpe.
—— (1995) Who will feed China? *The China Quarterly*, 143(Sept.): 801–13.
—— (1997) "China shoulders the cost of environmental change." *Environment*, 39(6): 6–9, 33–7.
—— (1999) "China's agricultural land." *The China Quarterly* 58: 414–29.
Smith, Neil (1987) "Academic war over the field of geography: the elimination of geography at Harvard, 1947–52." *Annals of the Association of American Geographers*, 77(2): 155–72.
—— (1989) "Uneven development and location theory: towards a synthesis,"

pp. 142–63 in Richard Peet and Nidel Thrift (eds), *New Models in Geography: The Political Economy Perspective, volume 1*. Boston: Unwin Hyman.

—— (1993) "Homeless/global: scaling places," pp. 87–119 in Jon Bird, Barry Curtis, Tim Putnam, George Robertson, and Lisa Tickner (eds), *Mapping the Futures: Local Cultures, Global Change*. London and New York: Routledge.

—— (1997) "The satanic geographies of globalization: uneven development in the 1990s." *Public Culture*, 10(1): 169–89.

Smith, Richard J. (1996) *Chinese Maps: Images of "All under Heaven."* Hong Kong: Oxford University Press.

So, Kwan-wai (1975) *Japanese Piracy in Ming China during the Sixteenth Century*. East Lansing: Michigan State University Press.

Soils of China (1990) English edition edited by Li Chingkwei and Sun Ou. Beijing: Science Press.

Solinger, Dorothy (1993) *China's Transformation from Socialism: Statist Legacies and Market Reforms*. Armonk, NY: M. E. Sharpe.

Spence, Jonathan D. (1990) *The Search for Modern China*. New York: W.W. Norton.

—— (1996) *God's Chinese Son: The Taiping Heavenly Kingdom of Hong Xiuquan*. New York: W. W. Norton.

Spring Bud Plan (1997) http://www.chinaplaza.com/springbud/

Stacey, Judith (1983) *Patriarchy and Socialist Revolution in China*. Berkeley: University of California Press.

The Star (Petaling Jaya, Malaysia) (1984a) "Stop sabotaging govt's plans: CM." July10, p. 1.

—— (1984b) "Thousands say no to govt plan for hill." October 10, p. 2.

—— (1984c) "Joint memorandum submitted to the Chief Minister of Melaka by the major Chinese organisations in Malaysia for the preservation of Bukit China in entirety." October 18, p. 2–3.

—— (1984d) "CPM exploiting Bukit China issue, says CM." November 23, p. 2.

—— (1984e) "Bukit China: Dr. M. urged to step in." December 3, p. 2.

—— (1985) "DAP didn't apply for license for posters, Court told." January 29, p. 4.

Stewart, Lynn (1995) "Bodies, visions, and spatial politics: a review essay on Henri Lefebvre's *The Production of Space*." *Environment and Planning D: Society and Space*, 13(5): 609–18.

Stockard, Janice E. (1989) *Daughters of the Canton Delta: Marriage Patterns and Economic Strategies in South China, 1860–1930*. Stanford, CA: Stanford University Press.

Storper, Michael (1997) *The Regional World: Territorial Development in a Global Economy*. New York: The Guilford Press.

The Straits Times (Singapore) (1997) "Need for Chinese-proficient elite." August 25, pp. 1, 28–30.

—— (1999) "Injustice to Hong Kong,." January 9, p. 56.

Strassberg, Richard E. (1994) *Inscribed Landscapes: Travel Writing from Imperial China*. Berkeley: University of California Press.

Struve, Lynn (1984) *The Southern Ming, 1644–1662*. New Haven, CT: Yale University Press.

Summerfield, Gale (1994a) "Chinese women and the post-Mao economic reforms," pp. 113–28 in N. Aslanbeigui, S. Pressman and G. Summerfield (eds), *Women in the Age of Economic Transformation: Gender Impact of Reforms in Post-socialist and Developing Countries*. London and New York: Routledge.

—— (1994b) "Economic reform and the employment of Chinese women." *Journal of Economic Issues* 28(3): 715–32.

—— (1997) "Economic transition in China and Vietnam: crossing the poverty line is just the first step for women and their families." *Review of Social Economy*, 55(2): 201–14.

Sun Ching-chih (ed.) (1962) *Economic Geography of South China: Kwangtung, Kwangsi, Fukien*. Washington, DC: Joint Publications Research Service.

Sun Han (1994) *Zhongguo nongye ziran ziyuan yu quyu fazhan* (*Agricultural Resources and Regional Development in China*). Nanjing: Jiangsu kexue jishu chubanshe.

Sung, Yun-Wing, Pak-Wai Liu, Yue-Chim Richard Wong, and Pui-King Lau (1995) *The Fifth Dragon: The Emergence of the Pearl River Delta*. Singapore and Reading, MA: Addison-Wesley.

Sung, Yun-wing (1996) "'Dragon head' of China's economy?," pp. 171–98 in Y. M. Yeung and Sung Yun-wing (eds), *Shanghai: Transformation and Modernization under China's Open Policy*. Hong Kong: Chinese University Press.

Suryadinata, Leo (1992) *Pribumi Indonesians, the Chinese Minority, and China*, 3rd edn. Singapore: Heinemann Asia.

Swyngedouw, Erik (1997a) "Excluding the other: the production of scale and scaled politics," in Roger Lee and Jane Wills (eds), pp. 167–76, *Geographies of Economies*. London: Arnold.

—— (1997b) "Neither global nor local: 'glocalization' and the politics of scale," pp. 137–66 in Kevin Cox (ed.), *Spaces of Globalization: Reasserting the Power of the Local*. New York: The Guilford Press.

Szymanski, Richard and John A. Agnew (1981) *Order and Skepticism: Human Geography and the Dialectic of Science*. Washington, DC: Association of American Geographers.

Tacey, Elisabeth (1996) "Tower drawings too complicated for public, says MTR." *South China Morning Post* (Hong Kong), December 6, p. 3.

Tam, Siumi Maria (1996) "Normalization of 'second wives': gender contestation in Hong Kong." *Asian Journal of Women's Studies*, 2: 113–132.

Tai, En-Sai (1918) *Treaty Ports in China*. New York: Columbia University.

Tan Chee Beng (1983) "Acculturation and the Chinese in Melaka: the expression of Baba identity today," pp. 56–78 in Peter Gosling and Linda Lim (eds), *The Chinese in Southeast Asia, volume 2*. Singapore: Maruzen Asia.

—— (1988) *The Baba of Melaka: Culture and Identity of a Chinese Peranakan Community in Malaysia*. Petaling Jaya, Malaysia: Pelanduk.

Tan Chee Koon (1984) "Bukit China – a hill steeped in history" and "What they say . . ." *The Star* (Petaling Jaya, Malaysia), October 3, p. 9.

Tan Hsueh Yun (1999) "PM outlines vision of S'pore, the cosmopolis." *Straits Times* (Singapore), April 20, p. 34.

Tan Liok Ee (2000) "Chinese schools in Malaysia: a case of cultural resilience," pp. 228–54 in Lee Kam Hing and Tan Chee-Beng (eds), *The Chinese in*

Malaysia. Shah Alam, Malaysia, and New York: Oxford University Press.

Tan Pek Leng (1984) "Bukit China: its past and present." *The Star* (Petaling Jaya, Malaysia), July 18, p. 2.

Taylor, Jay (1974) *China and Southeast Asia: Peking's Relations with Revolutionary Movements*. New York: Praeger.

Taylor, M. (1991) "Boom with a queue: Hong Kong has too much cash chasing too few assets." *Far Eastern Economic Review*, August 22, pp. 48–51.

TDCM (Tourist Development Corporation of Malaysia) (1991) *Melaka Map and Guide*. Kuala Lumpur: Tourist Development Corporation of Malaysia.

Teng, Jinhua Emma (1996) "The construction of the 'traditional Chinese woman' in the Western academy: a critical review." *Signs*, 22(1): 115–51.

Teng, Ssu-yü and John K. Fairbank (1954) *China's Response to the West: A Documentary Survey, 1839–1923*. Cambridge, MA: Harvard University Press.

Thrift, Nigel (1986) "The internationalisation of producer services and the integration of the Pacific Basin property market," pp. 142–92 in Michael Taylor and Nigel Thrift (eds), *Multinationals and the Restructuring of the World Economy*. London: Croom Helm.

T'ien Ju-K'ang (1988) *Male Anxiety and Female Chastity: A Comparative Study of Chinese Ethical Values in Ming-Ch'ing Times*. Leiden: E. J. Brill.

—— (1990) "The decadence of Buddhist temples in Fukien in late Ming and early Ch'ing," pp. 83–100 in Eduard B. Vermeer (ed.), *Development and Decline of Fukien Province in the Seventeenth and Eighteenth Centuries*. Leiden: E. J. Brill.

Tomaney, John and Neil Ward (2000) "Debates and surveys: England and the 'new regionalism'." *Regional Studies*, 34(5): 471–8.

Topley, Marjorie (1975) "Marriage resistance in rural Kwangtung," pp. 67–88 in Margery Wolf and Roxane Witke (eds), *Women in Chinese Society*. Stanford, CA: Stanford University Press.

Tsai, Shih-shan Henry (1996) *The Eunuchs in the Ming Dynasty*. Albany: State University of New York Press.

Tu Wei-ming (1984) *Confucian Ethics Today: The Singapore Challenge*. Singapore: Federal Publications.

—— (1991) "Cultural China: the periphery as the center." *Daedalus*, 120(2): 1–32.

Tuan, Yi-fu (1996) *Cosmos and Hearth: A Cosmopolite's Viewpoint*. Minneapolis: University of Minnesota Press.

Uhalley, Stephen Jr. (1994) "'Greater China': the contest of a term." *Positions*, 2(2): 274–93.

Ukers, William H. (1935) *All about Tea, volume 1*. New York: The Tea and Coffee Trade Journal Company.

United Nations (1995). *Human Development Report*. New York: Oxford University Press.

USDS, Despatches from US Consuls in Amoy (1844–1906) Records of the Department of State, United States National Archive, Washington, DC.

USDS, Numerical and Minor Files of the Department of State (1906–10) Records of the Department of State, United States National Archive, Washington, D.C.

Van Slyke, Lyman P. (1967) *Enemies and Friends: The United Front in Chinese Communist History*. Stanford, CA: Stanford University Press.

Vermeer, Eduard B. (1990) "Introduction: historical background and major issues," pp. 5–34 in E. B. Vermeer (ed.), *Development and Decline of Fukien Province in the Seventeenth and Eighteenth Centuries*. Leiden: E. J. Brill.

Viraphol, Sarasin (1977) *Tribute and Profit: Sino-Siamese Trade, 1652–1853*. Cambridge, MA: Council on East Asian Studies, Harvard University.

Vo Nhan Tri (1990) *Vietnam's Economic Policy since 1975*. Singapore: Institute of Southeast Asian Studies.

Vogel, Ezra F. (1989) *One Step Ahead in China: Guangdong under Reform*. Cambridge, MA: Harvard University Press.

Wade, Robert (1990) *Governing the Market: Economic Theory and the Role of Government in East Asian Industrialization*. Princeton, NJ: Princeton University Press.

—— (1993a) "Managing trade: Taiwan and South Korea as challenges to economics and political science." *Comparative Politics*, 25(2): 147–68.

—— (1993b) "The visible hand: the state and East Asia's economic growth." *Current History*, 92(578): 531–41.

—— (1998) "The Asian debt-and-development crisis of 1997–? Causes and consequences." *World Development* 26(8): 1535–53.

Wakeman, Frederic Jr (1978) "The Canton trade and the Opium War," pp. 163–213 in Denis Twitchett and John K. Fairbank (eds), *The Cambridge History of China, volume 10, part 1*. Cambridge: Cambridge University Press.

—— (1989) "All the rage in China." *The New York Review of Books*, March 2, pp. 19–21.

Wakeman, Frederic Jr and Wen-hsin Yeh (eds), (1992) "Introduction," pp. 1–14 in F. Wakeman and W. Yeh (eds), *Shanghai Sojourners*. Berkeley: University of California Press.

Walder, Andrew G. (1996) "China's transitional economy: interpreting its significance," pp. 1–17 in Andrew G. Walder, (ed.), *China's Transitional Economy* Oxford: Oxford University Press.

Waley, Arthur (1958) *The Opium War through Chinese Eyes*. Stanford, CA: Stanford University Press.

Wan, Cynthia (1996) "Anger at skyscraper reaches new heights." *Hong Kong Standard*, September 24, p. 3.

Wang Gungwu (1958) "The Nanhai trade: a study of the early history of Chinese trade in the South China Sea." *Journal of the Malayan Branch of the Royal Asiatic Society*, 31(2): 1–135.

—— (1959) *A Short History of the Nanyang Chinese*. Singapore: Donald Moore/ Eastern Universities Press.

—— (1976) "The limits of Nanyang Chinese nationalism, 1912–37," pp. 405–23 in C. D. Cowan and O. W. Wolters (eds), *Southeast Asian History and Historiography: Essays Presented to D. G. E. Hall*. Ithaca, NY: Cornell University Press.

—— (1981) "A note on the origins of *hua-ch'iao*," pp. 118–27 in *Community and Nation: Essays on Southeast Asia and the Chinese*. Singapore: Heinemann; Sydney: George Allen & Unwin.

—— (1985) "External China as a new policy area." *Pacific Affairs*, 58(1): 28–43.

—— (1988) "The study of Chinese identities in Southeast Asia," pp. 1–21 in Jennifer Cushman and Wang Gungwu (eds), *The Changing Identities of the Chinese in Southeast Asia*. Hong Kong: Hong Kong University Press.

—— (1996) "Sojourning: the Chinese experience in Southeast Asia," pp. 1–14 in Anthony Reid (ed.), *Sojourners and Settlers: Histories of Southeast Asia and the Chinese*. St. Leonards, Australia: Allen & Unwin.

Wang, Jing (1996) *High Culture Fever*. Berkeley: University of California Press.

Wang, Mingming (1995) "Place, administration, and territorial cults in late imperial China: a case study from south Fujian." *Late Imperial China*, 16(1): 33–78.

Wang Qi (1999) "State-society relations and women's political participation," pp. 19–44 in Jackie West, Zhao Minghua, Chang Xiangqun, and Cheng Yuan, (eds), *Women of China: Economic and Social Transformation*. London: Macmillan Press; New York: St Martin's Press.

Wang Xiangwei (1999) "Wang Qishan clout grows on GDE deal." *South China Morning Post*, October 5, Business Post sec., p. 12.

Wang, Ya Ping and Alan Murie (1996) "The process of commercialization of urban housing in China." *Urban Studies*, 33(6): 971–89.

Wasserstrom, Jeffrey N. (1998) "Are you now or have you ever been . . . postmodern?" *The Chronicle of Higher Education*, 45, pp. B4–5.

Watson, James L. (1985) "Standardizing the gods: the promotion of T'ien Hou ("Empress of Heaven") along the South China coast, 960–1960," pp. 292–324 in David Johnson, Andrew Nathan, and Evelyn S. Rawski (eds), *Popular Culture in Late Imperial China*. Berkeley: University of California Press.

Weidenbaum, Murray and Samuel Hughes (1996) *The Bamboo Network: How Expatriate Chinese Entrepreneurs Are Creating a New Economic Superpower in Asia*. New York: Martin Kessler Books.

Wheatley, Paul (1959) "Geographical notes on some commodities invovled in Sung maritime trade." *Journal of the Malayan Branch of the Royal Asiatic Society*, 32(2): 1–140.

White, Lynn T. III (1989) *Shanghai Shanghaied: Uneven Taxes in Reform China*. Hong Kong: University of Hong Kong Centre of Asian Studies.

—— (1998a) *Unstately Power. Volume 1, Local Causes of China's Economic Reforms*. Armonk, NY: M. E. Sharpe.

—— (1998b) *Unstately Power. Volume 2, Local Causes of China's Intellectual, Legal and Governmental Reforms*. Armonk, NY: M. E. Sharpe.

White, Lynn T. III and Li Cheng (1993) "China coast identities: regional, national, and global," pp. 155–93 in Lowell Ditmer and Samuel S. Kim (eds), *China's Quest for National Identity*. Ithaca, NY: Cornell University Press.

Whyte, Martin King (2000) "The perils of assessing trends in gender inequality in China," pp. 157–67 in Barbara Entwisle and Gail E. Henderson (eds), *Redrawing Boundaries: Work, Households, and Gender in China*. Berkeley: University of California Press.

Wiens, Harold (1954) *China's March toward the Tropics*. Hamden, CT: Shoe String Press.

Wigen, Kären (1999) "Culture, power, and place: the new landscapes of East Asian regionalism." *American Historical Review*, 104(4): 1183–201.

Williams, Raymond (1973) *The Country and the City*. New York: Oxford University Press.

Wills, John E., Jr. (1968) "Ch'ing relations with the Dutch, 1662–1690," pp. 225–56 in John K. Fairbank (ed.), *The Chinese World Order: Traditional China's Foreign Relations*. Cambridge, MA: Harvard University Press.

—— (1988) "Tribute, defensiveness, and dependency; uses and limits of some basic ideas about Mid-Qing dynasty foreign relations." *American Neptune*, 48(4): 225–9.

Wong, John (1986) "Promoting Confucianism for socioeconomic development: the Singapore experience," pp. 277–93 in Tu Wei-ming, (ed.), *Confucian Traditions in East Asian Modernity: Moral Education and Economic Culture in Japan and the Four Mini-dragons*. Cambridge, MA: Harvard University Press.

Wong, K. K. and X. B. Zhao (1999) "The influence of bureaucratic behavior on land apportionment in China: the informal process." *Environment and Planning C: Government and Policy*, 17(1): 113–26.

Wong, Linda and Huen Wai-Po (1998) "Reforming the household registration system: a preliminary glimpse of the blue chop household registration system in Shanghai and Shenzhen." *International Migration Review*, 32(4): 974–94.

Wong Siu-lun (1988) *Emigrant Entrepreneurs: Shanghai Industrialists in Hong Kong*. Hong Kong: Oxford University Press.

—— (1996) "The entrepreneurial spirit: Shanghai and Hong Kong compared," pp. 25–48 in Y. M. Yeung and Sung Yun-wing (eds), *Shanghai: Transformation and Modernization under China's Open Policy*. Hong Kong: Chinese University Press.

Wong, Yuk-lin Renita (1997) "Dispersing the 'public' and the 'private': gender and the state in the birth planning policy of China." *Gender and Society*, 11(4): 509–25.

Worcester, G. R. G. (1966) *Sail and Sweep in China*. London: Her Majesty's Stationery Office.

The World Bank (1993) *The East Asian Miracle: Economic Growth and Public Policy*. Oxford: Oxford University Press.

World Wide Fund for Nature Hong Kong (1996) *Wetland Conservation*. Discussion pack for environmental education. Hong Kong: World Wide Fund for Nature Hong Kong.

Wright, Mary Clabaugh (1968) "Introduction: the rising tide of change," pp. 1–63 in M. C. Wright (ed.), *China in Revolution: The First Phase 1900–1913*. New Haven, CT: Yale University Press.

Wright, Stanley F. (1927). *The Collection and Disposal of the Maritime and Native Customs Revenue since the Revolution of 1911*. Shanghai: Statistical Department of the Inspectorate General of Customs.

Wu Chengxi (1937) "Xiamen de huaqiao huikuan yu jinrong zucheng (Overseas Chinese from Xiamen and the organization of remittances and banking)." *Shehui kexue zazhi* (*Quarterly Review of Social Sciences*), 8(2): 193–252.

Wu Chuanqun and Guo Huancheng (1994) *Zhongguo tudi liyong* (*Land Use in*

China). Beijing: Kexue chubanshe.

Wu Chün-hsi (1967) *Dollars, Dependents, and Dogma: Overseas Chinese Remittances to Communist China.* Stanford: Hoover Institution.

Wu Jiayu and Lin Jiazhen (1983) "Fujian jinyan yundong 'Qudushe' (The Fujian anti-opium movement and the 'Anti-Opium League')." *Fuzhou wenshi ziliao xuanji* (*Select Materials on Fuzhou Culture and History*), 2: 15–18.

Wu, Linda and Bo Xiong (1998) "Women's federation warns government against forced maternity leave," *China News Digest,* March 6. http://www.cnd.org/CND-Global/CND-Global.98.1st/CND-Global.98–03–06a.html

Wu, Silas H. L. (1979) *Passage to Power: K'ang-hsi and His Heir Apparent, 1661–1722.* Cambridge, MA: Harvard University Press.

Wu, Sofia (1999) "Mainland men obtain residency by marrying elderly ROC women," Central News Agency (Taiwan), October 13. http://www.lexis-nexis.com/universe

Xie Xuanjun and Yuan Zhiming (1991/2) "Azure (Blue Sky)." *Chinese Sociology and Anthropology,* 24(2): 78–90.

Xiamen huaqiao zhi (*Xiamen overseas Chinese gazetteer*) (1991) Xiamen: Lujiang chubanshe.

Xiamen zhi (*Xiamen gazetteer*) (1839) Reprint, Taipei, 1967.

Xinghua fu zhi (*Xinghua Prefecture Gazetteer*) (1503) Reprint, Futian, 1871.

Xinhua (Beijing) (1987) "Deng Xiaoping meets US professors." May 16, item no. 0516083. http://www.lexis-nexis.com/universe

—— (1989) "International Confucius symposium held." October 5, item no. 1005215. http://www.lexis-nexis.com/universe

—— (1993a) "Rising city in Fujian stresses efficiency in land use." June 5, item no. 0605118.

—— (1993b) "Land development interests in Xiamen." June10, item no. 0610050. http://www.lexis-nexis.com/universe

—— (1995a) "Fujian steps up land protection efforts." February 20, item no. 0220030.

—— (1995b) "Lippo investment project proceeds smoothly." June 23, trans. in FBIS-CHI-95–121.

—— (1996) "Fujian's arable land shrinking." September 10, item no. 0910073.

—— (1997a) "Protection of cultivated land increasing." May 19, trans. in FBIS-CHI-97–096.

—— (1997b) "Preferential policies for overseas investors in southeast China." May 23, item no. 0523150.

—— (1998a) "PRC officials debate amendments to draft of land-use law." April 27, trans. in FBIS-CHI-98–117.

—— (1998b) "Fujian province to revise land-use plan." May 18, trans. in FBIS-CHI-98–089.

—— (1998c) "*Renmin Ribao* urges land protection." May 18, trans. in FBIS-CHI-97–097.

—— (1998d) "Public opinions incorporated into law." August 29, trans. in FBIS-CHI-98–241.

—— (1998e) "Labor disputes in southern Chinese province." October 14, item no. 1014222. http://www.lexis-nexis.com/universe

—— (1999a) "Tung Chee-hwa wants Hong Kong to become cosmopolitan Asian city." October 6. http://www.lexis-nexis.com/universe

—— (1999b) "Unsafe sex may worsen aids epidemic in China." November 15. http://www.lexis-nexis.com/universe

Xu Dixin and Wu Chengming (eds) (2000) *Chinese Capitalism, 1522–1840.* Basingstoke and London: Macmillan Press; New York: St Martin's Press.

Xu Fengxian, Mao Zhichong, and Yuan Juying (1993) "Su'nan moshi de xin fazhan (New developments in the Su'nan model)." *Jingji yanjiu* (*Economic Research*), 2: 49–55.

XZQH (*Zhonghua renmin gongheguo xingzheng quhua jiance*) (Brief guide to China's administrative divisions) (1978, 1995) Beijing: Ditu chubanshe.

Yabuki, Susumu (1995) *China's New Political Economy: The Giant Awakes.* Boulder, CO: Westview.

Yahuda, Michael (1993) "The foreign relations of Greater China." *The China Quarterly*, 136: 687–710.

Yao Xitang (ed.) (1990) *Shanghai Xianggang bijiao yanjiu* (*A Comparative Study of Shanghai and Hong Kong*), Shanghai: Shanghai renmin chubanshe.

Yang, Dali L. (1994) "Reform and restructuring of central-local relations," pp. 59–98 in David S. G. Goodman and Gerald Segal (eds), *China Deconstructs: Politics, Trade and Regionalism.* London and New York: Routledge.

—— (1997) *Beyond Beijing: Liberalization and the Regions in China.* London and New York: Routledge.

Yang Dongping (1994) *Chengshi jifeng: Beijing he Shanghai de wenhua jingshen* (*City Monsoon: The Cultural Spirit of Beijing and Shanghai*). Beijing: Dongfang chubanshe.

Yang, Mayfair Mei-hui (1994) *Gifts, Favors, and Banquets: The Art of Social Relationships in China.* Ithaca, NY: Cornell University Press.

—— (1999a) "Introduction," pp. 1–31 in Mayfair Mei-hui Yang (ed.), *Spaces of Their Own: Women's Public Sphere in Transnational China.* Minneapolis: University of Minnesota Press.

—— (1999b) "From gender erasure to gender difference: state feminism, consumer sexuality, and women's public sphere in China," pp. 35–67 in Mayfair Mei-hui Yang (ed.), *Spaces of their Own: Women's Public Sphere in Transnational China.* Minneapolis: University of Minnesota Press.

Yang You (1985) "Zhenghe xia xi yang suoyou de chuanbo: cong hanghai yu zuachuan de jiaodu kaolü (The ships of Zhenghe of the Western Ocean: perspectives on maritime navigation and boatbuilding)," pp. 108–18 in *Zhenghe xia xi yang lunwen ji* (*Collected Papers on Zhenghe of the Western Ocean*). Beijing: Remin jiaotong chubanshe.

Ye Wenzhen and Lin Qingguo (1996) "Fujiansheng shewai hunyin zhuangkuang yanjiu (Research on marriage with non-mainlanders in Fujian province)," *Renkou yu jingji* (*Population and Economics*), 2: 21–7.

Yeh, Anthony Gar-on and Fulong Wu (1996) "The new land development process and urban development in Chinese cities." *International Journal of Urban and Regional Research*, 20(2): 330.

Yeh, Anthony Gar-On and Xia Li (1997) "An integrated remote sensing and GIS approach in the monitoring and evaluation of rapid urban growth for

sustainable development in the Pearl River Delta, China." *International Planning Studies*, 2(2): 193–210.
Yeh, H. F. (1936) "The Chinese of Malacca." *Historical Guide of Malacca*. Malacca: Malacca Historical Society.
Yeh, Wen-Hsin (1996) *Provincial Passages: Culture, Space, and the Origins of Chinese Communism*. Berkeley: University of California Press.
Yeoh, Brenda S. A. (1991) "The control of sacred space: conflicts over the Chinese burial grounds in colonial Singapore, 1880–1930." *Journal of Southeast Asian Studies*, 22(2): 282–311.
Yeoh, Brenda S. A. and Katie Willis (1999) "'Heart' and 'wing', nation and diaspora: gendered discourses in Singapore's regionalisation process." *Gender, Place and Culture*, 6(4): 355–72.
Yeung, Henry Wai-chung and Kris Olds (eds) (2000) *Globalization of Chinese Business Firms*. New York: St Martin's Press.
Yi Fu (1998) "Toyland inferno: a journey through the ruins: the story of Shenzhen's 'Black Friday'." *Chinese Sociology and Anthropology*, 30(4): 8–34.
Yü, Chün-fang (1992) "P'u-t'o shan: pilgrimage and the creation of the Chinese potalaka," pp. 193–245 in Susan Naquin and Chün-fang Yü (eds), *Pilgrims and Sacred Sites in China*. Berkeley: University of California Press.
Yü, Ying-shih (1997) "Business culture and Chinese traditions: toward a study of the evolution of merchant culture in Chinese history," pp. 1–84 in Wang Gungwu and Wong Siu-lun (eds), *Dynamic Hong Kong: Business and Culture*. Hong Kong: University of Hong Kong.
Yu Zhisen (1997) "The relationship of guilds to government in the Shanghai and Suzhou area," pp. 62–79 in Brian H. A. Ranson (ed.), *Guild-hall and Government: An Exploration of Power, Control and Resistance in Britain and China. Volume 1, A Preliminary Study of the Social Organisation of Guilds in China*. Hong Kong: David C. Lam Institute for East–West Studies, Hong Kong Baptist University.
Yuan Bingling (1992) "Haishang maoyi yu Song Yuan Quanzhou shangye jingji (Maritime trade and commercial economy in Quanzhou during the Song and Yuan dynasties)." *Nanyang wenti yanjiu* (*Southeast Asian Affairs*), 3: 100–09.
ZGTJNJ (*Zhongguo tongji nianjian*) (China Statistical Yearbook) (various years) Beijing: Zhongguo tongji chubanshe.
ZGYHQZ (*Zhongguo yinhang Quanzhou fenhang hangshi bianweihui*) (Editorial Board, History of the Bank of China, Quanzhou Branch) (1996) *Minnan qiao pi shi jishu* (*Record of the History of Overseas Chinese Mail Remittances in South Fujian*). Xiamen: Xiamen daxue chubanshe.
Zha, Jiangying (1995) *China Pop: How Soap Operas, Tabloids, and Bestsellers Are Transforming a Culture*. New York: The New Press.
Zhang Junzuo (1994) "Development in a Chinese reality: rural women's organizations in China." *Journal of Communist Studies and Transition Politics*, 10(4): 71–91.
Zhangpu xian zhi (*Zhangpu County Gazetteer*) (1700).
Zhangzhou fu zhi (*Zhangzhou Prefecture Gazetteer*) (1573, 1628).
Zhao Jianhong (1994) "Xiangzhen gongye xiaoqu he qiye jituan jianshe (Construction of small industrial zones and enterprise groups in townships and

villages)." *Zhongguo nongye nianjian, 1993 (Agricultural Yearbook of China, 1993)*. Beijing: Nongye chubanshe, pp. 54–5.
Zhao Songqiao (1986) *The Physical Geography of China*. Beijing: Science Press; New York: John Wiley & Sons.
Zheng Linkuan (1940) *Fujian huaqiao huikuan (Fujian and Overseas Chinese Remittances)*. Fuzhou: Fujian sheng zhengfu mishu chu tongji shi.
Zheng Zhaojing (1993) *Zhongguo shuili shi (History of China's Rivers)*. Beijing: Shangwu Yinshu Guan.
Zhongguo duiwai jingji maoyi nianjian (Almanac of China's Foreign Economic Relations and Trade) (1996/7) Beijing: Zhongguo jingji chubanshe.
Zhongguo jingji tequ kaifa qu nianjian (Yearbook of China's Special Economic and Development Zones) (1996) Beijing: Gaige chubanshe.
Zhongguo lishi dituji (The Historical Atlas of China), (1982), 8 vols., Beijing: ditu chubanshe.
Zhou Yongming (1999) *Anti-Drug Crusades in Twentieth-century China: Nationalism, History, and State-building*. Lanham, MD: Rowman and Littlefield.
Zhu Hanqiang (1995) "Futian ziran baohuqu can bian baoshuiqu; gaifang xiulu shulin jian zao huangtu yanmo (Futian natural protected area tragically changed into a tax protection area; building houses, constructing roads, the forest gradually suffers inundation by yellow earth)." *Xianggang lianhebao (United Daily News*, Hong Kong), April 25, p. 2.
Zhu, Jieming (1994) "Changing land policy and its impact on local growth: the experience of the Shenzhen Special Economic Zone, China, in the 1980s." *Urban Studies*, 31(10): 1611–23.
Zhu Rongji (1998) "Special issue on Zhu Rongji." *China News Digest*, March 20. http://www.cnd.org/CND-Global/
Zhu Tonghua and Sun Bin (1994) *Su'nan moshi fazhan yanjiu (A Study of Development under the Su'nan Model)*. Nanjing: Nanjing daxue chubanshe.
Zimmerman, Rachel (1997) "Chinese village swells with pride as Washington governor seeks his roots on a pilgrimage." *The New York Times*, October 13, sec. A, p. 12.
Zito, Angela (1997) *Of Body and Brush: Grand Sacrifice as Text/Performance in Eighteenth-century China*. Chicago: Chicago University Press.

Index

Printed and bound by CPI Group (UK) Ltd, Croydon, CR0 4YY

23/04/2025

14660945-0003